LONDON MATHEMATICAL SOCIETY LECTURE NOTE SERIES

Managing Editor: Professor M. Reid, Mathematics Institute,
University of Warwick, Coventry CV4 7AL, United Kingdom

The titles below are available from booksellers, or from Cambridge University Press at
www.cambridge.org/mathematics

London Mathematical Society Lecture Note Series: 452

Partial Differential Equations in Fluid Mechanics

Edited by

CHARLES L. FEFFERMAN
Princeton University

JAMES C. ROBINSON
University of Warwick

JOSÉ L. RODRIGO
University of Warwick

CAMBRIDGE
UNIVERSITY PRESS

CAMBRIDGE
UNIVERSITY PRESS

University Printing House, Cambridge CB2 8BS, United Kingdom

One Liberty Plaza, 20th Floor, New York, NY 10006, USA

477 Williamstown Road, Port Melbourne, VIC 3207, Australia

314-321, 3rd Floor, Plot 3, Splendor Forum, Jasola District Centre, New Delhi - 110025, India

79 Anson Road, #06-04/06, Singapore 079906

Cambridge University Press is part of the University of Cambridge.

It furthers the University's mission by disseminating knowledge in the pursuit of education, learning and research at the highest international levels of excellence.

www.cambridge.org
Information on this title: www.cambridge.org/9781108460965
DOI: 10.1017/9781108610575

© Cambridge University Press 2018

First published 2018

A catalogue record for this publication is available from the British Library

Library of Congress Cataloging in Publication data
Names: Fefferman, Charles, 1949– editor. |
Robinson, James C. (James Cooper), 1969– editor. | Rodrigo Diez, Jose Luis, 1977– editor.
Title: Partial differential equations in fluid mechanics / edited by Charles L. Fefferman (Princeton University, New Jersey), James. C. Robinson (University of Warwick), José L. Rodrigo (University of Warwick).
Description: Cambridge ; New York, NY : Cambridge University Press, 2019. |
Series: London Mathematical Society lecture note series ; 452 |
Includes bibliographical references.
Identifiers: LCCN 2018027151 | ISBN 9781108460965 (hardback : alk. paper)
Subjects: LCSH: Fluid mechanics. | Differential equations, Partial.
Classification: LCC QA374 .P3484 2019 | DDC 532–dc23
LC record available at https://lccn.loc.gov/2018027151

ISBN 978-1-108-46096-5 Paperback

Contents

Contributors

Claude Bardos
Laboratoire J.-L. Lions, Université Denis Diderot,
5 Rue Thomas Mann 75205, Paris CEDEX 13. France.
claude.bardos@gmail.com

Guher Camliyurt
Department of Mathematics, University of Southern California,
Los Angeles, CA 90089. USA.
camliyur@usc.edu

Colm-cille P. Caulfield
BP Institute, University of Cambridge,
Madingley Rise, Madingley Road, Cambridge, CB3 0EZ. UK.
Department of Applied Mathematics & Theoretical Physics,
University of Cambridge, Centre for Mathematical Sciences,
Wilberforce Road, Cambridge, CB3 0WA. UK.
c.p.caulfield@damtp.cam.ac.uk

Charles L. Fefferman
Department of Mathematics, Princeton University,
Princeton, NJ, 08544. USA.
cf@math.princeton.edu

Giovanni P. Galdi
Department of Mechanical Engineering and Materials Science,
University of Pittsburgh, Pittsburgh, PA 15261. USA.
galdi@pitt.edu

John D. Gibbon
Department of Mathematics, Imperial College London,
London, SW7 2AZ. UK.
j.d.gibbon@ic.ac.uk

Igor Kukavica
Department of Mathematics, University of Southern California,
Los Angeles, CA 90089. USA.
kukavica@usc.edu

Mads Kyed
Fachbereich Mathematik, Technische Universität Darmstadt,
Schlossgartenstr. 7, 64289 Darmstadt. Germany.
kyed@mathematik.tu-darmstadt.de

Koji Ohkitani
School of Mathematics and Statistics, The University of Sheffield,
Hicks Building, Hounsfield Road, Sheffield, S3 7RH. UK.
k.ohkitani@sheffield.ac.uk

Wojciech S. Ożański
Mathematics Institute, University of Warwick,
Coventry, CV4 7AL. UK.
w.s.ozanski@warwick.ac.uk

Pooja Rao
Department of Aerospace Engineering,
University of Illinois at Urbana-Champaign,
319E Talbot Laboratory MC 236, 104 S. Wright, Urbana, IL 61801.
USA.
poojarao@illinois.edu

Benjamin C. Pooley
Mathematics Institute, University of Warwick,
Coventry, CV4 7AL. UK.
b.pooley@warwick.ac.uk

Reimund Rautmann
Institut für Mathematik der Universität Paderborn,
D33095 Paderborn. Germany.
rautmann@math.uni-paderborn.de

James C. Robinson
Mathematics Institute, University of Warwick,
Coventry, CV4 7AL. UK.
j.c.robinson@warwick.ac.uk

José L. Rodrigo
Mathematics Institute, University of Warwick,
Coventry, CV4 7AL. UK.
j.rodrigo@warwick.ac.uk

Jack W.D. Skipper
Institute of Applied Mathematics, Leibniz University Hannover,
Welfengarten 1, 30167 Hannover. Germany.
skipper@ifam.uni-hannover.de

Chuong V. Tran
School of Mathematics and Statistics, University of St Andrews,
St Andrews, KY16 9SS. UK.
cvt1@st-andrews.ac.uk

Vlad Vicol
Department of Mathematics, Princeton University,
Princeton, NJ 08544. USA.
vvicol@math.princeton.edu

Fei Wang
Department of Mathematics, University of Maryland,
College Park, MD 20742. USA.
fwang256@umd.edu

Emil Wiedemann
Institute of Applied Mathematics, Leibniz University Hannover,
Welfengarten 1, 30167 Hannover. Germany.
wiedemann@ifam.uni-hannover.de

Xinwei Yu
Department of Mathematical and Statistical Sciences,
University of Alberta, Edmonton, AB, T6G 2G1. Canada.
xinweiyu@math.ualberta.ca

Preface

This volume is the result of a workshop, "PDEs in Fluid Mechanics", which took place in the Mathematics Institute at the University of Warwick, September 26th–30th, 2016. This was the opening meeting of the Warwick EPSRC Symposium on *Partial Differential Equations and their Applications*, which continued throughout the 2016/17 academic year.

Several of the speakers agreed to write review papers related to their contributions to the workshop, while others have written more traditional research papers. We believe that this volume therefore provides an accessible summary of a wide range of active research topics, along with some exciting new results, and we hope that it will prove a useful resource both for graduate students new to the area and for more established researchers. We have also included a survey of the pioneering 1934 paper by Leray, rewritten in updated terminology and notation by Wojciech Ożański & Benjamin Pooley, which should make the many insights contained in Leray's work more accessible to the modern reader.

We would like to express our gratitude to the following sponsors of the workshop and the writing of this volume of proceedings: the meeting itself was supported (as part of the 2016/17 Symposium) by the EPSRC, grant EP/N003039/1. JLR is currently supported by the European Research Council (ERC grant agreement no. 616797).

Finally, it is a pleasure to thank Hazel Higgens and Yvonne Collins of Warwick's Mathematics Research Centre for all their help with the organisation of our workshop.

Charles L. Fefferman
James C. Robinson
José L. Rodrigo

1

Remarks on recent advances concerning boundary effects and the vanishing viscosity limit of the Navier–Stokes equations

Claude Bardos

Laboratoire J.-L. Lions,
Université Denis Diderot,
5 rue Thomas Mann 75205 Paris CEDEX 13,
and Pauli Fellow W.P.I. Vienna.
claude.bardos@gmail.com

Abstract

This contribution covers the topic of my talk at the 2016-17 Warwick-EPSRC Symposium: "PDEs and their applications". As such it contains some already classical material and some new observations. The main purpose is to compare several avatars of the Kato criterion for the convergence of a Navier–Stokes solution, to a regular solution of the Euler equations, with numerical or physical issues like the presence (or absence) of anomalous energy dissipation, the Kolmogorov $\frac{1}{3}$ law or the Onsager $C^{0,\frac{1}{3}}$ conjecture. Comparison with results obtained after September 2016 and an extended list of references have also been added.

1.1 Introduction and uniform estimates.

In this contribution I will describe the main topics of my talk at the 2016-17 Warwick-EPSRC Symposium: *PDEs in Fluid Mechanics* in September 2016. Most of these issues are the results of a long term collaboration with Edriss Titi. I will also comment on some more recent (after September 2016) results (also collaboration with Edriss Titi and several other coworkers). In the same way I am going to include (mostly with no details) some recent results of other researchers and an extended list of references whenever they contribute to the understanding of the problems. Eventually one of the guidelines is the comparison between the use of weak convergence and the use of a statistical theory of turbulence. Hence the paper is organized as follows. After introducing some basic and well-known estimates, the zero viscosity limit of solutions of the Navier–

Published as part of *Partial Differential Equations in Fluid Mechanics*, edited by Charles L. Fefferman, James C. Robinson, & José L. Rodrigo © Cambridge University Press 2018.

Stokes equations is considered with no-slip boundary condition but in the presence, for the same initial data, of a Lipschitz solution of the Euler equations. This leads to an extension of Kato's theorem and to the introduction of several (equivalent) criteria for convergence to a smooth solution and for the absence of anomalous energy dissipation. Comparison of these criteria with physical observations or classical ansatz are made. In particular emphasis is given to the issue of the anomalous energy dissipation which leads to the comparison with the Kolmogorov $\frac{1}{3}$ law in the statistical theory of turbulence. Then this leads to the issue of the Onsager $C^{0,\frac{1}{3}}$ conjecture.

As a starting point consider solutions of the Euler equations and of the Navier–Stokes equations in a space-time domain

$$\Omega \times [0,T] \subset \mathbb{R}^d \times \mathbb{R}_t^+, \quad d = 2,3.$$

We assume that the boundary $\partial\Omega$ is a C^1 manifold with $\vec{n}(x)$ denoting the outward normal at any point x in $\partial\Omega$. Then we introduce the function

$$d(x) = d(x,\partial\Omega) = \inf_{y\in\partial\Omega} |x - y| \geq 0$$

and the set

$$\mathcal{U}_\eta = \{x \in \Omega \,, d(x) < \eta\},$$

which have the following classical geometrical properties.

Proposition 1.1 *For $0 < \eta < \eta_0$ small enough $d(x)_{|\mathcal{U}_\eta} \in C^1(\overline{\mathcal{U}_\eta})$ and for any $x \in \mathcal{U}_\eta$ there exists a unique point $\sigma(x) \in \partial\Omega$ such that $d(x) = |x - \sigma(x)|$. Moreover for every $x \in \mathcal{U}_\eta$ we have*

$$x = \sigma(x) - d(x)\vec{n}(\sigma(x)) \quad and \quad \nabla_x d(x) = -\vec{n}(\sigma(x)). \tag{1.1}$$

To focus on the boundary effects, first, we consider a smooth (Lipschitz) solution $u(x,t)$ of the incompressible Euler equations with the impermeability condition:

$$\nabla \cdot u = 0 \quad and \quad \partial_t u + u \cdot \nabla u + \nabla p = 0 \text{ in } \Omega \times [0,T]$$
$$and \quad u \cdot \vec{n} = 0 \quad on \ \partial\Omega \times [0,T]. \tag{1.2}$$

The value of such solution for $t = 0$ is denoted by $u_0(x) = u(x,0)$. For the same initial data $u_\nu(x,0) = u_0(x)$ and for any $\nu > 0$ one considers a family $u_\nu(x,t)$ of Leray–Hopf solutions of the Navier–Stokes equations with the no-slip boundary condition:

$$\partial_t u_\nu + u_\nu \cdot \nabla u_\nu - \nu \Delta u_\nu + \nabla p_\nu = 0 \quad \text{and} \quad \nabla \cdot u_\nu = 0 \text{ in } \Omega \times [0, T]$$

$$\text{with } u_\nu = 0 \text{ on } \partial\Omega \times [0, T].$$

$$(1.3)$$

For Lipschitz solutions of the Euler equations we have the obvious energy balance relation

$$\int_\Omega \frac{|u(x, t)|^2}{2} \, dx = \int_\Omega \frac{|u(x, 0)|^2}{2} \, dx, \quad \text{for all } t \in [0, T], \qquad (1.4)$$

while for any Leray–Hopf solution of the Navier–Stokes equations we obtain

$$\int_\Omega \frac{|u_\nu(x, t)|^2}{2} \, dx + \nu \int_0^t \int_\Omega |\nabla u_\nu(x, s)|^2 \, dx \, ds \le \int_\Omega \frac{|u_\nu(x, 0)|^2}{2} \, dx, \quad (1.5)$$

for all $t \in [0, T]$.

It is well known that in dimension two the solution u_ν is smooth, unique and (1.5) is actually an equality instead of an inequality. The issue of the regularity of the solutions of (1.3) plays no role in the present contribution which focuses on the zero viscosity limit. It turns out there are no other estimates uniformly valid for all positive ν, and in particular for ν going to zero, other that the one that follows from (1.3). It implies the existence of (may be not unique) limits, in the weak star $L^\infty(0, T; L^2(\Omega))$ topology, of subsequence of solutions u_ν of (1.3). Any such limit is denoted by $\overline{u_\nu}$, and the main question is whether or not we have

$$\overline{u_\nu} = u \text{ in } \Omega \times [0, T].$$

As shown in Kato (1972) and Constantin (2005), in the absence of physical boundaries (torus or the whole space) u_ν converges to u.

In the presence of physical boundaries, this is much more subtle. The obvious difficulty comes from the fact that when $\nu \to 0$ only the impermeability boundary condition remains while (here τ denotes the tangential component at the boundary) the relation $(u_\nu)_\tau = 0$ does not persist. Therefore the solution has to become singular near the boundary. It creates a shear flow near the boundary, in solutions of (1.3), which generates vorticity that may propagate inside the domain by the advection term and by the effect of the pressure. This turns out to be the most natural way to generate turbulence (even homogeneous turbulence far from the boundary).

For any Lipschitz vector field w we denote by $S(w)$ its symmetric stress tensor

$$S(w) = \frac{\nabla w + (\nabla w)^{\perp}}{2}.$$

Denote by $(\cdot, \cdot)$ the $L^2(\Omega)$ scalar product. Then since both u_ν and u are divergence-free Lipschitz vector fields and since u is tangent to the boundary of Ω we obtain, by integration by parts, the classical formula

$$(u_\nu \cdot \nabla u_\nu - u \cdot \nabla u, u_\nu - u) = (u_\nu - u, S(u)(u_\nu - u)). \quad (1.6)$$

From (1.2) and (1.3) we also have

$$\partial_t(u_\nu - u) + u_\nu \cdot \nabla u_\nu - u \cdot \nabla u - \nu \Delta u_\nu + \nabla p_\nu - \nabla p = 0. \quad (1.7)$$

Taking the $L^2(\Omega)$ inner product of (1.7) with $(u_\nu - u)$ and observing that on the boundary Ω we have $u_\nu = 0$ and $u \cdot \vec{n} = 0$, thanks to (1.6), we obtain

$$\frac{d}{dt} \frac{1}{2} |u_\nu - u|^2_{L^2(\Omega)} + \nu \int_\Omega |\nabla u_\nu|^2 \, dx$$

$$\leq |(u_\nu - u, S(u)(u_\nu - u))| + \nu \int_\Omega (\nabla u_\nu : \nabla u) \, dx$$

$$- \nu \int_{\partial\Omega} (\partial_{\vec{n}} u_\nu)_\tau \cdot u \, d\sigma.$$

$$(1.8)$$

The analysis of the term

$$-\nu \int_{\partial\Omega} (\partial_{\vec{n}} u_\nu)_\tau \cdot u \, d\sigma,$$

which appears in the right-hand side of (1.8) is, in this section and in the next one, the cornerstone of this contribution. We observe that $(\partial_{\vec{n}} u_\nu)_\tau$ is the tangential component of the stress at the boundary. It creates a shear flow near the boundary and generates vorticity. In order to see this more clearly notice that since $(u_\nu)_\tau = 0$ on the boundary of Ω we obtain the following equality

$$-(\partial_{\vec{n}} u_\nu)_\tau \cdot u = (\nabla \wedge u_\nu) \cdot (\vec{n} \wedge u). \quad (1.9)$$

Therefore all the considerations concerning the left hand-side of (1.9) do have their counterpart on the right-hand side, i.e. in terms of the trace of the vorticity of u_ν on $\partial\Omega$.

Moreover, from (1.8) it follows the very easy, but essential result.

Proposition 1.2 *Let u be a Lipschitz solution of the Euler equations (1.2) and u_ν the solutions of the Navier–Stokes equations (1.3) with initial data $u_\nu(x, 0) = u(x, 0) = u_0(x)$. Then under the hypothesis*

$$\limsup_{\nu \to 0} \int_0^T -\nu \int_{\partial\Omega} (\partial_{\vec{n}} \cdot u_\nu)_\tau u \, d\sigma \, dt$$

$$= \limsup_{\nu \to 0} \nu \int_0^T \int_{\partial\Omega} -((\partial_{\vec{n}} u_\nu)_\tau \cdot u_\tau)_- \, d\sigma \, dt \le 0 \tag{1.10}$$

any weak limit $\overline{u_\nu}$ coincides with u in $\Omega \times [0, T]$.

In the proposition and throughout the paper we use $(X)_- = \inf(0, X)$.

Proof From (1.8), using the Cauchy–Schwarz and Young inequalities we deduce that

$$|u_\nu - u|^2_{L^2(\Omega)}(t) + \nu \int_0^t \int_\Omega |\nabla u_\nu(x, s)|^2 \, dx \, ds$$

$$\le \nu \int_0^t \int_\Omega |\nabla u|^2 \, dx \, ds$$

$$+ 2|S(u)|_{L^\infty(\Omega\times[0,T])} \int_0^t |(u_\nu - u)(s)|^2_{L^2(\Omega)} \, ds \tag{1.11}$$

$$+ 2 \int_0^t -\nu \int_{\partial\Omega} (\partial_{\vec{n}} u_\nu)_\tau \cdot u \, d\sigma \, ds.$$

Then, under the hypothesis (1.10), we have

$$\limsup_{\nu \to 0} |(u_\nu - u)(t)|^2_{L^2(\Omega)}$$

$$\le |S(u)|_{L^\infty(\Omega\times[0,T])} \int_0^t \limsup_{\nu \to 0} |(u_\nu - u)(s)|^2_{L^2(\Omega)} \, ds, \tag{1.12}$$

which implies, by Gronwall's inequality, that

$$\limsup_{\nu \to 0} |(u_\nu - u)(t)|^2_{L^2(\Omega)} = 0, \quad \text{for all } t \in [0, T],$$

and consequently, the relation

$$|\overline{u_\nu} - u|^2_{L^2(\Omega)}(t) \le \limsup_{\nu \to 0} |(u_\nu - u)(t)|^2_{L^2(\Omega)} \tag{1.13}$$

implies $\overline{u_\nu} = u$ in $\Omega \times [0, T]$. $\qquad \square$

1.2 Kato criterion for convergence to the regular solution.

In a remarkable paper Kato (1984) related the convergence to the smooth solution of the Euler equations to the absence of anomalous energy dissipation in a boundary layer of size ν. At present it turns out that this criterion (this is the object of the Theorem 1.3 below) has several equivalent forms (see Theorem 4.1 in Bardos & Titi (2013) and Constantin et al (2018) for more references). Some of these equivalent forms (in particular the above hypothesis (1.10)) have natural physical interpretations.

Theorem 1.3 *Assume the existence of a Lipschitz solution $u(x,t)$ of the incompressible Euler equations in $\Omega \times]0,T[$. Let $u_\nu(x,t)$ be a Leray–Hopf weak solution of the Navier–Stokes equations (1.3) with no slip boundary condition, that coincides with u at the time $t = 0$. Define the region*

$$\mathcal{U}_\nu = \Omega \cap \{d(x, \partial\Omega) < \nu\}.$$

Then the following facts are equivalent:

$$\lim_{\nu \to 0} \nu \int_0^T \int_{\partial\Omega} ((\partial_{\vec{n}} u_\nu)_\tau \cdot u_\tau)_- \, d\sigma \, dt = 0, \tag{1.14a}$$

$$u_\nu(t) \to u(t) \text{ in } L^2(\Omega) \text{ uniformly in } t \in [0,T], \tag{1.14b}$$

$$u_\nu(t) \to u(t) \text{ weakly in } L^2(\Omega) \text{ for each } t \in [0,T], \tag{1.14c}$$

$$\lim_{\nu \to 0} \nu \int_0^T \int_\Omega |\nabla u_\nu(x,t)|^2 \, dx \, dt = 0, \tag{1.14d}$$

$$\lim_{\nu \to 0} \nu \int_0^T \int_{\mathcal{U}_\nu} |\nabla u_\nu(x,t)|^2 \, dx \, dt = 0, \tag{1.14e}$$

$$\lim_{\nu \to 0} \frac{1}{\nu} \int_0^T \int_{\mathcal{U}_\nu} |(u_\nu(x,t))_\tau|^2 \, dx \, dt = 0, \text{ and} \tag{1.14f}$$

$$\lim_{\nu \to 0} \nu \int_0^T \int_{\partial\Omega} (\frac{\partial u_\nu}{\partial \vec{n}}(\sigma,t))_\tau \cdot w(\sigma,t) \, d\sigma \, dt = 0 \tag{1.14g}$$
for all $w(x,t) \in Lip(\partial\Omega \times [0,T])$ tangent to $\partial\Omega$.

Proof The proof is an updated version (cf. Bardos & Titi, 2013) of the basic result of Kato (1984). First observe that (1.14a) is (with $w = u$) a direct consequence of (1.14g).

The fact that (1.14a) implies (1.14b) was observed in the previous section, while (1.14c) clearly follows from (1.14b).

From (1.14c), for any $0 < t < T$, we deduce

$$\lim_{\nu \to 0} 2\nu \int_0^t \int_\Omega |\nabla u_\nu(x,s)|^2 \, \mathrm{d}x \, \mathrm{d}s$$

$$\leq \int_\Omega |u(x,0)|^2 \, \mathrm{d}x - \liminf_{\nu \to 0} \int_\Omega |u_\nu(x,t)|^2 \, \mathrm{d}x \qquad (1.15)$$

$$\leq \int_\Omega |u(x,0)|^2 \, \mathrm{d}x - \int_\Omega |u(x,t)|^2 \, \mathrm{d}x \leq 0,$$

which gives (1.14d) from which (1.14e) easily follows, as $\mathcal{U}_\nu \subset \Omega$.

Since $u_\nu = 0$ on $\partial\Omega \times]0, T[$ the estimate (1.14f) is deduced from (1.14e) using the Poincaré inequality.

The only non trivial statement is the fact that (1.14f) implies (1.14g) and its proof is inspired by the construction of Kato (1984). We introduce a cut-off function

$$\Theta \in C^\infty(\mathbb{R}), \text{ with } \Theta(0) = 1 \quad \text{and} \quad \Theta(s) = 0 \text{ for } s > 1. \qquad (1.16)$$

Then, with $\nu < \eta_0$, use Proposition 1.1 to extend w to a Lipschitz, divergence-free, tangent to the boundary vector field $\hat{w}_\nu$ according to the formula:

$$\hat{w}_\nu(x,t) = 0, \text{ for } x \notin \mathcal{U}_\nu,$$

$$\hat{w}_\nu(x,t) = \nabla \wedge (\vec{n}(\sigma) \wedge w(\sigma,t)d(x,\partial\Omega)\Theta(\frac{d(x,\partial\Omega)}{\nu})), \qquad (1.17)$$

$$\text{for } x = \sigma(x) - d(x,\partial\Omega)\vec{n}(\sigma(x)) \in \mathcal{U}_\nu.$$

Multiplication of the Navier–Stokes equation satisfied by u_ν and integrating by part gives

$$\nu \int_{\partial\Omega} (\frac{\partial u_\nu}{\partial \vec{n}}(\sigma,t))_\tau w(\sigma,t) \, \mathrm{d}\sigma$$

$$= \nu(\nabla u_\nu, \nabla \hat{w}_\nu)_{L^2(\Omega)} - (u_\nu \otimes u_\nu, \nabla \hat{w}_\nu)_{L^2(\Omega)} + (\partial_t u_\nu, \hat{w}_\nu)_{L^2(\Omega)}. \qquad (1.18)$$

To show that the right-hand side of (1.18) goes to 0 with ν observe that, the only non trivial terms to consider are those that contain the highest power of ν^{-1}.

We have the following estimates, where C denotes any constant which depends on the geometry and on the Jacobian of the transformation defined on $\mathcal{U}_\nu$ by the relation $x = \sigma(x) - d(x, \partial\Omega)\vec{n}(\sigma(x))$.

$$\nu\left|\int_0^T \int_{\mathcal{U}_\nu} |(\nabla u_\nu, \nabla \hat{w}_\nu)|\,dx\,dt\right|$$

$$= -\nu\left|\int_0^T \int_{\mathcal{U}_\nu} u_\nu : \Delta\hat{w}_\nu\,dx\,dt\right|$$

$$\leq \nu C \int_0^T \int_0^\nu \int_{\sigma\in\partial\Omega} |(u_\nu)_\tau(\sigma, s)||w(\sigma)|\frac{s}{\nu^3}|\Theta'''(\frac{s}{\nu})|\,ds\,d\sigma dt$$

$$+ o(\nu)$$

$$(1.19)$$

and

$$\left|\int_0^T \int_{\mathcal{U}_\nu} (u_\nu \otimes u_\nu, \nabla\hat{w}_\nu)_{L^2(\Omega)}\,dt\right|$$

$$\leq \left|\int_0^T \int_{\mathcal{U}^\nu} ((u_\nu)_\tau(u_\nu)_n\partial_n(\hat{w}_\tau)\,dx\,dt\right| + o(\nu)$$

$$\leq C \int_0^T \int_0^\nu \int_{\sigma\in\partial\Omega} |(u_\nu)_\tau(\sigma, s)||(u_\nu)_n(\sigma, s)||w(\sigma, t)|\frac{s}{\nu^2}\Theta''(\frac{s}{\nu})\,ds\,d\sigma\,dt$$

$$+ o(\nu).$$

$$(1.20)$$

Therefore using Cauchy–Schwarz we obtain from (1.19)

$$\left|\nu\int_0^T \int_{\mathcal{U}_\nu} (\nabla u_\nu, \nabla\hat{w}_\nu)\,dx\,dt\right|$$

$$\leq C\frac{1}{\nu^2}\left(\int_0^T \int_0^\nu \int_{\sigma\in\partial\Omega} |(u_\nu)_\tau(\sigma, s)|^2\,ds\,d\sigma\,dt\right)^{\frac{1}{2}}$$

$$\times \left(\int_0^T \int_{\partial\Omega} \int_0^\nu s^2\,ds\,d\sigma\,dt\right)^{\frac{1}{2}}$$

$$\leq C\left(\frac{1}{\nu}\int_0^T \int_0^\nu \int_{\sigma\in\partial\Omega} |(u_\nu)_\tau(\sigma, s)|^2\,ds\,d\sigma\,dt\right)^{\frac{1}{2}}$$

$$(1.21)$$

and similarly for (1.20) we have

$$\left| \int_0^T \int_{\mathcal{U}_\nu} |(u_\nu \otimes u_\nu, \nabla \hat{w}_\nu)| \, dx \, dt \right|$$

$$\leq \left| \int_0^T \int_{\mathcal{U}^\nu} ((u_\nu)_\tau (u_\nu)_n \partial_n (\hat{w}_\tau) \, dx \, dt \right| + o(\nu)$$

$$\leq C \int_0^T \int_0^\nu \int_{\sigma \in \partial\Omega} |(u_\nu)_\tau(\sigma, s)| |(u_\nu)_n(\sigma, s)| |w(\sigma, t)| \frac{s}{\nu^2} \Theta''\left(\frac{s}{\nu}\right) ds \, d\sigma \, dt$$

$$+ o(\nu)$$

$$\leq \frac{C}{\nu} \int_0^T \int_0^\nu \int_{\sigma \in \partial\Omega} |(u_\nu)_\tau(\sigma, s)| |(u_\nu)_n(\sigma, s)| |w(\sigma, t)| \, ds \, d\sigma \, dt + o(\nu)$$

$$\leq C \left(\frac{1}{\nu} \int_0^T \int_0^\nu \int_{\sigma \in \partial\Omega} |(u_\nu)_\tau(\sigma, s)|^2 \, ds \, d\sigma dt \right)^{\frac{1}{2}}$$

$$\times \left(\frac{1}{\nu} \int_0^T \int_0^\nu \int_{\sigma \in \partial\Omega} |(u_\nu)_n(\sigma, s)|^2 \, ds \, d\sigma \, dt \right)^{\frac{1}{2}} + o(\nu).$$

$$(1.22)$$

Moreover, since $u_\nu = 0$ on $\partial\Omega$, with the Poincaré inequality, we have

$$\int_0^T \int_0^\nu \int_{\partial\Omega} |(u_\nu)_n(\sigma, s, t)|^2 \, ds \, d\sigma \, dt$$

$$(1.23)$$

$$\leq \nu^2 \int_0^T \int_0^\nu \int_{\partial\Omega} |(u_\nu)_n|^2 \, ds \, d\sigma \, dt \leq C \|u_0(x)\|_{L^2(\Omega)}^2.$$

Therefore the last term of both (1.21) and (1.22) is uniformly bounded by

$$C \frac{1}{\nu} \int_0^T \int_{\mathcal{U}_\nu} |(u_\nu(x, t))_\tau|^2 \, dx \, dt + o(\nu)$$

and this shows that (1.14f) implies (1.14g), completing the proof. $\quad\square$

1.3 Mathematical and physical interpretation of Theorem 1.3

1.3.1 Recirculation

Since $u_\nu = 0$ on $\partial\Omega$ and u is tangent to the boundary, the fact that

$$\left(\frac{\partial u_\nu}{\partial \vec{n}}(\sigma, t) \right)_\tau u_\tau = ((\nabla \wedge u_\nu) \wedge \vec{n}) \cdot u < 0$$

Laminar regime

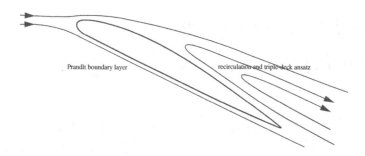

Figure 1.1 Laminar flow with recirculation around an airfoil.

for ν small enough, indicates that somewhere near the boundary the viscous flows u_ν go in the opposite direction to the base flow u that solves the Euler equations, or equivalently that this flow exhibits some backward vorticity. This configuration is known as "recirculation" and does not prevent the fluid from remaining laminar or from having an asymptotic behavior given by the Euler equations, as long as such recirculation is not too big. And this moderate recirculation, shown in Figure 1.1, corresponds to the hypothesis (1.10).

1.3.2 The Prandtl equations and the Stewartson triple-deck ansatz.

As already observed, in the zero-viscosity limit, the boundary condition $(u_\nu)_\tau = 0$ may not persist; hence some type of singularity has to appear near the boundary. However, for linear parabolic problems of the form

$$\partial_t u_\nu - \nu \Delta u_\nu = 0, \qquad u_\nu(x,0) = u_0(x), \qquad u_\nu(x,t)_{|\partial\Omega} = 0 \qquad (1.24)$$

and also, for the linearised Navier–Stokes equations (cf. Ding & Jiang, 2018), the solution converges strongly away from the boundary and near the boundary, in a layer $B_{\sqrt{\nu}} = \{x \in \Omega, d(x,\partial\Omega) < \sqrt{\nu}\}$ of size $\sqrt{\nu}$. It can be described in the laminar regime by a "parabolic scaling", that is, by a smooth function of the form

$$x = x_\tau - d(x, \partial\Omega)\vec{n}(x_\tau) \mapsto U\left(\frac{d(x, \partial\Omega)}{\sqrt{\nu}}, x_\tau\right).$$

In fact, Prandtl (1904) proposed to represent the solution of the Navier–Stokes equation according to the formula

$$u_\nu(x, t) \simeq U_\tau\left(\frac{d(x, \partial\Omega)}{\sqrt{\nu}}, x_\tau, t\right) + \sqrt{\nu}U_{\vec{n}}\left(\frac{d(x, \partial\Omega)}{\sqrt{\nu}}, x_\tau, t\right), \quad (1.25)$$

where the indices τ and $\vec{n}$ refer to tangent and normal components of the fluid velocity $U(x, t)$. To describe the behaviour near the boundary one requires that

$$U_\tau(0, x_\tau, t) + \sqrt{\nu}U_{\vec{n}}(0, x_\tau, t) = 0 \quad \text{and}$$
$$\lim_{y \to \infty} U_\tau\left(\frac{y}{\sqrt{\nu}}, x_\tau, t\right) + \sqrt{\nu}U_{\vec{n}}\left(\frac{y}{\sqrt{\nu}}, x_\tau, t\right) = u_\tau(x_\tau, t). \quad (1.26)$$

Much later, Stewartson (1974) proposed an ansatz that would incorporate a certain amount of recirculation. To do so he considered three layers of fluid near the boundary and hence called this ansatz "the triple deck".

1. In the Upper Deck $\{x \in \Omega |\ \sqrt{\nu} < d(x, \partial\Omega)\}$ the solution is described by the Euler flow.

2 In the Lower Deck $\{x \in \Omega |\ 0 < d(x, \partial\Omega) < \nu^{\frac{5}{8}}\}$ the solution is described by the above Prandtl boundary layer ansatz.

3 In Middle Deck $\{x \in \Omega |\ \nu^{\frac{5}{8}} < d(x, \partial\Omega) < \sqrt{\nu}\}$, which connects the two above layers the following scaling is proposed:

$$u_\nu(x, t) \simeq (\nu^{\frac{1}{8}}U_\tau(\frac{d(x, \partial\Omega)}{\nu^{\frac{5}{8}}}, \frac{x_\tau}{\nu^{\frac{3}{8}}}, t), \nu^{\frac{3}{8}}U_{\vec{n}}(\frac{d(x, \partial\Omega)}{\nu^{\frac{5}{8}}}, \frac{x_\tau}{\nu^{\frac{3}{8}}}, t)). \quad (1.27)$$

Observe that the Prandtl and the triple Deck ansatzs share in common the following property. If they have, for $0 < t < T$, a smooth solution and if this solution gives an accurate description of the genuine behaviour of the solution u_ν of the Navier–Stokes equation then u_ν satisfies any (and of course all) the criteria of Theorem 1.3.

However, to the best of our knowledge there are no mathematical results concerning the validity of the triple deck ansatz.

Additionally, results concerning the Prandtl equations have been established near a flat boundary. The problem in general is ill posed. It is well posed for analytic initial data (see Asano, 1991 and Sammartino & Caflisch, 1998) or for Gevrey initial data (see Gerard-Varet & Masmoudi, 2015). However even with such data the solution may blow up in a finite time, see E & Enquist (1997). Eventually one can construct examples where both the Navier–Stokes and Prandtl equations have, with the same initial data, a smooth solution for a finite time. But in this special case the Prandtl equations do not provide a correct approximation of Navier–Stokes equations as $\nu \to 0$.

Since a boundary layer of size ν is much smaller (as $\nu \to 0$) than a boundary layer of size $\sqrt{\nu}$ it follows from the above considerations that if the criteria (1.14e) is not satisfied asymptotic ansatz like the Prandtl equation or the triple deck cannot be uniformly valid for the description of the zero viscosity limit.

1.3.3 Von Karman turbulent Layer

From the above considerations on boundary layers one deduces that turbulence generation in a region of size ν around the boundary is the basic cause for the non convergence to the smooth solution of the Euler equations. Despite the absence of any type of proof I think that it may be useful to compare this issue with the empirical rule for turbulence at the boundary. Since convergence in a strong norm is not expected, a turbulent boundary layer for $\overline{u_\nu}$ should be present, in general, around some part of the boundary.

To the best of my knowledge, the only practical thing available is a description based on experiment, numerical analysis and dimensional analysis: the Prandtl–Von Karman turbulent layer (1932). It provides an ansatz for the tangential component of the velocity $u_\tau(x_{\vec{n}}, x_\tau, t)$ in the layer

$$B_{\text{turbulent}} = \{x \in \Omega, d(x, \Omega) < \nu\} \cap \mathcal{W},$$

with $\mathcal{W}$ denoting a neighborhood of some part of the boundary.

On $\partial\Omega \cap \overline{\mathcal{W}}$ the quantity

$$u^* = \sqrt{\nu \partial_{\vec{n}} u_\tau}, \tag{1.28}$$

which has the dimension of a velocity, is assumed to be of the order of one.

Then in $B_{\text{turbulent}}$ we have

$$u_\tau(x_{\vec{n}}, x_\tau) = u^* U_\tau(s), \quad s = u^* \frac{x_{\vec{n}}}{\nu}, \tag{1.29}$$

with $U_\tau(s)$ an intrinsic function of the quantity s. With phenomenological arguments this function is almost linear for $0 < s < 20$ and given by a Prandtl–Von Karman wall law of the form

$$U_\tau(s) = \kappa \log s + \beta \quad \text{for} \quad 20 < s < 100. \tag{1.30}$$

However, either with (1.28), which implies that

$$\nu \partial_{\vec{n}}(u_\tau)_{|\partial\Omega} \geq \alpha > 0$$

or with (1.29), which implies that

$$\nu \int_{\{x\in\Omega \backslash d(x,\partial\Omega) < \nu^{\frac{1}{2}}\}} |\nabla u_\nu(x,,t)|^2 \, dx \geq \varepsilon > 0 \tag{1.31}$$

we observe that the existence of such a boundary layer is consistent with the fact that $\overline{u_\nu}$ does not converge to the smooth solution u. This is necessary for the appearance of a turbulent wake.

1.3.4 Energy limit and d'Alembert paradox.

Eventually the avatar

$$\lim_{\nu\to 0} \frac{1}{\nu} \int_0^T \int_{\{x\in\Omega | d(x,\partial\Omega) < \nu\}} |(u_\nu(x,t))_\tau|^2 \, dx \, dt = 0 \tag{1.32}$$

of the Kato criteria represents the asymptotic (for $\nu \to 0$) amount energy exerted on the boundary of the vessel (or on the obstacle, such as an airfoil for instance) by the flow. This may contribute to the explanation of the d'Alembert paradox.

In a modern setting such a paradox would concern a fluid defined in a domain $\Omega \subset \mathbb{R}^3$ which is the complement of a bounded obstacle K with a divergence-free velocity u constant at infinity.

$$\lim_{|x|\to\infty} u(x) = (u_\infty, 0, 0). \tag{1.33}$$

If the evolution of the fluid is described by the Euler equations, with initial data $u_0(x)$, satisfying condition (1.33) and $\nabla \wedge u_0 = 0$, then the vector field u would remain irrotational and satisfy the boundary condition

$$u(x,t) \cdot \vec{n} = 0 \quad \text{on} \quad \partial\Omega.$$

Therefore u would be a potential flow $u = \nabla\varphi$. And from the relation

$$(u \cdot \nabla u)_i = \sum_{1\leq j\leq d} u_j \partial_{x_j} u_i = \sum_{1\leq j\leq d} \partial_{x_j}\varphi \partial_{x_j}\partial_{x_i}\varphi = \frac{1}{2}\partial_{x_i}|\nabla\varphi|^2 \tag{1.34}$$

we deduce the equation

$$u \cdot \nabla u - \nabla(\frac{|u|^2}{2}) = 0, \qquad (1.35)$$

which implies that $u(x,t)$ is a stationary solution $u(x,t) = U(x)$ satisfying:

$$\nabla \cdot U = \Delta\varphi = 0 \quad \text{in} \quad \Omega. \qquad (1.36)$$

The force exerted on the obstacle $K = \mathbb{R}^3 \backslash \partial\Omega$ can be computed using Green's formula and we have (see Marchioro & Pulvirenti, 1994 page 54):

$$
\begin{aligned}
F &= \int_{\partial\Omega} p\vec{n} \, d\sigma = \int_{\partial\Omega} (p\vec{n} + (\vec{n} \cdot U)U) \, d\sigma \\
&= \int_{\Omega \cap \{|x| < R\}} (\nabla p + U \cdot \nabla U) \, dx - \int_{|x|=R} (\frac{\vec{x}}{|x|}p + (\frac{\vec{x}}{|x|} \cdot U)U) \, d\sigma \\
&= -\lim_{R \to \infty} \int_{|x|=R} (\frac{\vec{x}}{|x|}p + (\frac{\vec{x}}{|x|} \cdot U)U) \, d\sigma = 0,
\end{aligned}
$$

$$(1.37)$$

where we have taken R large enough so that K is contained in $|x| < R$. This fact leads to the following conclusion.

If such stationary flows are described by the Euler equations birds and planes cannot glide! To resolve the paradox one observes that the flow around the obstacle cannot be a regular solution of the Euler equation but on the other hand may well be the zero-viscosity limit of a flow described by the Navier–Stokes equations with no slip boundary condition. In such case the non-zero limit of the right-hand side of (1.32) would represent the amount of energy needed to sustain the flight. This also indicates that the case where the Kato criterion is satisfied seems to be the exception rather than the general rule.

1.4 Kato's criterion, anomalous energy dissipation, and turbulence

Statistical theory of turbulence is based on a notion of an average $\langle . \rangle$. It may denote the ensemble average, average of different quantities of the same fluid at some point, time average, spatial average, etc. It involves in particular two objects:

- the mean energy dissipation denoted by $\varepsilon(\nu) = \nu\langle|\nabla u_\nu|^2\rangle$, and

- the mean flow velocity increment denoted by $\langle |u_\nu(x+r) - u_\nu(x)| \rangle$.

With the idea that the notion of weak convergence is a deterministic counterpart of the notion of average, as used in the statistical theory of turbulence, among other things to justify the Prandtl–Von Karman law mentioned above it is interesting to compare what is presently known for the relation between anomalous energy dissipation and loss of regularity.

Under the universality assumption about homogeneous isotropic turbulence, these quantities are related by the Kolmogorov 1/3 law:

$$\left\langle \frac{|u_\nu(x+r) - u_\nu(x)|}{|r|^{\frac{1}{3}}} \right\rangle \simeq \varepsilon^{\frac{1}{3}} = \left(\nu \langle |\nabla u_\nu|^2 \rangle \right)^{\frac{1}{3}}. \qquad (1.38)$$

One of the starting observations, object of the Kato criterion, is the fact that if a weak limit $\overline{u_\nu}$ of a sequence of solutions of the Navier–Stokes equation is of constant energy, that is,

$$\forall t \in [0, T] \quad \int_\Omega |\overline{u_\nu}(x, t)|^2 \, \mathrm{d}x = \int_\Omega |\overline{u_\nu}(0, t)|^2 \, \mathrm{d}x, \qquad (1.39)$$

then there is no anomalous energy dissipation, i.e.

$$\lim_{\nu \to 0} \int_0^T \int_\Omega |\nabla u_\nu(x, t)|^2 \, \mathrm{d}x \, \mathrm{d}t = 0. \qquad (1.40)$$

In particular, in passing to the limit, some regularity implies the absence of anomalous energy dissipation. On the other hand, to prove that (1.40) implies the convergence to a smooth solution of the Euler equations Kato uses, in an essential way, an extra hypothesis: the existence of such a smooth solution. To the best of our knowledge the Kato criterion provides the only deterministic (without involving any statistical theory of turbulence) configuration where there is a complete equivalence between regularity and absence of anomalous energy dissipation. In particular, with simple examples (cf. Bardos & Titi, 2010), one can show the existence of solutions $u_\nu(x, t)$ of Navier–Stokes which converge weakly in $L^\infty(0, T; L2(\Omega))$ to a non regular solution $\overline{u_\nu}$ of the Euler equation that conserves energy and hence without anomalous energy dissipation.

Then, from the Kolmogorov relation (1.38), one may infer that solutions of the Euler equations which belong to $L^3((0, T); C^{0,\alpha}(\Omega))$ with $\alpha > \frac{1}{3}$ are of constant total energy. This is the so called Onsager conjecture. Mathematical proofs of this conjecture were given by Eyink (1994) and by Constantin, E, & Titi (1994).

The pertinence of such a threshold was confirmed by the recent conclusion by Buckmaster et al. (2018a) of the program initiated by C. De Lellis and L. Székelyhidi about convex integration and wild solutions. In Buckmaster et al. (2018a) the following fact is proved: given any $\beta < 1/3$, a time interval $[0, T]$ and any smooth energy profile $e : [0, T] \mapsto (0, \infty)$, there exists a weak solution $v \in C^\beta([0, T] \times \mathbb{T}^3)$ of the three-dimensional Euler equations with

$$e(t) = \int_{\mathbb{T}^3} |v(x, t)|^2 \, \mathrm{d}x.$$

Extension of some of these results to domains with boundary have also been recently obtained. The conservation of energy for solutions in $L^3((0, T); W^{\alpha, 3}(\Omega)$ with $\alpha > \frac{1}{3}$ has been proven by Robinson, Rodrigo, & Skipper (2018) when $\Omega = \mathbb{T}^2 \times \mathbb{R}^+$ using a symmetry argument. The case of a general domain has been treated by Bardos & Titi (2018) and by Bardos, Titi, & Wiedemann (2018) where first one proves the validity of the local energy conservation:

$$\partial_t \frac{|u|^2}{2} + \nabla_x \cdot \left(\left(\frac{|u|^2}{2} + p \right) u \right) = 0 \quad \text{in } \mathcal{D}'(0, T) \times \Omega \qquad (1.41)$$

under a "local" $C^{0,\alpha}$ (with $\alpha > \frac{1}{3}$) hypothesis, using an approach inspired by a paper of Duchon & Robert (2000) and then extending the property up to the boundary.

In contrast to what is done in Robinson et al. (2018) for the symmetric case, some weak hypotheses on the pressure are requested. This can be related to the following facts:

1 In the absence of boundary ($\Omega = \mathbb{R}^3$ or $\Omega = \mathbb{T}^3$) the presence of a smooth solution u means that any admissible weak solution $\tilde{u}$ that is a solution in the sense of distributions and that satisfies the relation

$$\forall 0 < t < T \int_\Omega \frac{|\tilde{u}(x, t)|^2}{2} \, \mathrm{d}x \leq \int_\Omega \frac{|u(x, 0)|^2}{2} \, \mathrm{d}x \quad \text{with } \tilde{u}(x, 0) = u(x, 0)$$

$$(1.42)$$

coincides with $u(x, t)$ for $t \in (0, T)$ as proven in De Lellis & Székelyhidi (2010).

2 However in the presence of boundary effects the above hypothesis are not enough to show the coincidence of $\tilde{u}$ and u and some extra hypothesis on $\tilde{u}$ and the corresponding pressure $\tilde{p}$ are compulsory (cf. Bardos, Székelyhidi, & Wieidemann, 2014). With these "natural hypothesis" one obtains (cf. Theorem 5.1. in Bardos et al., 2014) sufficient conditions for the convergence to a smooth solution of the Euler equation and for the absence of anomalous energy dissipation. It should be emphasized that the above conditions are compatible with the appearance

of a parabolic boundary layer in the tangent component of the velocity in the neighborhood of the boundary, as described by the Prandtl ansatz.

As a final remark, observe that the above results concerning the energy conservation and the zero viscosity limit can be extended (under some convenient hypothesis and basically before the formation of shocks) to the solutions of the "compressible" Navier–Stokes and Euler equations, see for instance Carrillo, Feireisl, & Gwiazda (2017), Feireisl et al. (2107), and Bardos & Toan (2016).

References

Asano, A. (1991) Zero-viscosity limit of the incompressible Navier–Stokes equations: *Conference at the IVth Workshop on Mathematical Aspects of Fluid and Plasma dynamics*, Kyoto.

Bardos, C., Székelyhidi, L., Jr., & Wieidemann, E. (2014) On the absence of uniqueness for the Euler equations: the effect of the boundary. (Russian) *Uspekhi Mat. Nauk* **69**, no. 2(416), 3–22; translation in *Russian Math. Surveys* **69**, no. 2, 189–207.

Bardos, C. & Titi, E. (2010) Loss of smoothness and energy conserving rough weak solutions for the 3d Euler equations. *Discrete Contin. Dyn. Syst. Ser. S* **3** no. 2, 185–197.

Bardos, C. & Titi, E. (2013) Mathematics and turbulence: where do we stand? *Journal of Turbulence* **14**, no. 3, 42–76.

Bardos, C. & Titi, E. (2018) Onsager's Conjecture for the Incompressible Euler Equations in Bounded Domains. *Arch. Rational Mech. Anal.*, **228**, no. 1, 191–207.

Bardos C., Titi, E., & Wiedemann, E. (2012) The vanishing viscosity as a selection principle for the Euler equations: the case of 3D shear flow.*C. R. Math. Acad. Sci. Paris* **350**, no. 15-16, 757–760.

Bardos C., Titi, E., & Wiedemann, E. (2018) Onsager's Conjecture with Physical Boundaries and an Application to the Viscosity Limit. *Submitted*, arXiv:1803.04939.

Bardos, C. & Toan, N. (2016) Remarks on the inviscid limit for the compressible flows *Recent Advances in Partial Differential Equations and Applications* Contemporary Mathematics **666** (Edited by Radulescu, D., Sequeira, A., & Solonnikov, V.A.) 66–76.

Buckmaster, T., De Lellis. C, & Székelyhidi Jr, L. (2016) Dissipative Euler flows with Onsager-critical spatial regularity. *Comm. Pure and Applied Math.* **69**, no. 9, 1613–1670.

Buckmaster, T., De Lellis, C., Székelyhidi Jr, L., & Vicol, V. (2018) Onssger's conjecture for admissible weak solutions. Preprint arXiv:1701.08678.

Carrillo, J., Feireisl, E. Gwiazda, P., & Świerczewska-Gwiazda, A. (2017) A Weak solutions for Euler systems with non-local interactions. *J. Lond. Math. Soc* (2) **95**, no. 3, 705–724.

Cheskidov, A., Constantin, P., Friedlander, S., & Shvydkoy, R. (2008) Energy conservation and Onsager's conjecture for the Euler equations. *Nonlinearity* **21**, no. 6, 1233–1252.

Constantin, P. (2005) Euler equations, Navier–Stokes equations and turbulenc, in *Mathematical Foundation of Turbulent Viscous Flows*. Lectures given at the C.I.M.E. Summer School, Martina Franca, Italy. Editors M. Cannone & T. Miyakawa, Springer Lecture Notes in Mathematics **1871**, 1–43.

Constantin, P., E, W., & Titi, E. (1994) Onsager's Conjecture on the Energy Conservation for Solutions of Euler's equation. *Comm. Math. Phys.* **165**, 207–209.

Constantin, P., Elgindi, T., Ignatova, M., & Vicol, V. (2018) Remarks on the inviscid limit for the Navier–Stokes equations for uniformly bounded velocity fields. Preprint arXiv:1512.05674.

Constantin, P., Kukavica, I., & Vicol, V. (2015) On the inviscid limit of the Navier–Stokes equations. *Proc. Amer. Math. Soc.* **143**, no. 7, 3075–3090.

Ding, Y. & Jiang, N. (2018) Zero viscosity and thermal diffusivity limit of the linearized compressible Navier–Stokes-Fourier equations in the half plane. Preprint arXiv:1402.1390.

De Lellis, C. & Székelyhidi, Jr., L. (2009) The Euler equations as a differential inclusion. *Ann. of Math.* *(2)* **170**, no. 3, 1417–1436.

De Lellis, C., & Székelyhidi, Jr., L. (2010) On admissibility criteria for weak solutions of the Euler equations. *Arch. Ration. Mech. Anal.* **195** no. 1, 225–260.

Duchon, J. & Robert, R. (2000) Inertial energy dissipation for weak solutions of incompressible Euler and Navier–Stokes equations. *Nonlinearity* **13**, 249–255.

E, W. & Engquist, B. (1997) Blowup of solutions of the unsteady Prandtl's equation. *Comm. Pure Appl. Math.* **50** , no. 12, 1287–1293.

Eyink, G.L. (1994) Energy dissipation without viscosity in ideal hydrodynamics, I. Fourier analysis and local energy transfer. *Phys. D* **78**, 222–240.

Feireisl, E., Gwiazda, P., Świerczewska-Gwiazda, A., & Wiedemann, E. (2017) Regularity and energy conservation for the compressible Euler equations, *Arch. Ration. Mech. Anal.* **223**, 1375–139

Frisch, U. (1995) *Turbulence*. Cambridge University Press, Cambridge.

Gerard-Varet, D. & Masmoudi, N. (2015) Well-posedness for the Prandtl system without analyticity or monotonicity. *Ann. Sci. Ec. Norm. Supér.* **(4)** 48, no. 6, 1273–1325

Kato, T, (1972) Nonstationary flows of viscous and ideal fluids in $\mathbb{R}^3$. *J. Functional Analysis* **9**, 296–305.

Kato, T. (1984) Remarks on zero viscosity limit for nonstationary Navier–Stokes flows. with boundary, in *Seminar on Nonlinear Partial Differential Equations*, (Berkeley, Calif., 1983), Math. Sci. Res. Inst. Publ., **2**, Springer, New York, 85–98.

Kukavica, I. & Vicol, V. (2013) On the local existence of analytic solutions to the Prandtl boundary layer equations. *Communications in Mathematical Sciences* **11**, no. 1, 269–292.

Landau, L.D. & Lifshitz, E.M. (1959) *Fluid Mechanics*. Pergamon Press, New York.

Marchioro, C. & Pulvirenti, M. (1994) *Mathematical Theory of Incompressible Nonviscous Fluids*. Applied Mathematical Sciences **96**, Springer–Verlag, New York.

Mohammadi, B., Pironneau , O. (1994) *Analysis of the K-Epsilon Turbulence Model*. Wiley–Masson, Paris.

Onsager, L. (1949) Statistical Hydrodynamics. *Nuovo Cimento* **6** 2, 279–287.

Prandtl, L. (1904) Uber flussigkeits-bewegung bei sehr kleiner reibung, in Actes du 3me Congrès international des Mathematiciens, Heidelberg. Teubner, Leipzig, 484–491.

Robinson, J.C., Rodrigo, J.L., & Skipper J.W.D. (2018) Energy conservation in the 3D Euler equations on $T^2 \times \mathbb{R}^+$, in *Partial Differential Equations in Fluid Mechanics*. Eds. C.L. Fefferman, J.C. Robinson, & J.L. Rodrigo. Cambridge University Press.

Sammartino, M. & Caflisch, R. (1998) Zero viscosity limit for analytic solutions, of the Navier–Stokes equation on a half-space. I. Existence for Euler and Prandtl equations. *Comm. Math. Phys.* **192**, no. 2, 433–461.

Scheffer, V. (1993) An inviscid flow with compact support in space-time. *J. of Geom. Anal.* **3** , 343–401.

Shnirelman, A. (1997) On the nonuniqueness of weak solutions of the Euler equations. *Comm. Pure Appl. Math.* **50**, 1261–1286.

Stewartson, K. (1974) Multistructured boundary layer on flat plates and related bodies. *Advances Appl. Mech.* **14**, no. 1, 45–239.

2

Time-periodic flow of a viscous liquid past a body

Giovanni P. Galdi

Department of Mechanical Engineering and Materials Science,
University of Pittsburgh, Pittsburgh, PA 15261. USA.
galdi@pitt.edu

Mads Kyed

Fachbereich Mathematik, Technische Universität Darmstadt,
Schlossgartenstr. 7, 64289 Darmstadt. Germany.
kyed@mathematik.tu-darmstadt.de

Abstract

We investigate the existence, uniqueness and regularity of time-periodic solutions to the Navier–Stokes equations governing the flow of a viscous liquid past a three-dimensional body moving with a time-periodic translational velocity. The net motion of the body over a full time-period is assumed to be non-zero. In this case, the appropriate linearization is the time-periodic Oseen system in a three-dimensional exterior domain. A priori L^q estimates are established for this linearization. Based on these "maximal regularity" estimates, the existence and uniqueness of smooth solutions to the fully nonlinear Navier–Stokes problem is obtained by the contraction mapping principle.

2.1 Introduction

Consider a three-dimensional body moving with a prescribed time-periodic translational velocity $v_b(t)$ in a viscous liquid whose motion is governed by the Navier–Stokes equations. If the body occupies a bounded, simply connected domain $\mathcal{B} \subset \mathbb{R}^3$, the liquid flowing past it occupies the corresponding exterior domain $\Omega := \mathbb{R}^3 \setminus \mathcal{B}$. The motion of the liquid can then be described, in a coordinate system attached to the body, by

Published as part of *Partial Differential Equations in Fluid Mechanics*, edited by C.L. Fefferman, J.C. Robinson, & J.L. Rodrigo. © Cambridge University Press 2018.

the following system of equations:

$$
\begin{cases}
\partial_t u + u \cdot \nabla u = \nu \Delta u - \nabla \mathfrak{p} + f & \text{in } \mathbb{R} \times \Omega, \\
\operatorname{div} u = 0 & \text{in } \mathbb{R} \times \Omega, \\
u = u_* & \text{on } \mathbb{R} \times \partial\Omega, \\
\lim_{|x| \to \infty} u(t, x) = -v_b(t).
\end{cases}
\tag{2.1}
$$

Here, $u \colon \mathbb{R} \times \Omega \to \mathbb{R}^3$ denotes the Eulerian velocity field of the liquid and $\mathfrak{p} \colon \mathbb{R} \times \Omega \to \mathbb{R}$ the corresponding pressure field. As is natural for time-periodic problems, the time axis is taken to be the whole of $\mathbb{R}$, and so $(t, x) \in \mathbb{R} \times \Omega$ denotes the time variable t and spatial variable x of the system, respectively. A body force $f \colon \mathbb{R} \times \Omega \to \mathbb{R}^3$ and velocity distribution $u_* \colon \mathbb{R} \times \partial\Omega \to \mathbb{R}^3$ of the liquid on the surface of the body have been included. The constant coefficient of kinematic viscosity of the liquid is denoted by ν. An investigation of time-periodic solutions, that is, solutions $(u, \mathfrak{p})$ satisfying for some fixed $\mathcal{T} > 0$

$$
u(t + \mathcal{T}, x) = u(t, x), \quad \mathfrak{p}(t + \mathcal{T}, x) = \mathfrak{p}(t, x),
$$

corresponding to time-periodic data of the same period,

$$
v_b(t + \mathcal{T}) = v_b(t), \quad f(t + \mathcal{T}, x) = f(t, x), \quad u_*(t + \mathcal{T}, x) = u_*(t, x),
$$

will be carried out.

We assume the net motion of the body over a full time-period is non-zero, that is,

$$
\int_0^{\mathcal{T}} v_b(t) \, dt \neq 0.
\tag{2.2}
$$

We shall not treat the case of a vanishing net motion. The distinction between the two cases is justified by the physics of the problem. In the former case, the body performs a nonzero translatory motion, which induces a wake in the region behind it. In the latter case, the motion of the body would be purely oscillatory without a wake. The different properties of the solutions in the two cases also influence the mathematical analysis of the problem. If the net motion of the body has a nonzero translatory component, the appropriate linearization of (2.1) is a time-periodic Oseen system. If the net motion over a period is zero, the linearization is a time-periodic Stokes system. The analysis performed in this paper is based on suitable L^q estimates for solutions to the time-periodic Oseen system. Similar estimates do not hold for the corresponding Stokes system, in which case a different approach is needed. Without loss of generality, we shall assume that the translational motion of the body is directed along the e_1 axis:

$$v_b(t) = u_\infty(t)e_1, \quad u_\infty(t) \in \mathbb{R}. \tag{2.3}$$

Our main result is contained in Theorem 2.7. There we establish existence and uniqueness of a solution to (2.1) for data f, v_b and u_* sufficiently restricted in "size". The solution is strong both in the sense of local regularity and global summability. The proof is based on the contraction mapping principle and suitable L^q estimates for solutions to a linearization of (2.1). More specifically, we linearize (2.1) around v_b and obtain, due to (2.2), a time-periodic Oseen system. In Theorem 2.2 and Corollary 2.6 we identify a time-periodic Sobolev-type space that is mapped homeomorphically onto a time-periodic L^p space by the Oseen operator ("maximal regularity" spaces). We then employ embedding properties of the Sobolev-type space to show existence of a solution to the fully nonlinear problem (2.1) by a fixed-point argument. A similar result was obtained for a two-dimensional exterior domain in Galdi (2013b).

We wish now to give a general overview of the main literature related to the topic addressed in this paper. The study of time-periodic solutions to the Navier–Stokes equations was originally suggested in Serrin (1959). However, the first rigorous and complete investigations of the classical time-periodic Navier–Stokes problem in bounded domains are due to Prodi (1960), Yudovich (1960) and Prouse (1963). Further properties and extensions to other types of domains and related issues have been studied by a number of authors over the years: Kaniel & Shinbrot (1967), Takeshita (1970), Morimoto (1972), Miyakawa & Teramoto (1982), Teramoto (1983), Maremonti (1991a), Maremonti (1991b), Maremonti & Padula (1996), Kozono & Nakao (1996), Yamazaki (2000), Galdi & Sohr (2004), Galdi & Silvestre (2006, 2009), Taniuchi (2009), Van Baalen & Wittwer (2011), Silvestre (2012), Galdi (2013a), Kyed (2014a), Kyed (2014b), Kyed (2016), Nguyen (2014), and Geissert, Hieber, & Nguyen (2016).

Among all the above articles, we would like to point out, in particular, the contributions of Galdi & Silvestre (2006), Galdi & Silvestre (2009), Nguyen (2014), Geissert et al. (2016), Kyed (2014b), Kyed (2014a), Galdi (2013a), and Kyed (2016), where the authors provide results for the same type of flow past a body described by (2.1). However, these results either lack of the uniqueness property (Galdi & Silvestre, 2006, 2009) or can ensure very little regularity on the solutions (Nguyen, 2014; Geissert et al., 2016), or else are restricted to flow in the whole space (Kyed, 2014a,b; Galdi, 2013a; Kyed, 2016). Therefore, it is worth emphasizing that the findings of this paper resolve all these issues. As mentioned earlier on, the crucial tool in achieving this goal is contained in Corollary 2.6 ensuring maximal regularity properties for the linearized (Oseen) operator, which is also of independent interest.

2.2 Notation

Constants in capital letters in the proofs and theorems are global, while constants in small letters are local to the proof in which they appear. The notation $C(\xi)$ is used to emphasize the dependence of a constant on a parameter ξ.

The notation B_R is used to denote balls in $\mathbb{R}^3$ centered at 0 with radius $R > 0$.

The symbol Ω denotes an exterior domain of $\mathbb{R}^3$, that is, an open connected set that is the complement of the closure of a simply connected bounded domain $\mathcal{B} \subset \mathbb{R}^3$. Without loss of generality, it is assumed that $0 \in \mathcal{B}$. Two constants $R_0 > R_* > 0$ with $\mathcal{B} \subset\subset B_{R_*}$ remain fixed. Moreover, we employ the notation $\Omega_R := \Omega \cap B_R$, $\Omega_{R_1,R_2} := \Omega \cap B_{R_2} \setminus B_{R_1}$ and $\Omega^R := \Omega \setminus B_R$.

Points in $\mathbb{R} \times \mathbb{R}^3$ are denoted by (t, x). Throughout, t is referred to as the time and x as the spatial variable. For a sufficiently regular function $u : \mathbb{R} \times \mathbb{R}^3 \to \mathbb{R}$, $\partial_i u$ denotes the spatial derivative $\partial_{x_i} u$.

2.3 Preliminaries

A framework will be employed based on the function space of smooth and compactly supported $\mathcal{T}$-time-periodic functions:

$$C_{0,\mathrm{per}}^\infty(\mathbb{R} \times \overline{\Omega}) := \big\{ f \in C^\infty(\mathbb{R} \times \Omega) : \; f(t + \mathcal{T}, x) = f(t, x),$$
$$f \in C_0^\infty([0, \mathcal{T}] \times \overline{\Omega}) \big\}.$$

For simplicity, the interval $(0, \mathcal{T})$ of one time period is denoted by $\mathbb{T}$. An L^q norm on the time-space domain $\mathbb{T} \times \Omega$ is given by

$$\|f\|_q := \|f\|_{q, \mathbb{T} \times \Omega} := \left(\frac{1}{\mathcal{T}} \int_0^{\mathcal{T}} \int_\Omega |f(t, x)|^q \, dx \, dt \right)^{\frac{1}{q}}, \quad q \in [1, \infty).$$

Lebesgue spaces of time-periodic functions are defined by

$$L^q_{\mathrm{per}}(\mathbb{R} \times \Omega) := \overline{C^\infty_{0,\mathrm{per}}(\mathbb{R} \times \overline{\Omega})}^{\|\cdot\|_q}.$$

It is easy to see that the elements of $L^q_{\mathrm{per}}(\mathbb{R} \times \Omega)$ coincide with the $\mathcal{T}$-time-periodic extension of functions in $L^q((0,\mathcal{T}) \times \Omega)$. Observe that the Lebesgue space $L^q(\Omega)$ can be considered as the subspace of functions in $L^q_{\mathrm{per}}(\mathbb{R} \times \Omega)$ that are time-independent. For such functions, the $L^q(\Omega)$ norm coincides with the norm $\|f\|_q$ introduced above.

Sobolev spaces of $\mathcal{T}$-time-periodic functions are also introduced as completions of $C^\infty_{0,\mathrm{per}}(\mathbb{R} \times \overline{\Omega})$ in appropriate norms:

$$W^{1,2,q}_{\mathrm{per}}(\mathbb{R} \times \Omega) := \overline{C^\infty_{0,\mathrm{per}}(\mathbb{R} \times \overline{\Omega})}^{\|\cdot\|_{1,2,q}},$$

$$\|u\|_{1,2,q} := \left(\sum_{|\beta| \leq 1} \|\partial^\beta_t u\|^q_q + \sum_{0 < |\alpha| \leq 2} \|\partial^\alpha_x u\|^q_q \right)^{\frac{1}{q}}.$$

These Sobolev spaces are clearly subspaces of the classical anisotropic Sobolev spaces $W^{1,2,q}((0,\mathcal{T}) \times \Omega)$. Homogeneous Sobolev spaces of time-periodic functions are defined analogously:

$$D^{1,q}_{\mathrm{per}}(\mathbb{R} \times \Omega) := \overline{C^\infty_{0,\mathrm{per}}(\mathbb{R} \times \overline{\Omega})}^{\langle \cdot \rangle_{1,q}},$$

$$\langle \mathfrak{p} \rangle_{1,q} := \|\nabla \mathfrak{p}\|_q + \frac{1}{\mathcal{T}} \int_0^{\mathcal{T}} \left| \int_{\Omega_{R_0}} \mathfrak{p}(t,x) \, \mathrm{d}x \right| \, \mathrm{d}t.$$

The space $D^{1,q}_{\mathrm{per}}(\mathbb{R} \times \Omega)$ can be identified with $L^q((0,\mathcal{T}); D^{1,q}(\Omega))$, where $D^{1,q}(\Omega)$ is the classical homogeneous Sobolev space.

In a similar manner, Lebesgue and Sobolev spaces of time-periodic vector-valued functions are defined for any Banach space X respectively as

$$L^q_{\mathrm{per}}(\mathbb{R}; X) := \overline{C^\infty_{\mathrm{per}}(\mathbb{R}; X)}^{\|\cdot\|_{L^q((0,\mathcal{T});X)}},$$

$$W^{m,q}_{\mathrm{per}}(\mathbb{R}; X) := \overline{C^\infty_{\mathrm{per}}(\mathbb{R}; X)}^{\|\cdot\|_{W^{m,q}((0,\mathcal{T});X)}}.$$

One readily verifies that $L^q_{\mathrm{per}}(\mathbb{R}; X)$ coincides with the $\mathcal{T}$-periodic extensions of functions in the Lebesgue space $L^q_{\mathrm{per}}((0,\mathcal{T}); X)$. One may further verify for sufficiently regular domains, say Ω of class C^1, that

$$W^{1,2,q}_{\mathrm{per}}(\mathbb{R} \times \Omega) = W^{1,q}_{\mathrm{per}}(\mathbb{R}; L^q(\Omega)) \cap L^q_{\mathrm{per}}(\mathbb{R}; W^{2,q}(\Omega))$$

$$= \{u \in L^q_{\mathrm{per}}(\mathbb{R} \times \Omega) : \|u\|_{1,2,q} < \infty\}.$$

Sufficiently regular $\mathcal{T}$-time-periodic functions $u \colon \mathbb{R} \times \mathbb{R}^n \to \mathbb{R}$ can be decomposed into what we term a *steady-state* part $\mathcal{P}u\colon \mathbb{R}\times\mathbb{R}^n\to\mathbb{R}$ and *oscillatory* part $\mathcal{P}_\perp u\colon \mathbb{R}\times\mathbb{R}^n\to\mathbb{R}$ by

$$\mathcal{P}u(t,x) := \frac{1}{\mathcal{T}}\int_0^{\mathcal{T}} u(s,x)\,\mathrm{d}s,$$

$$\mathcal{P}_\perp u(t,x) := u(t,x) - \mathcal{P}u(t,x)$$

whenever these expressions are well defined. Note that the steady-state part of a $\mathcal{T}$-time-periodic function u is time independent, and the oscillatory part $\mathcal{P}_\perp u$ has vanishing time-average over the period. Also note that $\mathcal{P}$ and $\mathcal{P}_\perp$ are complementary projections, that is, $\mathcal{P}^2 = \mathcal{P}$ and $\mathcal{P}_\perp = \mathrm{Id} - \mathcal{P}$. Based on these projections, the following subspaces are defined:

$$C^\infty_{0,\mathrm{per},\perp}(\mathbb{R}\times\Omega) := \{f\in C^\infty_{0,\mathrm{per}}(\mathbb{R}\times\Omega): \ \mathcal{P}f = 0\},$$
$$L^q_{\mathrm{per},\perp}(\mathbb{R}\times\Omega) := \{f\in L^q_{\mathrm{per}}(\mathbb{R}\times\Omega): \ \mathcal{P}f = 0\},$$
$$W^{1,2,q}_{\mathrm{per},\perp}(\mathbb{R}\times\Omega) := \{u\in W^{1,2,q}_{\mathrm{per}}(\mathbb{R}\times\Omega): \ \mathcal{P}u = 0\}.$$

Intersections of these spaces are denoted by

$$L^{q,r}_{\mathrm{per},\perp}(\mathbb{R}\times\Omega) := L^q_{\mathrm{per},\perp}(\mathbb{R}\times\Omega)\cap L^r_{\mathrm{per},\perp}(\mathbb{R}\times\Omega),$$
$$W^{1,2,q,r}_{\mathrm{per},\perp}(\mathbb{R}\times\Omega) := W^{1,2,q}_{\mathrm{per},\perp}(\mathbb{R}\times\Omega)\cap W^{1,2,r}_{\mathrm{per},\perp}(\mathbb{R}\times\Omega)$$

and equipped with the canonical norms. Similar subspaces of homogeneous Sobolev spaces are defined by

$$D^{1,q}_{\mathrm{per},\perp}(\mathbb{R}\times\Omega) := \{\mathfrak{p}\in D^{1,q}_{\mathrm{per}}(\mathbb{R}\times\Omega): \ \mathcal{P}\mathfrak{p} = 0\},$$
$$D^{1,q,r}_{\mathrm{per},\perp}(\mathbb{R}\times\Omega) := D^{1,q}_{\mathrm{per},\perp}(\mathbb{R}\times\Omega)\cap D^{1,r}_{\mathrm{per},\perp}(\mathbb{R}\times\Omega),$$
$$D^{1,q}_{\mathrm{per},\perp,R_0}(\mathbb{R}\times\Omega) := \left\{\mathfrak{p}\in D^{1,q}_{\mathrm{per}}(\mathbb{R}\times\Omega): \ \mathcal{P}u = 0,\ \int_{\Omega_{R_0}}\mathfrak{p}(t,x)\,\mathrm{d}x = 0\right\},$$
$$D^{1,q,r}_{\mathrm{per},\perp,R_0}(\mathbb{R}\times\Omega) := D^{1,q}_{\mathrm{per},\perp,R_0}(\mathbb{R}\times\Omega)\cap D^{1,r}_{\mathrm{per},\perp,R_0}(\mathbb{R}\times\Omega).$$

All spaces above are clearly Banach spaces.

Finally, we introduce for $\lambda > 0$ and $q\in(1,2)$ the Sobolev-type space

$$X^q_\lambda(\Omega) := \{v\in L^q_{\mathrm{loc}}(\Omega): \ \|v\|_{X^q_\lambda} < \infty\},$$

$$\|v\|_{X^q_\lambda} := \lambda^{\frac{1}{2}}\|v\|_{\frac{2q}{2-q}} + \lambda^{\frac{1}{4}}\|\nabla v\|_{\frac{4q}{4-q}} + \lambda\|\partial_1 v\|_q + \|\nabla^2 v\|_q,$$

which is used to characterize the velocity field of a steady-state solution to the Oseen system in three-dimensional exterior domains. An appropriate function space for the corresponding pressure term is the homogeneous Sobolev space

$$D_{R_0}^{1,q}(\Omega) := \left\{ v \in D^{1,q}(\Omega) : \int_{B_{R_0}} p \, dx = 0 \right\}$$

equipped with the norm $\langle \cdot \rangle_{1,q} := \|\nabla \cdot\|_q$. As mentioned above, $D^{1,q}(\Omega)$ denotes the classical homogeneous Sobolev space.

2.4 An Embedding Theorem

Embedding properties of the Sobolev spaces of time-periodic functions defined in the previous section shall be established. As such properties may be of use in applications beyond the scope of this article, we consider in this section an exterior domain $\Omega \subset \mathbb{R}^n$ of arbitrary dimension $n \geq 2$.

Theorem 2.1 *Let $\Omega \subset \mathbb{R}^n$ ($n \geq 2$) be an exterior domain of class C^1 and $q \in (1, \infty)$. Assume that $\alpha \in [0, 2]$ and $p_0, r_0 \in [q, \infty]$ satisfy*

$$\begin{cases} r_0 \leq \dfrac{2q}{2 - \alpha q} & \text{if } \alpha q < 2, \\ r_0 < \infty & \text{if } \alpha q = 2, \\ r_0 \leq \infty & \text{if } \alpha q > 2, \end{cases} \qquad \begin{cases} p_0 \leq \dfrac{nq}{n - (2 - \alpha)q} & \text{if } (2 - \alpha)q < n, \\ p_0 < \infty & \text{if } (2 - \alpha)q = n, \\ p_0 \leq \infty & \text{if } (2 - \alpha)q > n, \end{cases}$$
$$\tag{2.4}$$

and that $\beta \in [0, 1]$ and $p_1, r_1 \in [q, \infty]$ satisfy

$$\begin{cases} r_1 \leq \dfrac{2q}{2 - \beta q} & \text{if } \beta q < 2, \\ r_1 < \infty & \text{if } \beta q = 2, \\ r_1 \leq \infty & \text{if } \beta q > 2, \end{cases} \qquad \begin{cases} p_1 \leq \dfrac{nq}{n - (1 - \beta)q} & \text{if } (1 - \beta)q < n, \\ p_1 < \infty & \text{if } (1 - \beta)q = n, \\ p_1 \leq \infty & \text{if } (1 - \beta)q > n. \end{cases}$$
$$\tag{2.5}$$

Then for all $u \in W_{\text{per}}^{1,2,q}(\mathbb{R} \times \Omega)$

$$\|u\|_{L_{\text{per}}^{r_0}(\mathbb{R}; L^{p_0}(\Omega))} + \|\nabla u\|_{L_{\text{per}}^{r_1}(\mathbb{R}; L^{p_1}(\Omega))} \leq C_1 \|u\|_{1,2,q}, \tag{2.6}$$

with $C_1 = C_1(\mathcal{T}, n, \Omega, r_0, p_0, r_1, p_1)$.

Proof The regularity of Ω suffices to ensure the existence of a continuous extension operator $E \colon W_{\text{per}}^{1,2,q}(\mathbb{R} \times \Omega) \to W_{\text{per}}^{1,2,q}(\mathbb{R} \times \mathbb{R}^n)$ as in the case of classical Sobolev spaces. Consequently, it suffices to show (2.6) for functions $u \in W_{\text{per}}^{1,2,q}(\mathbb{R} \times \mathbb{R}^n)$. For this purpose, we identify $W_{\text{per}}^{1,2,q}(\mathbb{R} \times \mathbb{R}^n)$ with a Sobolev space of functions defined on the group

$G := (\mathbb{R}/\mathcal{T}\mathbb{Z}) \times \mathbb{R}^n$. Endowed with the quotient topology induced by the quotient map $\pi \colon \mathbb{R} \times \mathbb{R}^n \to (\mathbb{R}/\mathcal{T}\mathbb{Z}) \times \mathbb{R}^n$, G becomes a locally compact abelian group. A Haar measure is given by the product of the Lebesgue measure on $\mathbb{R}^n$ and the normalized Lebesgue measure on the interval $[0, \mathcal{T}) \simeq \mathbb{R}/\mathcal{T}\mathbb{Z}$. Also via the quotient map, the space of smooth functions on G is defined as

$$C^\infty(G) := \{f \colon G \to \mathbb{R} \colon f \circ \pi \in C^\infty(\mathbb{R} \times \mathbb{R}^n)\}. \tag{2.7}$$

Sobolev spaces $W^{1,2,q}(G)$ can then be defined as the closure of $C_0^\infty(G)$, the subspace of compactly supported functions in $C^\infty(G)$, in the norms $\|\cdot\|_{1,2,q}$. It is easy to verify that $W^{1,2,q}(G)$ and $W_{\mathrm{per}}^{1,2,q}(\mathbb{R} \times \mathbb{R}^n)$ are homeomorphic. Further details can be found in Kyed (2012, 2014b). The availability of the Fourier transform $\mathscr{F}_G$ is an immediate advantage in the group setting. Denoting points in the dual group $\widehat{G} := \mathbb{Z} \times \mathbb{R}^n$ by (k, ξ), we obtain the representation

$$\partial_j \mathcal{P}_\perp u = \mathscr{F}_G^{-1}\left[\frac{\mathrm{i}\xi_j\big(1 - \delta_\mathbb{Z}(k)\big)}{|\xi|^2 + \mathrm{i}\frac{2\pi}{\mathcal{T}}k}\mathscr{F}_G\big[\partial_t u - \Delta u\big]\right],$$

where $\delta_\mathbb{Z}$ denotes the delta distribution on $\mathbb{Z}$, that is, $\delta_\mathbb{Z}(0) = 1$ and $\delta_\mathbb{Z}(k) = 0$ for $k \neq 0$. Since $\mathscr{F}_G = \mathscr{F}_{\mathbb{R}/\mathcal{T}\mathbb{Z}} \circ \mathscr{F}_{\mathbb{R}^n}$, it follows that

$$\partial_j \mathcal{P}_\perp u = \mathscr{F}_{\mathbb{R}/\mathcal{T}\mathbb{Z}}^{-1}\left[(1 - \delta_\mathbb{Z})|k|^{-\frac{1}{2}\beta}\right] *_{\mathbb{R}/\mathcal{T}\mathbb{Z}} \mathscr{F}_{\mathbb{R}^n}^{-1}\left[|\xi|^{\beta-1}\right] *_{\mathbb{R}^n} F \tag{2.8}$$

with

$$F := \mathscr{F}_G^{-1}\left[M(k,\xi)\,\mathscr{F}_G\big[\partial_t u - \Delta u\big]\right],$$

$$M(k,\xi) := \frac{|k|^{\frac{1}{2}\beta}\,|\xi|^{1-\beta}\,\mathrm{i}\xi_j\big(1 - \delta_\mathbb{Z}(k)\big)}{|\xi|^2 + \mathrm{i}\frac{2\pi}{\mathcal{T}}k}.$$

Owing to the fact that M has no singularities, one can utilize a so-called transference principle to verify that M is an $L^q(G)$ Fourier multiplier for all $q \in (1,\infty)$; see Kyed (2014b) for the details on such an approach. It follows that $F \in L^q(G)$ with $\|F\|_q \le C\|u\|_{1,2,q}$. It is standard to compute the inverse Fourier transform

$$\gamma_\beta := \mathscr{F}_{\mathbb{R}/\mathcal{T}\mathbb{Z}}^{-1}\left[(1 - \delta_\mathbb{Z})|k|^{-\frac{1}{2}\beta}\right]$$

appearing on the right-hand side in (2.8). Choosing for example the interval $[-\frac{1}{2}\mathcal{T}, \frac{1}{2}\mathcal{T})$ as a realization of $\mathbb{R}/\mathcal{T}\mathbb{Z}$, one has

$$\gamma_\beta(t) = Ct^{-1+\frac{1}{2}\beta} + h(t),$$

with $h \in C^\infty(\mathbb{R}/\mathcal{T}\mathbb{Z})$; see e.g. Example 3.1.19 in Grafakos (2008). Clearly $\gamma_\beta \in L^{\frac{1}{1-\frac{1}{2}\beta},\infty}(\mathbb{R}/\mathcal{T}\mathbb{Z})$. Thus, by Young's inequality, see Theorem 1.4.24 in Grafakos (2008) (for example), the mapping $\varphi \mapsto \gamma_\beta *_{\mathbb{R}/\mathcal{T}\mathbb{Z}} \varphi$ extends to a bounded operator from $L^q(\mathbb{R}/\mathcal{T}\mathbb{Z})$ into $L^r(\mathbb{R}/\mathcal{T}\mathbb{Z})$ for $r \in (1,\infty)$ satisfying

$$\frac{1}{r} = \left(1 - \frac{1}{2}\beta\right) + \frac{1}{q} - 1. \tag{2.9}$$

The mapping $\varphi \mapsto \mathscr{F}_{\mathbb{R}^n}^{-1}\left[|\xi|^{\beta-1}\right] *_{\mathbb{R}^n} \varphi$ can be identified with the operator $\Delta_{\mathbb{R}^n}^{\frac{\beta-1}{2}}$. It is well known (see for example Theorem 6.1.13 in Grafakos, 2009) that this operator is bounded from $L^q(\mathbb{R}^n)$ into $L^p(\mathbb{R}^n)$ for any $p \in (1,\infty)$ satisfying

$$\frac{1}{p} = \frac{1}{q} - \frac{1-\beta}{n}. \tag{2.10}$$

We now consider $r_1, p_1 \in (1,\infty)$ that satisfy (2.9) and (2.10). We then recall (2.8) to estimate

$$\|\partial_j \mathcal{P}_\perp u\|_{L^{r_1}(\mathbb{R}/\mathcal{T}\mathbb{Z};L^{p_1}(\Omega))}$$

$$= \left(\int_{\mathbb{R}/\mathcal{T}\mathbb{Z}} \|\mathscr{F}_{\mathbb{R}^n}^{-1}\left[|\xi|^{\beta-1}\right] *_{\mathbb{R}^n} \gamma_\beta *_{\mathbb{R}/\mathcal{T}\mathbb{Z}} F(t,\cdot)\|_{p_1}^{r_1} \, dt\right)^{\frac{1}{r_1}}$$

$$\leq c_1 \left(\int_{\mathbb{R}/\mathcal{T}\mathbb{Z}} \|\gamma_\beta *_{\mathbb{R}/\mathcal{T}\mathbb{Z}} F(t,\cdot)\|_q^{r_1} \, dt\right)^{\frac{1}{r_1}}$$

$$\leq c_2 \left(\int_{\mathbb{R}^n} \|\gamma_\beta *_{\mathbb{R}/\mathcal{T}\mathbb{Z}} F(\cdot,x)\|_{r_1}^q \, dx\right)^{\frac{1}{q}} \leq c_3 \|F\|_q \leq c_4 \|u\|_{1,2,q},$$

where Minkowski's integral inequality is employed to conclude the second inequality above. Classical Sobolev embedding yields $\nabla \mathcal{P} u \in L^{p_1}(\mathbb{R}^n)$ with $\|\nabla \mathcal{P} u\|_{p_1} \leq c_5 \|u\|_{1,2,q}$. By the above, it thus follows that

$$\|\nabla u\|_{L^{r_1}(\mathbb{R}/\mathcal{T}\mathbb{Z};L^{p_1}(\Omega))} \leq c_6 \|u\|_{1,2,q}.$$

By interpolation, the same estimate follows for all $r_1, p_1 \in [q,\infty)$ satisfying (2.5). In a similar manner, the estimate

$$\|u\|_{L^{r_0}(\mathbb{R}/\mathcal{T}\mathbb{Z};L^{p_0}(\Omega))} \leq c_7 \|u\|_{1,2,q}$$

can be shown for parameters $r_0, p_0 \in [q,\infty)$ satisfying (2.4). This concludes the proof. $\square$

2.5 Linearized Problem

A suitable linearization of (2.1) is given by the time-periodic Oseen system

$$
\begin{cases}
\partial_t u - \nu \Delta u + \lambda \partial_1 u + \nabla \mathfrak{p} = F & \text{in } \mathbb{R} \times \Omega, \\
\operatorname{div} u = 0 & \text{in } \mathbb{R} \times \Omega, \\
u = 0 & \text{on } \mathbb{R} \times \partial\Omega, \\
\lim_{|x|\to\infty} u(t,x) = 0, \quad u(t+\mathcal{T},x) = u(t), \quad \mathfrak{p}(t+\mathcal{T},x) = \mathfrak{p}(t),
\end{cases}
\tag{2.11}
$$

where $\lambda > 0$. The goal in this section is to identify a Banach space X of functions $(u, \mathfrak{p})$ satisfying $(2.11)_{2-4}$ such that the differential operator on the left-hand side in $(2.11)_1$

$$
\mathcal{L}(u, \mathfrak{p}) := \partial_t u + \lambda \partial_1 u - \nu \Delta u + \nabla \mathfrak{p}
$$

becomes a homeomorphism $\mathcal{L} \colon X \to L^q_{\mathrm{per}}(\mathbb{R} \times \Omega)^3$. In other words, what can be referred to as "maximal L^q regularity" of the time-periodic system (2.11) will be established.

The projections $\mathcal{P}$ and $\mathcal{P}_\perp$ will be used to decompose (2.11) into two problems. More specifically, for data $F \in L^q_{\mathrm{per}}(\mathbb{R} \times \Omega)^3$ a solution to (2.11) is investigated as the sum of a solution corresponding to the steady-state part of the data $\mathcal{P}F$, and a solution corresponding to the oscillatory part $\mathcal{P}_\perp F$. We start with the latter and consider data in the space $L^q_{\mathrm{per},\perp}(\mathbb{R} \times \Omega)^3$. In this case, appropriate L^q estimates can be established whether λ vanishes or not. The case $\lambda = 0$ is therefore included in the theorem below.

Theorem 2.2 *Let $\Omega \subset \mathbb{R}^3$ be an exterior domain of class C^2, and take $q \in (1, \infty)$ and $\lambda \in [0, \lambda_0]$. For any vector field $F \in L^q_{\mathrm{per},\perp}(\mathbb{R} \times \Omega)^3$ there is a solution*

$$
(u, \mathfrak{p}) \in W^{1,2,q}_{\mathrm{per},\perp}(\mathbb{R} \times \Omega)^3 \times D^{1,q}_{\mathrm{per},\perp}(\mathbb{R} \times \Omega)
\tag{2.12}
$$

to (2.11) which satisfies

$$
\|u\|_{1,2,q} + \|\nabla \mathfrak{p}\|_q \leq C_2 \|F\|_q
\tag{2.13}
$$

with $C_2 = C_2(q, \Omega, \nu, \lambda_0)$. If $r \in (1, \infty)$ and

$$
(\widetilde{u}, \widetilde{\mathfrak{p}}) \in W^{1,2,r}_{\mathrm{per},\perp}(\mathbb{R} \times \Omega)^3 \times D^{1,r}_{\mathrm{per},\perp}(\mathbb{R} \times \Omega)
$$

is another solution, then $\widetilde{u} = u$ and $\widetilde{\mathfrak{p}} = \mathfrak{p} + d(t)$ for some $\mathcal{T}$-periodic function $d : \mathbb{R} \to \mathbb{R}$.

The proof of Theorem 2.2 will be based on three lemmas. The first lemma states that the theorem holds in the case $q = 2$.

Lemma 2.3 *Let Ω and λ be as in Theorem 2.4. For any vector field $F \in L^2_{\mathrm{per},\perp}(\mathbb{R} \times \Omega)^3$ there is a solution*

$$(u, \mathfrak{p}) \in W^{1,2,2}_{\mathrm{per},\perp}(\mathbb{R} \times \Omega)^3 \times D^{1,2}_{\mathrm{per},\perp}(\mathbb{R} \times \Omega)$$

to (2.11). Moreover, the solution obeys the estimate

$$\|u\|_{1,2,2} + \|\nabla\mathfrak{p}\|_2 \leq C_3 \|F\|_2 \tag{2.14}$$

with $C_3 = C_3(\lambda_0, \mathcal{T}, \Omega)$.

Proof The proof in Lemma 5 of Galdi (2013b) can easily be adapted to establish the desired statement. For the sake of completeness, a sketch of the proof is given here. For a Hilbert space H, we introduce the function space $L^2_{\mathrm{per}}(\mathbb{R}; H)$ whose elements are the $\mathcal{T}$-periodic extensions of the functions in $L^2((0, \mathcal{T}); H)$. Classical theory for Fourier series, in particular the theorem of Parseval, is available in the setting of such spaces. Since clearly $L^2_{\mathrm{per}}(\mathbb{R} \times \Omega) = L^2_{\mathrm{per}}(\mathbb{R}; L^2(\Omega))$, we may express the data F as a Fourier series $F = \sum_{k \in \mathbb{Z}} F_k e^{i\frac{2\pi}{\mathcal{T}}kt}$ with Fourier coefficients $F_k \in L^2(\Omega)^3$. Since $\mathcal{P}F = 0$, it follows that $F_0 = 0$. Consider for each $k \in \mathbb{Z} \setminus \{0\}$ the system

$$\begin{cases} ik\dfrac{2\pi}{\mathcal{T}}u_k - \nu\Delta u_k + \lambda\partial_1 u_k + \nabla\mathfrak{p}_k = F_k & \text{in } \Omega, \\[2mm] \mathrm{div}\, u_k = 0 & \text{in } \Omega, \\[2mm] u_k = 0 & \text{on } \partial\Omega. \end{cases}$$

Standard methods from the theory of elliptic systems can be employed to investigate this problem. As a result, the existence of a solution $(u_k, \mathfrak{p}_k) \in W^{2,2}(\Omega) \times (D^{1,2}(\Omega) \cap L^6(\Omega))$ which satisfies

$$\frac{2\pi}{\mathcal{T}}|k|\|u_k\|_2 + \|\nabla^2 u_k\|_2 + \|\nabla\mathfrak{p}_k\|_2 \leq c_8 \|F_k\|_2, \tag{2.15}$$

with c_8 independent on k, can be established. Observe that the space $D^{1,2}(\Omega) \cap L^6(\Omega)$ is a Hilbert space in the norm $\|\nabla\cdot\|_2$. Thus, by (2.15) and Parseval's theorem, the vector fields

$$u := \sum_{k \in \mathbb{Z} \setminus \{0\}} u_k\, e^{i\frac{2\pi}{\mathcal{T}}kt}, \qquad \mathfrak{p} := \sum_{k \in \mathbb{Z} \setminus \{0\}} \mathfrak{p}_k\, e^{i\frac{2\pi}{\mathcal{T}}kt}$$

are well defined in $L^2_{\mathrm{per}}(\mathbb{R}; W^{2,2}(\Omega))$ and $L^2_{\mathrm{per}}(\mathbb{R}; D^{1,2}(\Omega))$, respectively, with $\partial_t u \in L^2_{\text{per}}(\mathbb{R}; L^2(\Omega))$. Parseval's theorem yields

$$\|\partial_t u\|_{L^2_{\text{per}}(\mathbb{R};L^2(\Omega))} + \|u\|_{L^2_{\text{per}}(\mathbb{R};W^{2,2}(\Omega))} + \|\nabla \mathfrak{p}\|_{L^2_{\text{per}}(\mathbb{R};L^2(\Omega))}$$
$$\leq c_9 \|F\|_{L^2_{\text{per}}(\mathbb{R};L^2(\Omega))}.$$

Finally observe that $\mathcal{P}u = \mathcal{P}\mathfrak{p} = 0$ as both Fourier coefficients u_0 and $\mathfrak{p}_0$ vanish by definition of u and $\mathfrak{p}$. We can therefore conclude that $(u, \mathfrak{p}) \in W^{1,2,2}_{\text{per},\perp}(\mathbb{R} \times \Omega)^3 \times D^{1,2}_{\text{per},\perp}(\mathbb{R} \times \Omega)$ and satisfies (2.14). By construction, $(u, \mathfrak{p})$ is a solution to (2.11). □

The next lemma states that the assertions in Theorem 2.2 are valid if Ω is replaced with a bounded domain.

Lemma 2.4 *Let $\mathfrak{B} \subset \mathbb{R}^3$ be a bounded domain of class C^2, $q \in (1, \infty)$ and $\lambda \in [0, \lambda_0]$. For any vector field $F \in L^q_{\text{per},\perp}(\mathbb{R} \times \mathfrak{B})^3$ there is a solution*

$$(u, \mathfrak{p}) \in W^{1,2,q}_{\text{per},\perp}(\mathbb{R} \times \mathfrak{B})^3 \times D^{1,q}_{\text{per},\perp}(\mathbb{R} \times \mathfrak{B}) \tag{2.16}$$

to

$$\begin{cases} \partial_t u - \nu \Delta u + \lambda \partial_1 u + \nabla \mathfrak{p} = F & \text{in } \mathbb{R} \times \mathfrak{B}, \\ \text{div} u = 0 & \text{in } \mathbb{R} \times \mathfrak{B}, \\ u = 0 & \text{on } \mathbb{R} \times \partial \mathfrak{B}, \end{cases} \tag{2.17}$$

which satisfies

$$\|u\|_{1,2,q} + \|\nabla \mathfrak{p}\|_q \leq C_4 \|F\|_q, \tag{2.18}$$

with $C_4 = C_4(q, \mathfrak{B}, \nu, \lambda_0)$. If $r \in (1, \infty)$ and

$$(\widetilde{u}, \widetilde{\mathfrak{p}}) \in W^{1,2,r}_{\text{per},\perp}(\mathbb{R} \times \mathfrak{B})^3 \times D^{1,r}_{\text{per},\perp}(\mathbb{R} \times \mathfrak{B})$$

is another solution, then $\widetilde{u} = u$ and $\widetilde{\mathfrak{p}} = \mathfrak{p} + d(t)$ for some $\mathcal{T}$-periodic function $d : \mathbb{R} \to \mathbb{R}$.

Proof One may verify that the proof of Lemma 9 in Galdi (2013b) for a two-dimensional domain also holds for a three-dimensional domain. For the sake of completeness, a sketch is given here when $\lambda = 0$. By density of $C^\infty_{0,\text{per},\perp}(\mathfrak{B})$ in $L^q_{\text{per},\perp}(\mathbb{R} \times \mathfrak{B})$, it is sufficient to consider only $F \in C^\infty_{0,\text{per},\perp}(\mathfrak{B})^3$. The existence of a solution $(u, \mathfrak{p})$ that belongs to the space $W^{1,2,2}_{\text{per},\perp}(\mathbb{R} \times \mathfrak{B})^3 \times D^{1,2}_{\text{per},\perp}(\mathbb{R} \times \mathfrak{B})$ to (2.17) can be shown with the same argument used in the proof of Lemma 2.3. By Sobolev's embedding theorem, it may be assumed that this solution is continuous in the sense that $u \in C_{\text{per}}(\mathbb{R}; L^2_\sigma(\mathfrak{B}))$. Clearly, for this solution a $t_0 \in [0, \mathcal{T}]$ can be chosen such that $u(t_0) \in W^{2,2}(\mathfrak{B})$ and $\mathrm{div}u(t_0) = 0$. Using Sobolev's embedding theorem, it follows that $u(t_0) \in L^q_\sigma(\mathfrak{B})$. Consider now the initial-value problem

$$
\begin{cases}
\partial_t v = \nu\Delta v - \nabla p & \text{in } (t_0, \infty) \times \mathfrak{B}, \\
\mathrm{div}v = 0 & \text{in } (t_0, \infty) \times \mathfrak{B}, \\
v = 0 & \text{on } (t_0, \infty) \times \partial\mathfrak{B}, \\
v(t_0, \cdot) = u(t_0, \cdot) & \text{in } \mathfrak{B}.
\end{cases}
\tag{2.19}
$$

It is well known that the Stokes operator $A := \mathcal{P}_H\Delta$ generates a bounded analytic semi-group on $L^q_\sigma(\mathfrak{B})$; see Giga (1981). Consequently, the solution to the initial-value problem (2.19) given by $v := \exp\big(A(t-t_0)\big)u(t_0)$ satisfies

$$
\|\partial_t v(t)\|_q + \|Av(t)\|_q \le c_{10}\,(t - t_0)^{-1} \qquad \text{for every } t > t_0; \tag{2.20}
$$

see for example Theorem II.4.6 in Engel & Nagel (2000). Also by classical results, see for example Theorem 2.8 in Giga & Sohr (1991), there exists a solution $(w, \pi) \in W^{1,2,q}\big((t_0, \infty) \times \mathfrak{B}\big) \times W^{0,1,q}\big((t_0, \infty) \times \mathfrak{B}\big)$ to

$$
\begin{cases}
\partial_t w = \nu\Delta w - \nabla\pi + F & \text{in } (t_0, \infty) \times \mathfrak{B}, \\
\mathrm{div}w = 0 & \text{in } (t_0, \infty) \times \mathfrak{B}, \\
w = 0 & \text{on } (t_0, \infty) \times \partial\mathfrak{B}, \\
w(t_0, \cdot) = 0 & \text{in } \mathfrak{B}
\end{cases}
\tag{2.21}
$$

that is continuous in the sense that $w \in C\big([t_0, \infty); L^q_\sigma(\mathfrak{B})\big)$ and for all $\tau \in (t_0, \infty)$ satisfies the estimate

$$
\|w\|_{W^{1,2,q}((t_0,\tau)\times\mathfrak{B})} + \|\pi\|_{W^{0,1,q}((t_0,\tau)\times\mathfrak{B})} \le c_{11}\,\|F\|_{L^q((t_0,\tau)\times\mathfrak{B})}, \tag{2.22}
$$

with c_{11} independent of τ. Since u and $v+w$ solve the same initial-value problem, a standard uniqueness argument implies $u = v+w$. Due to the $\mathcal{T}$-periodicity of u and F, it follows for all $m \in \mathbb{N}$ that

$$
\begin{aligned}
\int_0^T &\|\partial_t u(t)\|_q^q + \|Au(t)\|_q^q \, dt \\
&= \frac{1}{m}\int_{2T}^{(m+2)T} \|\partial_t u(t)\|_q^q + \|Au(t)\|_q^q \, dt \\
&\le c_{12}\frac{1}{m}\int_T^\infty t^{-q}\, dt + c_{13}\frac{1}{m}\|F\|_{L^q((0,(m+1)T)\times\mathfrak{B})}^q \\
&\le c_{14}\frac{1}{m}\frac{1}{q-1}T^{1-q} + c_{13}\frac{m+1}{m}\|F\|_{L^q((0,T)\times\mathfrak{B})}^q.
\end{aligned}
\tag{2.23}
$$

We now let $m \to \infty$ and conclude that

$$\|\partial_t u\|_{L^q_{\mathrm{per}}(\mathbb{R} \times \mathfrak{B})} + \|Au\|_{L^q_{\mathrm{per}}(\mathbb{R} \times \mathfrak{B})} \leq c_{15} \|F\|_{L^q_{\mathrm{per}}(\mathbb{R} \times \mathfrak{B})}.$$

The estimate $\|\nabla^2 u\|_{L^q_{\mathrm{per}}(\mathbb{R} \times \mathfrak{B})} \leq c_{16} \|Au\|_{L^q_{\mathrm{per}}(\mathbb{R} \times \mathfrak{B})}$ is a consequence of well-known L^q theory for the Stokes problem in bounded domains; see for example Theorem IV.6.1 in Galdi (2011). Consequently, the estimate $\|u\|_{1,2,q} \leq c_{17} \|F\|_q$ follows by the Poincaré inequality $\|u\|_q \leq c_{18} \|\partial_t u\|_q$. Now modify the pressure $\mathfrak{p}$ by adding a function depending only on t such that $\int_{\mathfrak{B}} \mathfrak{p}(t, x)\, \mathrm{d}x = 0$, which ensures the validity of Poincaré's inequality for $\mathfrak{p}$. A similar estimate is then obtained for $\mathfrak{p}$ by isolating $\nabla \mathfrak{p}$ in (2.17)$_1$. This establishes (2.18). To show the assertion on uniqueness, a duality argument can be employed. For this purpose, let $\varphi \in C^\infty_{0,\mathrm{per}}(\mathbb{R} \times \mathfrak{B})^3$ and let (ψ, η) be a solution to the problem

$$
\begin{cases}
\partial_t \psi = -\nu \Delta \psi - \nabla \mathfrak{p} + \varphi & \text{in } \mathbb{R} \times \mathfrak{B}, \\
\mathrm{div}\psi = 0 & \text{in } \mathbb{R} \times \mathfrak{B}, \\
\psi = 0 & \text{on } \mathbb{R} \times \partial\mathfrak{B}, \\
\psi(t + \mathcal{T}, x) = \psi(t, x)
\end{cases}
\tag{2.24}
$$

adjoint to (2.17). The existence of a solution (ψ, η) follows by the same arguments that yield a solution to (2.17). Since $\varphi \in C^\infty_{0,\mathrm{per}}(\mathbb{R} \times \mathfrak{B})^3$, the solution satisfies $(\psi, \eta) \in W^{1,2,s}_{\mathrm{per}}(\mathbb{R} \times \mathfrak{B}) \times W^{1,s}_{\mathrm{per}}(\mathbb{R} \times \mathfrak{B})$ for all $s \in (1, \infty)$. The regularity of (ψ, η) ensures validity of the following computation:

$$\int_0^{\mathcal{T}} \int_{\mathfrak{B}} (w - \widetilde{w}) \cdot \varphi \, \mathrm{d}x\mathrm{d}t = \int_0^{\mathcal{T}} \int_{\mathfrak{B}} (w - \widetilde{w}) \cdot (\partial_t \psi + \nu \Delta \psi + \nabla \mathfrak{p}) \, \mathrm{d}x\mathrm{d}t$$

$$= \int_0^{\mathcal{T}} \int_{\mathfrak{B}} \left(\partial_t [w - \widetilde{w}] - \nu \Delta [w - \widetilde{w}] + \nabla[\pi - \widetilde{\pi}] \right) \cdot \psi \, \mathrm{d}x\mathrm{d}t = 0.$$

Since $\varphi \in C^\infty_{0,\mathrm{per}}(\mathbb{R} \times \mathfrak{B})^3$ was arbitrary, $\widetilde{w} - w = 0$ follows. In turn, $\nabla \pi = \nabla \widetilde{\pi}$ and thus $\widetilde{\pi} = \pi + d(t)$ follows. $\qquad\square$

The final lemma concerns estimates of the pressure term in (2.11). The following lemma was originally proved for a two-dimensional exterior domain in Galdi (2013b). We employ the ideas from the proof of Lemma 6 in Galdi (2013b) and establish the lemma in a slightly modified form for a three-dimensional exterior domain.

Lemma 2.5 *Let Ω and λ be as in Theorem 2.4 and $s \in (1, \infty)$. There is a constant $C_5 = C_5(R_0, \Omega, s)$ such that a solution*

$$(u, \mathfrak{p}) \in W^{1,2,r}_{\mathrm{per},\perp}(\mathbb{R} \times \Omega)^3 \times D^{1,r}_{\mathrm{per},\perp,R_0}(\mathbb{R} \times \Omega)$$

to (2.11) *corresponding to data* $F \in L^r_{\text{per},\perp}(\mathbb{R} \times \Omega)^3$ *for some* $r \in (1, \infty)$ *satisfies*

$$\|\mathfrak{p}(t, \cdot)\|_{\frac{3}{2}s, \Omega_{R_0}} \leq C_5 \left(\|F(t, \cdot)\|_s + \|\nabla u(t, \cdot)\|_{s, \Omega_{R_0}} \right.$$
$$\left. + \|\nabla u(t, \cdot)\|_{s, \Omega_{R_0}}^{\frac{s-1}{s}} \|\nabla u(t, \cdot)\|_{1, s, \Omega_{R_0}}^{\frac{1}{s}} \right) \tag{2.25}$$

for a.e. $t \in \mathbb{R}$. *Moreover, for every* $\rho > R_*$ *there is a constant* $C_6 = C_6(\rho, \Omega, s)$ *such that*

$$\|\nabla \mathfrak{p}(t, \cdot)\|_{s, \Omega^\rho} \leq C_6 \left(\|F(t, \cdot)\|_s + \|\mathfrak{p}(t, \cdot)\|_{s, \Omega_\rho} \right) \tag{2.26}$$

for a.e. $t \in \mathbb{R}$.

Proof For the sake of simplicity, the t-dependency of functions is not indicated. All norms are taken with respect to the spatial variables only. Consider an arbitrary $\varphi \in C_0^\infty(\overline{\Omega})$. Observe that for any $\psi \in C_{\text{per}}^\infty(\mathbb{R})$

$$\int_0^T \int_\Omega \partial_t u \cdot \nabla \varphi \, \mathrm{d}x \, \psi \, \mathrm{d}t = \int_0^T \int_\Omega \operatorname{div} u \, \varphi \, \partial_t \psi \, \mathrm{d}x \, \mathrm{d}t = 0,$$

which implies that $\int_\Omega \partial_t u \cdot \nabla \varphi \, \mathrm{d}x = 0$ for a.e. t. Moreover,

$$\int_\Omega \partial_1 u \cdot \nabla \varphi \, \mathrm{d}x = -\int_\Omega \operatorname{div} u \, \partial_1 \varphi \, \mathrm{d}x = 0.$$

Hence it follows from (2.11) that $\mathfrak{p}$ is a solution to the weak Neumann problem for the Laplacian:

$$\int_\Omega \nabla \mathfrak{p} \cdot \nabla \varphi \, \mathrm{d}x = \int_\Omega F \cdot \nabla \varphi + \Delta u \cdot \nabla \varphi \, \mathrm{d}x \qquad \forall \varphi \in C_0^\infty(\overline{\Omega}).$$

Recall that

$$\mathfrak{p} \in L_{\text{loc}}^1(\Omega), \qquad \nabla \mathfrak{p} \in L^r(\Omega)^3, \quad \text{and} \quad \int_{\Omega_{R_0}} \mathfrak{p} \, \mathrm{d}x = 0. \tag{2.27}$$

It is well known that the weak Neumann problem for the Laplacian in an exterior domain is uniquely solvable in the class (2.27); see for example Section III.1 in Galdi (2011) or Simader (1990). We can thus write $\mathfrak{p}$ as a sum $\mathfrak{p} = \mathfrak{p}_1 + \mathfrak{p}_2$ of two solutions (in the class above) to the weak Neumann problems

$$\int_\Omega \nabla \mathfrak{p}_1 \cdot \nabla \varphi \, \mathrm{d}x = \int_\Omega F \cdot \nabla \varphi \, \mathrm{d}x \qquad \forall \varphi \in C_0^\infty(\overline{\Omega})$$

and

$$\int_\Omega \nabla \mathfrak{p}_2 \cdot \nabla \varphi \, dx = \int_\Omega \Delta u \cdot \nabla \varphi \, dx \qquad \forall \varphi \in C_0^\infty(\overline{\Omega}),$$

respectively. The a priori estimate

$$\|\nabla \mathfrak{p}_1\|_q \le c_{19} \|F\|_q \qquad \forall q \in (1, \infty) \tag{2.28}$$

is well known. An estimate for $\mathfrak{p}_2$ will now be established. Consider for this purpose an arbitrary function $g \in C_0^\infty(\Omega_{R_0})$ with $\int_{\Omega_{R_0}} g \, dx = 0$. Existence of a vector field $h \in C_0^\infty(\Omega_{R_0})^3$ with $\operatorname{div} h = g$ and

$$\forall q \in (1, \infty) : \quad \|h\|_{1,q} \le c_{20} \|g\|_q$$

is well known; see for example Theorem III.3.3 in Galdi (2011). Let Φ be a solution to the following weak Neumann problem for the Laplacian:

$$\int_\Omega \nabla \Phi \cdot \nabla \varphi \, dx = \int_\Omega h \cdot \nabla \varphi \, dx \qquad \forall \varphi \in C_0^\infty(\overline{\Omega}).$$

By classical theory, such a solution exists with

$$\Phi \in C^\infty(\overline{\Omega}) \quad \text{and} \quad \|\nabla \Phi\|_{1,q} \le c_{21} \|h\|_{1,q} \le c_{22} \|g\|_q \quad \forall q \in (1, \infty).$$

Since Φ is harmonic in $\mathbb{R}^3 \setminus \overline{B_{R_0}}$, the following asymptotic expansion as $|x| \to \infty$ is valid:

$$\partial^\alpha \Phi(x) = \partial^\alpha c_{23} + \partial^\alpha \Gamma_{\mathrm{L}}(x) \cdot \int_{\partial B_\rho} \frac{\partial \Phi}{\partial n} \, dS + O\big(|x|^{-2-|\alpha|}\big),$$

where c_{23} is a constant and $\Gamma_{\mathrm{L}} : \mathbb{R}^3 \setminus \{0\} \to \mathbb{R}$, $\Gamma_{\mathrm{L}}(x) := (4\pi|x|)^{-1}$ is the fundamental solution of the Laplacian in $\mathbb{R}^3$. Observing that

$$\int_{\partial B_\rho} \frac{\partial \Phi}{\partial n} \, dS = \int_{\partial \Omega} \frac{\partial \Phi}{\partial n} \, dS + \int_{\Omega_\rho} \Delta \Phi \, dx$$

$$= 0 + \int_{\Omega_\rho} \operatorname{div} h \, dx = \int_{\Omega_\rho} g \, dx = 0,$$

we thus deduce that $\nabla \Phi(x) = O\big(|x|^{-3}\big)$ as $|x| \to \infty$. Similarly, we see that $\mathfrak{p}_2 = c_{24} + O\big(|x|^{-1}\big)$. We therefore conclude that

$$\lim_{\rho \to \infty} \int_{\partial B_\rho} \mathfrak{p}_2 \frac{\partial \Phi}{\partial n} \, dx = 0.$$

We can thus compute

$$
\int_\Omega \mathfrak{p}_2\, g\, \mathrm{d}x = \int_\Omega \mathfrak{p}_2\, \Delta\Phi\, \mathrm{d}x = \lim_{\rho\to\infty} \int_{\Omega_\rho} \mathfrak{p}_2\, \Delta\Phi\, \mathrm{d}x
$$

$$
= \lim_{\rho\to\infty} \left(\int_{\partial\Omega_\rho} \mathfrak{p}_2\, \frac{\partial\Phi}{\partial n}\, \mathrm{d}S - \int_{\Omega_\rho} \nabla\mathfrak{p}_2 \cdot \nabla\Phi\, \mathrm{d}x \right) \qquad (2.29)
$$

$$
= -\int_\Omega \nabla\mathfrak{p}_2 \cdot \nabla\Phi\, \mathrm{d}x = -\int_\Omega \Delta u \cdot \nabla\Phi\, \mathrm{d}x.
$$

The decay of $\nabla\Phi$ further implies that $\partial_i u_j\, \partial_k \Phi \in L^1(B^{2R_0})$. We can thus find a sequence $\{\rho_k\}_{k=1}^\infty$ of positive numbers with $\lim_{k\to\infty}\rho_k = \infty$ such that

$$
\lim_{\rho_k\to\infty} \int_{\partial B_\rho} \partial_j u_i\, \partial_i \Phi\, n_j - \partial_j u_i\, \partial_j \Phi\, n_i\, \mathrm{d}S = 0.
$$

Returning to (2.29), we continue the computation and find that

$$
\int_\Omega \mathfrak{p}_2\, g\, \mathrm{d}x = -\lim_{k\to\infty} \int_{\Omega_{\rho_k}} \Delta u \cdot \nabla\Phi\, \mathrm{d}x
$$

$$
= -\lim_{k\to\infty} \int_{\partial\Omega_{\rho_k}} \partial_j u_i\, \partial_i \Phi\, n_j - \partial_j u_i\, \partial_j \Phi\, n_i \mathrm{d}S
$$

$$
= -\int_{\partial\Omega} \partial_j u_i\, \partial_i \Phi\, n_j - \partial_j u_i\, \partial_j \Phi\, n_i \mathrm{d}S
$$

$$
= -\int_{\partial\Omega} \nabla u : \left(\nabla\Phi \otimes n - n \otimes \nabla\Phi \right) \mathrm{d}S.
$$

Applying first the Hölder and then a trace inequality (see Theorem II.4.1 in Galdi, 2011) we deduce

$$
\left| \int_\Omega \mathfrak{p}_2\, g\, \mathrm{d}x \right| \le c_{25} \|\nabla u\|_{s,\partial\Omega} \|\nabla\Phi\|_{\frac{s}{s-1},\partial\Omega}
$$

$$
\le c_{26} \|\nabla u\|_{s,\partial\Omega} \|\nabla\Phi\|_{1,\frac{3s}{3s-2},\Omega_{R_0}} \le c_{27} \|\nabla u\|_{s,\partial\Omega} \|g\|_{\frac{3s}{3s-2},\Omega_{R_0}}.
$$

Recalling that $\int_{\Omega_{R_0}} \mathfrak{p}_2\, \mathrm{d}x = 0$, we thus obtain

$$
\|\mathfrak{p}_2\|_{\frac{3}{2}s,\Omega_{R_0}} = \sup_{\substack{g\in C_0^\infty(\Omega_{R_0}),\, \|g\|_{\frac{3s}{3s-2}}=1 \\ \int_{\Omega_{R_0}} g\, \mathrm{d}x=0}} \left| \int_\Omega \mathfrak{p}_2\, g\, \mathrm{d}x \right| \le c_{28} \|\nabla u\|_{s,\partial\Omega}.
$$

Another application of a trace inequality (see again Theorem II.4.1 in Galdi, 2011) now yields

$$
\|\mathfrak{p}_2\|_{\frac{3}{2}s,\Omega_{R_0}} \le c_{29} \left(\|\nabla u\|_{s,\Omega_{R_0}} + \|\nabla u\|_{s,\Omega_{R_0}}^{\frac{s-1}{s}} \|\nabla u\|_{1,s,\Omega_{R_0}}^{\frac{1}{s}} \right).
$$

Recalling (2.28), we employ Sobolev's embedding theorem to conclude finally that

$$\|\mathfrak{p}\|_{\frac{3}{2}s,\Omega_{R_0}} \leq c_{30} \|\nabla \mathfrak{p}_1\|_{s,\Omega_{R_0}} + \|\mathfrak{p}_2\|_{\frac{3}{2}s,\Omega_{R_0}}$$

$$\leq c_{31} \left(\|F\|_s + \|\nabla u\|_{s,\Omega_{R_0}} + \|\nabla u\|_{s,\Omega_{R_0}}^{\frac{s-1}{s}} \|\nabla u\|_{1,s,\Omega_{R_0}}^{\frac{1}{s}} \right)$$

and thus (2.25). To show (2.26), we introduce $R \in (R_*, \rho)$ and a "cutoff" function $\chi \in C^\infty(\mathbb{R}^3; \mathbb{R})$ with $\chi = 1$ on Ω^R and $\chi = 0$ on Ω_{R_*}. We then put $\pi := \chi\mathfrak{p}$ and observe from (2.11) that π is a solution to the weak Neumann problem for the Laplacian

$$\forall \varphi \in C_0^\infty(\overline{\Omega}) : \int_\Omega \nabla \pi \cdot \nabla \varphi \, dx = \langle \mathcal{F}_1, \varphi \rangle + \langle \mathcal{F}_2, \varphi \rangle \qquad (2.30)$$

with

$$\langle \mathcal{F}_1, \varphi \rangle := \int_\Omega \left(2\mathfrak{p} \nabla \chi + \chi F \right) \cdot \nabla \varphi \, dx,$$

$$\langle \mathcal{F}_2, \varphi \rangle := \int_\Omega \left(\nabla \chi \cdot F + \Delta \chi \, \mathfrak{p} \right) \varphi \, dx.$$

We clearly have

$$\sup_{\|\nabla \varphi\|_{s^*} = 1} |\langle \mathcal{F}_1, \varphi \rangle| \leq c_{32} \left(\|\mathfrak{p}\|_{s,\Omega_\rho} + \|F\|_s \right).$$

Since $\chi = 1$ on Ω^R, we further observe that

$$\sup_{\|\nabla \varphi\|_{s^*} = 1} |\langle \mathcal{F}_2, \varphi \rangle| = \sup_{\substack{\|\nabla \varphi\|_{s^*} = 1 \\ \mathrm{supp}\varphi \subset \Omega_\rho}} |\langle \mathcal{F}_2, \varphi \rangle| \leq c_{33} \left(\|F\|_s + \|\mathfrak{p}\|_{s,\Omega_\rho} \right),$$

where Poincaré's inequality is used to obtain the last estimate. A standard a priori estimate for the weak Neumann problem (2.30) now implies (2.26); see for example Section III.1 in Galdi (2011) or Simader (1990). $\qquad \square$

Proof of Theorem 2.2 By density of $C_{0,\mathrm{per},\perp}^\infty(\mathbb{R} \times \Omega)$ in $L_{\mathrm{per},\perp}^q(\mathbb{R} \times \Omega)$, it suffices to consider only $F \in C_{0,\mathrm{per},\perp}^\infty(\mathbb{R} \times \Omega)^3$. Let

$$(u, \mathfrak{p}) \in W_{\mathrm{per},\perp}^{1,2,2}(\mathbb{R} \times \Omega)^3 \times D_{\mathrm{per},\perp}^{1,2}(\mathbb{R} \times \Omega)$$

be the solution from Lemma 2.3. By adding to $\mathfrak{p}$ a function that only depends on time, we may assume that $\int_{\Omega_{R_0}} \mathfrak{p} \, dx = 0$. For the scope of the proof, we fix a constant ρ with $R_* < \rho < R_0$.

We shall establish two fundamental estimates. To show the first one, we introduce a "cutoff" function $\psi_1 \in C^\infty(\mathbb{R}^3; \mathbb{R})$ with $\psi_1(x) = 1$ for

$|x| \geq \rho$ and $\psi_1(x) = 0$ for $|x| \leq R_*$. We let $\Gamma_L \colon \mathbb{R}^3 \setminus \{0\} \to \mathbb{R}$ denote the fundamental solution to the Laplace operator $(\Gamma_L(x) := (4\pi|x|)^{-1})$ and define

$$
\begin{aligned}
&V \colon \mathbb{R} \times \mathbb{R}^3 \to \mathbb{R}^3, \quad V = \nabla\Gamma_L *_{\mathbb{R}^3} \left(\nabla\psi_1 \cdot u\right), \\
&P \colon \mathbb{R} \times \mathbb{R}^3 \to \mathbb{R}, \quad P = \Gamma_L *_{\mathbb{R}^3} \left([\partial_t - \Delta + \lambda\partial_1](\nabla\psi_1 \cdot u)\right), \\
&w \colon \mathbb{R} \times \mathbb{R}^3 \to \mathbb{R}^3, \quad w(t,x) := \psi_1(x)\, u(t,x) - V(t,x), \\
&\pi \colon \mathbb{R} \times \mathbb{R}^3 \to \mathbb{R}, \quad \pi(t,x) := \psi_1(x)\, \mathfrak{p}(t,x) - P(t,x).
\end{aligned}
\tag{2.31}
$$

Then (w, π) is a solution to the whole-space problem

$$
\begin{cases}
\partial_t w - \Delta w + \lambda\partial_1 w + \nabla\pi = \\
\qquad \psi_1 F - 2\nabla\psi_1 \cdot \nabla u - \Delta\psi_1 u + \lambda\partial_1\psi_1 u + \nabla\psi_1 \mathfrak{p} & \text{in } \mathbb{R} \times \mathbb{R}^3, \\
\operatorname{div} w = 0 & \text{in } \mathbb{R} \times \mathbb{R}^3.
\end{cases}
\tag{2.32}
$$

The precise regularity of (w, π) is not important at this point. It is enough to observe that w and π belong to the space of tempered time-periodic distributions $\mathscr{S}'_{per}(\mathbb{R} \times \mathbb{R}^3)$, which is easy to verify from the definition (2.31) and the regularity of u and $\mathfrak{p}$. It is not difficult to show, see Lemma 5.3 in Kyed (2014a), that a solution w to (2.32) is unique in the class of distributions in $\mathscr{S}'_{per}(\mathbb{R} \times \mathbb{R}^3)$ satisfying $\mathcal{P}w = 0$. Consequently, w coincides with the solution from Theorem 2.1 in Kyed (2014b) and therefore satisfies

$$
\begin{aligned}
\|w\|_{1,2,s} &\leq c_{34} \left\| \psi_1 F - 2\nabla\psi_1 \cdot \nabla u - \Delta\psi_1 u + \lambda\partial_1\psi_1 u + \nabla\psi_1 \mathfrak{p} \right\|_s \\
&\leq c_{35} \left(\|F\|_s + \|u\|_{s,\mathrm{T}\times\Omega_\rho} + \|\nabla u\|_{s,\mathrm{T}\times\Omega_\rho} + \|\mathfrak{p}\|_{s,\mathrm{T}\times\Omega_\rho} \right)
\end{aligned}
$$

for all $s \in (1, \infty)$. Clearly

$$
\|\nabla V\|_s + \|\nabla^2 V\|_s \leq c_{36}\left(\|u\|_{s,\Omega_\rho} + \|\nabla u\|_{s,\Omega_\rho}\right).
$$

Since $u = w + V$ for $x \in \Omega^\rho$, we conclude that

$$
\begin{aligned}
\|\nabla u\|_{s,\mathrm{T}\times\Omega^\rho} &+ \|\nabla^2 u\|_{s,\mathrm{T}\times\Omega^\rho} \\
&\leq c_{37}\left(\|F\|_s + \|u\|_{s,\mathrm{T}\times\Omega_\rho} + \|\nabla u\|_{s,\mathrm{T}\times\Omega_\rho} + \|\mathfrak{p}\|_{s,\mathrm{T}\times\Omega_\rho}\right)
\end{aligned}
$$

for all $s \in (1, \infty)$. For a similar estimate of u and $\partial_t u$, we turn first to (2.11) and then apply (2.26) to deduce

$$
\begin{aligned}
\|\partial_t u\|_{s,\mathrm{T}\times\Omega^\rho} &\leq c_{38}\left(\|F\|_s + \|\Delta u\|_{s,\mathrm{T}\times\Omega^\rho} + \|\lambda\partial_1 u\|_{s,\mathrm{T}\times\Omega^\rho} + \|\nabla\mathfrak{p}\|_{s,\mathrm{T}\times\Omega^\rho}\right) \\
&\leq c_{39}\left(\|F\|_s + \|u\|_{s,\mathrm{T}\times\Omega_\rho} + \|\nabla u\|_{s,\mathrm{T}\times\Omega_\rho} + \|\mathfrak{p}\|_{s,\mathrm{T}\times\Omega_\rho}\right).
\end{aligned}
$$

Since $\mathcal{P}u = 0$, Poincaré's inequality yields $\|u\|_{s,\mathbb{T}\times\Omega^\rho} \leq c_{40}\|\partial_t u\|_{s,\mathbb{T}\times\Omega^\rho}$. We have thus shown

$$\|u\|_{1,2,s,\mathbb{T}\times\Omega^\rho} \leq c_{41} \left(\|F\|_s + \|u\|_{s,\mathbb{T}\times\Omega_\rho} + \|\nabla u\|_{s,\mathbb{T}\times\Omega_\rho} + \|\mathfrak{p}\|_{s,\mathbb{T}\times\Omega_\rho} \right) \tag{2.33}$$

for all $s \in (1,\infty)$.

Next, we seek to establish a similar estimate for u over the bounded domain $\mathbb{T}\times\Omega_\rho$. For this we introduce a "cutoff" function $\psi_2 \in C^\infty(\mathbb{R}^3; \mathbb{R})$ with $\psi_2(x) = 1$ for $|x| \leq \rho$ and $\psi_2(x) = 0$ for $|x| \geq R_0$. We then introduce a vector field V with

$$V \in W^{1,2,2}_{\mathrm{per},\perp}(\mathbb{R}\times\mathbb{R}^3), \quad \mathrm{supp}V \subset \mathbb{R}\times\Omega_{\rho,R_0}, \quad \mathrm{div}V = \nabla\psi_2 \cdot u,$$

$$\|V\|_{1,2,s} \leq c_{42} \left(\|u\|_{s,\mathbb{T}\times\Omega_{\rho,R_0}} + \|\nabla u\|_{s,\mathbb{T}\times\Omega_{\rho,R_0}} + \|\partial_t u\|_{s,\mathbb{T}\times\Omega_{\rho,R_0}} \right) \tag{2.34}$$

for all $s \in (1,\infty)$. Since

$$\int_{\Omega_{\rho,R_0}} \nabla\psi_2 \cdot u \, dx = \int_{\Omega_{R_0}} \mathrm{div}\left(\psi_2 u\right) dx = \int_{\partial\Omega_{R_0}} u \cdot n \, dS = 0,$$

the existence of a vector field V with the properties above can be established by the same construction as the one used in Theorem III.3.3. in Galdi (2011); see also the proof of Lemma 3.2.1 in Kyed (2012). We now let

$$\begin{aligned} w\colon \mathbb{R}\times\mathbb{R}^3 &\to \mathbb{R}^3, \quad w(t,x) := \psi_2(x)\,u(t,x) - V(t,x), \\ \pi\colon \mathbb{R}\times\mathbb{R}^3 &\to \mathbb{R}, \quad \pi(t,x) := \psi_2(x)\,\mathfrak{p}(t,x). \end{aligned} \tag{2.35}$$

Then $(w,\pi) \in W^{1,2,2}_{\mathrm{per},\perp}(\mathbb{R}\times\Omega_{R_0}) \times D^{1,2}_{\mathrm{per},\perp}(\mathbb{R}\times\Omega_{R_0})$ is a solution to the problem

$$\begin{cases} \partial_t w - \Delta w + \lambda\partial_1 w + \nabla\pi \\ \quad = \psi_2 F - 2\nabla\psi_2 \cdot \nabla u - \Delta\psi_2 u + \lambda\partial_1\psi_2 u + \nabla\psi_2\mathfrak{p} \\ \quad\quad + [\partial_t - \Delta + \lambda\partial_1]V & \text{in } \mathbb{R}\times\Omega_{R_0}, \\ \mathrm{div}w = 0 & \text{in } \mathbb{R}\times\Omega_{R_0}, \\ w = 0 & \text{on } \mathbb{R}\times\partial\Omega_{R_0}. \end{cases}$$

By Lemma 2.4 we thus deduce

$$\begin{aligned} \|w\|_{1,2,s} &\leq c_{43} \, \| \psi_2 F - 2\nabla\psi_2 \cdot \nabla u - \Delta\psi_2 u + \lambda\partial_1\psi_2 u + \nabla\psi_2\mathfrak{p} \\ &\quad\quad + [\partial_t - \Delta + \lambda\partial_1]V\|_s \\ &\leq c_{44} \left(\|F\|_s + \|u\|_{s,\mathbb{T}\times\Omega_{\rho,R_0}} + \|\nabla u\|_{s,\mathbb{T}\times\Omega_{\rho,R_0}} \right. \\ &\quad\quad \left. + \|\mathfrak{p}\|_{s,\mathbb{T}\times\Omega_{\rho,R_0}} + \|V\|_{1,2,s} \right) \end{aligned}$$

for all $s \in (1, \infty)$. Since $\Omega_{\rho, R_0} \subset \Omega^\rho$, we can combine the estimate for w with (2.34) to obtain

$$\|u\|_{1,2,s,\Omega_\rho} \leq c_{45} \left(\|F\|_s + \|u\|_{s,\mathrm{T} \times \Omega_\rho} + \|\nabla u\|_{s,\mathrm{T} \times \Omega_\rho} + \|\mathfrak{p}\|_{s,\mathrm{T} \times \Omega_\rho} \right),$$

which was the intermediate goal at this stage. Combining the estimate above with (2.33), we have

$$\|u\|_{1,2,s} \leq c_{46} \left(\|F\|_s + \|u\|_{s,\mathrm{T} \times \Omega_{R_0}} + \|\nabla u\|_{s,\mathrm{T} \times \Omega_{R_0}} + \|\mathfrak{p}\|_{s,\mathrm{T} \times \Omega_{R_0}} \right)$$
$$(2.36)$$

for all $s \in (1, \infty)$.

We now move on to the final part of the proof. We emphasize that estimate (2.36) has been established for all $s \in (1, \infty)$, but we do not actually know whether the right-hand side is finite or not. At the outset, we only know the right-hand side is finite for $s \leq 2$. We shall now use a bootstrap argument to show that it is also finite when $s \in (2, \infty)$. For this purpose, we employ the embedding properties of $W_{\mathrm{per}}^{1,2,s}(\mathbb{R} \times \Omega)$ stated in Theorem 2.1. Choosing for example $\alpha = \beta = \frac{1}{2}$ in Theorem 2.1, we obtain the implication

$$u \in W_{\mathrm{per}}^{1,2,s}(\mathbb{R} \times \Omega) \ \Rightarrow \ u, \nabla u \in L_{\mathrm{per}}^{\frac{3}{2}s}(\mathbb{R} \times \Omega) \qquad \forall s \in [2, \infty). \quad (2.37)$$

We now turn to estimate (2.25) of the pressure term. By Hölder's inequality,

$$\int_0^T \left(\|\nabla u(t, \cdot)\|_{s,\Omega_{R_0}}^{\frac{s-1}{s}} \|\nabla u(t, \cdot)\|_{1,s,\Omega_{R_0}}^{\frac{1}{s}} \right)^{\frac{3}{2}s} dt$$
$$\leq \left(\int_0^T \|\nabla u(t, \cdot)\|_{s,\Omega_{R_0}}^{\frac{(s-1)s}{\frac{2}{3}s-1}} dt \right)^{\frac{s-\frac{3}{2}}{s}} \|u\|_{1,2,s}^{\frac{3}{2}}.$$

Utilizing again Theorem 2.1, this time with $\beta = 1$, we see that the right-hand side above is finite for all $s \in [2, \infty)$, provided $u \in W_{\mathrm{per}}^{1,2,s}(\mathbb{R} \times \Omega)$. Due to the normalization of the pressure $\mathfrak{p}$ carried out in the beginning of the proof, Lemma 2.5 can be applied to infer from (2.25) that

$$\forall s \in [2, \infty): \quad u \in W_{\mathrm{per}}^{1,2,s}(\mathbb{R} \times \Omega) \ \Rightarrow \ \mathfrak{p} \in L_{\mathrm{per}}^{\frac{3}{2}s}(\mathbb{R} \times \Omega_{R_0}). \quad (2.38)$$

Combining (2.36) with the implications (2.37) and (2.38), we find that

$$\forall s \in [2, \infty): \quad u \in W_{\mathrm{per}}^{1,2,s}(\mathbb{R} \times \Omega) \ \Rightarrow \ u \in W_{\mathrm{per}}^{1,2,\frac{3}{2}s}(\mathbb{R} \times \Omega). \quad (2.39)$$

Starting with $s = 2$, we can now bootstrap (2.39) a sufficient number of times to deduce that $u \in W_{\mathrm{per}}^{1,2,s}(\mathbb{R} \times \Omega)$ for any $s \in (2, \infty)$. Knowing now that the right-hand side of (2.36) is finite for all $s \in (1, \infty)$, we can use interpolation and (2.25) in combination with Young's inequality to deduce

$$\|u\|_{1,2,s} \leq c_{47} \left(\|F\|_s + \|u\|_{s, \mathbb{T} \times \Omega_{R_0}} \right)$$

for all $s \in (1, \infty)$. It then follows directly from (2.11) that

$$\|u\|_{1,2,s} + \|\nabla \mathfrak{p}\|_s \leq c_{47} \left(\|F\|_s + \|u\|_{s, \mathbb{T} \times \Omega_{R_0}} \right) \tag{2.40}$$

for all $s \in (1, \infty)$.

If $(U, \mathfrak{P}) \in W^{1,2,r}_{\mathrm{per},\perp}(\mathbb{R} \times \Omega) \times D^{1,r}_{\mathrm{per},\perp}(\mathbb{R} \times \Omega)$ with $r \in (1, \infty)$ is a solution to (2.11) with homogeneous right-hand side, then $U = \nabla \mathfrak{P} = 0$ follows by a duality argument. More specifically, since for an arbitrary $\varphi \in C^\infty_{0,\mathrm{per},\perp}(\mathbb{R} \times \Omega)^3$ the existence of a solution

$$(W, \Pi) \in W^{1,2,r'}_{\mathrm{per},\perp}(\mathbb{R} \times \Omega) \times D^{1,r'}_{\mathrm{per},\perp}(\mathbb{R} \times \Omega)$$

to (2.11) with φ as the right-hand side has just been established, the computation

$$
\begin{aligned}
0 &= \int_0^T \int_\Omega (\partial_t U - \Delta U + \lambda \partial_1 U + \nabla \mathfrak{P}) \cdot W \, dx dt \\
&= - \int_0^T \int_\Omega U \cdot (\partial_t W - \Delta W + \lambda \partial_1 W + \nabla \Pi) \, dx dt \\
&= \int_0^T \int_\Omega U \cdot \varphi \, dx dt
\end{aligned}
\tag{2.41}
$$

is valid. It follows that $U = 0$ and in turn, directly from (2.11), that also $\nabla \mathfrak{P} = 0$.

We now return to the estimate (2.40). Owing to the fact a solution to (2.11) with homogeneous right-hand is necessarily zero (which we have just shown above) a standard contradiction argument (see for example the proof of Proposition 2 in Galdi, 2013b) can be used to eliminate the lower order term on the right-hand side in (2.40) to conclude that

$$\|u\|_{1,2,q} + \|\nabla \mathfrak{p}\|_q \leq c_{48} \|F\|_q. \tag{2.42}$$

It is easy to verify that $C^\infty_{0,\mathrm{per},\perp}(\mathbb{T} \times \Omega)$ is dense in $L^q_{\mathrm{per},\perp}(\mathbb{T} \times \Omega)$. By a density argument, the existence of a solution

$$(u, \mathfrak{p}) \in W^{1,2,q}_{\mathrm{per},\perp}(\mathbb{R} \times \Omega)^3 \times D^{1,q}_{\mathrm{per},\perp}(\mathbb{T} \times \Omega)$$

to (2.11) that satisfies (2.42) follows for any $F \in L^q_{\mathrm{per},\perp}(\mathbb{T} \times \Omega)^3$.

Finally, assume that $(\tilde{u}, \tilde{\mathfrak{p}}) \in W^{1,2,r}_{\mathrm{per},\perp}(\mathbb{R} \times \Omega) \times D^{1,r}_{\mathrm{per},\perp}(\mathbb{R} \times \Omega)$ is another solution to (2.11) with $r \in (1, \infty)$. The duality argument used in (2.41) applied to the difference $(u - \widetilde{u}, \mathfrak{p} - \widetilde{\mathfrak{p}})$ yields $u = \widetilde{u}$ and $\nabla \mathfrak{p} = \nabla \widetilde{\mathfrak{p}}$. The proof of the theorem is thereby complete. $\qquad\square$

By combining Theorem 2.2 with well-known L^q estimates for the steady-state Oseen system, we can formulate what can be referred to as "maximal L^q regularity" for the time-periodic Oseen system (2.11).

Corollary 2.6 *Let $\Omega \subset \mathbb{R}^3$ be an exterior domain of class C^2, take $\lambda \in (0, \lambda_0]$ and $q \in (1, 2)$. Define*

$$\mathcal{X}_\lambda^q(\Omega) := \{ v \in X_\lambda^q(\Omega)^3 : \operatorname{div} v = 0, \ v = 0 \ on \ \partial\Omega \}.$$

Moreover, let $r \in (1, \infty)$ and

$$\mathcal{W}_{per,\perp}^{1,2,q,r}(\mathbb{R} \times \Omega) := \{ w \in W_{per,\perp}^{1,2,q,r}(\mathbb{R} \times \Omega)^3 : \operatorname{div} w = 0, \ w = 0 \ on \ \partial\Omega \}.$$

Then the $\mathcal{T}$-time-periodic Oseen operator

$$A_O \colon \left(\mathcal{X}_\lambda^q(\Omega) \oplus \mathcal{W}_{per,\perp}^{1,2,q,r}(\mathbb{R} \times \Omega) \right) \times \left(D_{R_0}^{1,q}(\Omega) \oplus D_{per,\perp,R_0}^{1,q,r}(\mathbb{R} \times \Omega) \right) \rightarrow$$

$$L^q(\Omega)^3 \oplus L_{per,\perp}^{q,r}(\mathbb{R} \times \Omega)^3,$$

$$A_O(v + w, p + \pi) := \partial_t w - \nu \Delta(v + w) + \lambda \partial_1(v + w) + \nabla(p + \pi)$$

$$(2.43)$$

is a homeomorphism with $\|A_O^{-1}\|$ depending only on q, r, Ω, ν and λ. If $q \in (1, \frac{3}{2})$, then $\|A_O^{-1}\|$ depends only on the upper bound λ_0 and not on λ itself.

Proof It is well known that the steady-state Oseen operator, that is, the Oseen operator from (2.43) restricted to time-independent functions, is a homeomorphism as a mapping $A_O \colon \mathcal{X}_\lambda^q(\Omega) \times D_{R_0}^{1,q}(\Omega) \rightarrow L^q(\Omega)^3$; see for example Theorem VII.7.1 in Galdi (2011). By Theorem 2.2, it follows that also the time-periodic Oseen operator is a homeomorphism as a mapping $A_O \colon W_{per,\perp}^{1,2,q,r}(\mathbb{R} \times \Omega) \times D_{per,\perp,R_0}^{1,q,r}(\mathbb{R} \times \Omega) \rightarrow L_{per,\perp}^{q,r}(\mathbb{R} \times \Omega)^3$. Since clearly $\mathcal{P}$ and $\mathcal{P}_\perp$ commute with A_O, it further follows that A_O is a homeomorphism as an operator in the setting (2.43). The dependence of $\|A_O^{-1}\|$ on the various parameters follows from Theorem VII.7.1 in Galdi (2011) and Theorem 2.2. $\qquad\square$

2.6 Fully Nonlinear Problem

Existence of a solution to the fully nonlinear problem (2.1) will now be established. We employ a fixed point argument based on the estimates established for the linearized system (2.11) in the previous section. We need to assume (2.2) to ensure that (2.11) is indeed a suitable linearization of (2.1). Moreover, we need to assume (2.3) to ensure that all terms appearing on the right-hand side after linearizing (2.1) are subordinate to the linear operator on the left-hand side.

For convenience, we rewrite (2.1) by replacing u with $u+v_b$ and obtain the following equivalent problem:

$$
\begin{cases}
\partial_t u + (u - v_b) \cdot \nabla u = \nu \Delta u - \nabla \mathfrak{p} + f & \text{in } \mathbb{R} \times \Omega, \\
\operatorname{div} u = 0 & \text{in } \mathbb{R} \times \Omega, \\
u = u_* & \text{on } \mathbb{R} \times \partial\Omega, \\
\lim_{|x|\to\infty} u(t,x) = 0.
\end{cases}
\tag{2.44}
$$

Theorem 2.7 *Let $\Omega \subset \mathbb{R}^3$ be an exterior domain of class C^2. Assume that $v_b(t) = -u_\infty(t)e_1$ for a $\mathcal{T}$-periodic function $u_\infty \colon \mathbb{R} \to \mathbb{R}$ with $\lambda := \mathcal{P}u_\infty > 0$. Moreover, let $q \in \left[\frac{6}{5}, \frac{4}{3}\right]$. There is an $\lambda_0 > 0$ such that for all $\lambda \in (0, \lambda_0]$ there is a $\varepsilon_0 > 0$ such that for all $f \in L^q_{\mathrm{per}}(\mathbb{R} \times \Omega)^3$, $u_\infty \in L^\infty_{\mathrm{per}}(\mathbb{R})$ and*

$$
u_* \in W^{1,q}_{\mathrm{per}}\left(\mathbb{R}; W^{2-\frac{3-q}{3q}, \frac{3q}{3-q}}(\partial\Omega)\right)^3
\tag{2.45}
$$

satisfying

$$
\|f\|_q + \|\mathcal{P}_\perp f\|_{\frac{3q}{3-q}} + \|\mathcal{P}_\perp u_\infty\|_\infty + \|u_*\|_{W^{1,q}_{\mathrm{per}}\left(\mathbb{R}; W^{2-\frac{3-q}{3q}, \frac{3q}{3-q}}(\partial\Omega)\right)} \leq \varepsilon_0
\tag{2.46}
$$

there is a solution $(u, \mathfrak{p})$ to (2.44) with

$$
u \in \left(X^q_\lambda(\Omega) \oplus W^{1,2,q,\frac{3q}{3-q}}_{\mathrm{per},\perp}(\mathbb{R} \times \Omega)\right)^3,
\tag{2.47}
$$

$$
\mathfrak{p} \in \left(D^{1,q}(\Omega) \oplus D^{1,q,\frac{3q}{3-q}}_{\mathrm{per},\perp}(\mathbb{R} \times \Omega)\right).
\tag{2.48}
$$

Proof In order to "lift" the boundary values in (2.44), that is, in order to rewrite the system as one with homogeneous boundary values, a solution $(\mathcal{W}, \Pi_\perp) \in W_{\mathrm{per},\perp}^{1,2,q,\frac{3q}{3-q}}(\mathbb{R} \times \Omega) \times D_{\mathrm{per},\perp}^{1,q,\frac{3q}{3-q}}(\mathbb{R} \times \Omega)$ to

$$
\begin{cases}
-\nu\Delta\mathcal{W} + \nabla\Pi_\perp = \mathcal{W} & \text{in } \mathbb{R} \times \Omega, \\
\mathrm{div}\,\mathcal{W} = 0 & \text{in } \mathbb{R} \times \Omega, \\
\mathcal{W} = \mathcal{P}_\perp u_* & \text{on } \mathbb{R} \times \partial\Omega,
\end{cases}
\tag{2.49}
$$

is introduced. Observe that (2.49) is a Stokes resolvent-type problem. One can therefore use standard methods to solve (2.49) in $\mathcal{T}$-time periodic function spaces and obtain a solution that satisfies

$$
\forall r \in \left(1, \frac{3q}{3-q}\right]:
$$

$$
\|\mathcal{W}\|_{1,2,r} + \|\nabla\Pi_\perp\|_r \le c_{49}\|u_*\|_{W_{\mathrm{per}}^{1,q}\left(\mathbb{R}; W^{2-\frac{3-q}{3q}, \frac{3q}{3-q}}(\partial\Omega)\right)},
\tag{2.50}
$$

where $c_{49} = c_{49}(r, q, \Omega, \nu)$. Furthermore, classical results for the steady-state Oseen problem (Galdi, 2011, Theorem VII.7.1) ensure existence of a solution $(\mathcal{V}, \Pi_s) \in X_\lambda^q(\Omega) \times D^{1,q}(\Omega)$ to

$$
\begin{cases}
-\nu\Delta\mathcal{V} + \lambda\partial_1\mathcal{V} + \nabla\Pi_s = 0 & \text{in } \Omega, \\
\mathrm{div}\,\mathcal{V} = 0 & \text{in } \Omega, \\
\mathcal{V} = \mathcal{P}u_* & \text{on } \partial\Omega,
\end{cases}
\tag{2.51}
$$

which satisfies, since $q \ge \frac{6}{5}$ implies $\frac{3q}{3-q} \ge 2$,

$$
\forall r \in (1,2): \ \|\mathcal{V}\|_{X_\lambda^r(\Omega)} + \|\nabla\Pi_s\|_r \le c_{50}\|\mathcal{P}u_*\|_{W^{2-\frac{3-q}{3q}, \frac{3q}{3-q}}(\partial\Omega)},
\tag{2.52}
$$

where $c_{50} = c_{50}(r, \Omega, \nu)$. We shall now establish the existence of a solution $(u, \mathfrak{p})$ to (2.44) of the form

$$
u = v + \mathcal{V} + w + \mathcal{W}, \quad \mathfrak{p} = p + \Pi_s + \pi + \Pi_\perp,
\tag{2.53}
$$

where $(v, p) \in X_\lambda^q(\Omega) \times D^{1,q}(\Omega)$ is a solution to the steady-state problem

$$
\begin{cases}
-\nu\Delta v + \lambda\partial_1 v + \nabla p = \mathfrak{R}_1(v, w, \mathcal{V}, \mathcal{W}) & \text{in } \Omega, \\
\mathrm{div}\,v = 0 & \text{in } \Omega, \\
v = 0 & \text{on } \partial\Omega,
\end{cases}
\tag{2.54}
$$

with

$$
\begin{aligned}
\mathfrak{R}_1(v, w, \mathcal{V}, \mathcal{W}) :=& \\
&- v \cdot \nabla v - v \cdot \nabla\mathcal{V} - \mathcal{V} \cdot \nabla v - \mathcal{V} \cdot \nabla\mathcal{V} \\
&- \mathcal{P}[w \cdot \nabla w] - \mathcal{P}[w \cdot \nabla\mathcal{W}] - \mathcal{P}[\mathcal{W} \cdot \nabla w] - \mathcal{P}[\mathcal{W} \cdot \nabla\mathcal{W}] \\
&- \mathcal{P}[\mathcal{P}_\perp u_\infty\, \partial_1 w] - \mathcal{P}[\mathcal{P}_\perp u_\infty\, \partial_1\mathcal{W}] + \mathcal{P}f,
\end{aligned}
$$

and $(w,\pi) \in W^{1,2,q,\frac{3q}{3-q}}_{\mathrm{per},\perp}(\mathbb{R} \times \Omega) \times D^{1,q,\frac{3q}{3-q}}_{\mathrm{per},\perp}(\mathbb{R} \times \Omega)$ a solution to

$$\begin{cases} \partial_t w - \nu\Delta w + \lambda\partial_1 w + \nabla\pi = \mathfrak{R}_2(v,w,\mathcal{V},\mathcal{W}) & \text{in } \mathbb{R} \times \Omega, \\ \operatorname{div}w = 0 & \text{in } \mathbb{R} \times \Omega, \qquad (2.55)\\ w = 0 & \text{on } \mathbb{R} \times \partial\Omega, \end{cases}$$

with

$$\begin{aligned} \mathfrak{R}_2(v,w,\mathcal{V},\mathcal{W}) := \\ &- \mathcal{P}_\perp[w\cdot\nabla w] - \mathcal{P}_\perp[w\cdot\nabla\mathcal{W}] - \mathcal{P}_\perp[\mathcal{W}\cdot\nabla w] - \mathcal{P}_\perp[\mathcal{W}\cdot\nabla\mathcal{W}] \\ &- v\cdot\nabla w - v\cdot\nabla\mathcal{W} - w\cdot\nabla v - w\cdot\nabla\mathcal{V} \\ &- \mathcal{V}\cdot\nabla w - \mathcal{V}\cdot\nabla\mathcal{W} - \mathcal{W}\cdot\nabla v - \mathcal{W}\cdot\nabla\mathcal{V} \\ &- \mathcal{P}_\perp u_\infty\,\partial_1 v - \mathcal{P}_\perp u_\infty\,\partial_1\mathcal{V} - \mathcal{P}_\perp[\mathcal{P}_\perp u_\infty\,\partial_1 w] - \mathcal{P}_\perp[\mathcal{P}_\perp u_\infty\,\partial_1\mathcal{W}] \\ &- \partial_t\mathcal{W} - \mathcal{W} + \lambda\partial_1\mathcal{W} + \mathcal{P}_\perp f. \end{aligned}$$

The systems (2.54) and (2.55) emerge as the result of inserting (2.53) into (2.44) and subsequently applying $\mathcal{P}$ and $\mathcal{P}_\perp$ to the equations. Recalling the function spaces introduced in Corollary 2.6, we define the Banach space

$$\begin{aligned} \mathcal{K}^q_\lambda(\mathbb{R} \times \Omega) := \\ \mathcal{X}^q_\lambda(\Omega) \oplus W^{1,2,q,\frac{3q}{3-q}}_{\mathrm{per},\perp}(\mathbb{R} \times \Omega) \times D^{1,q}_{R_0}(\Omega) \oplus D^{1,q,\frac{3q}{3-q}}_{\mathrm{per},\perp,R_0}(\mathbb{R} \times \Omega). \end{aligned} \qquad (2.56)$$

We can obtain solutions (v,p) and (w,π) to (2.54) and (2.55), respectively, as a fixed point of the mapping

$$\mathcal{N} \colon \mathcal{K}^q_\lambda(\mathbb{R} \times \Omega) \to \mathcal{K}^q_\lambda(\mathbb{R} \times \Omega),$$
$$\mathcal{N}(v+w,p+\pi) := A_O^{-1}\big(\mathfrak{R}_1(v,w,\mathcal{V},\mathcal{W}) + \mathfrak{R}_2(v,w,\mathcal{V},\mathcal{W})\big).$$

We shall show that $\mathcal{N}$ is a contracting self-mapping on a ball of sufficiently small radius. For this purpose, take $\rho > 0$ and consider some $(v+w,p+\pi) \in \mathcal{K}^q_\lambda \cap B_\rho$. Suitable estimates of $\mathfrak{R}_1$ and $\mathfrak{R}_2$ in combination with a smallness assumption on ε_0 from (2.46) are needed to guarantee that $\mathcal{N}$ has the desired properties.

We first estimate the terms of $\mathfrak{R}_1$. Since $q \in [\frac{6}{5},\frac{4}{3}]$ implies that we have $\frac{4q}{4-q} \leq 2 \leq \frac{3q}{3-q}$, we can employ first Hölder's inequality and then interpolation to estimate

$$\|v\cdot\nabla v\|_q \leq \|v\|_{\frac{2q}{2-q}}\|\nabla v\|_2 \leq \lambda^{-\frac{1}{2}}\|v\|_{X^q_\lambda}\|\nabla v\|^\theta_{\frac{4q}{4-q}}\|\nabla v\|^{1-\theta}_{\frac{3q}{3-q}}$$

with $\theta = \frac{10q-12}{q}$. It thus follows by the embedding $W^{1,q}(\Omega) \hookrightarrow L^{\frac{3q}{3-q}}(\Omega)$

that

$$\|v \cdot \nabla v\|_q \leq c_{51} \lambda^{-\frac{1}{2}-\frac{\theta}{4}} \|v\|_{X_\lambda^q}^{1+\theta} \|\nabla^2 v\|_q^{1-\theta}$$

$$\leq c_{51} \lambda^{-\frac{3q-3}{q}} \|v\|_{X_\lambda^q}^2 \leq c_{51} \lambda^{-\frac{3q-3}{q}} \rho^2. \tag{2.57}$$

The other terms in the definition of $\mathfrak{R}_1$ can be estimated in a similar fashion to conclude in combination with (2.46) that

$$\|\mathfrak{R}_1(v, w, \mathcal{V}, \mathcal{W})\|_{L^q(\Omega)} \leq c_{52}\left(\lambda^{-\frac{3q-3}{q}} \rho^2 + \lambda^{-\frac{1}{2}} \rho\varepsilon_0 + \rho\varepsilon_0 + \varepsilon_0{}^2 + \varepsilon_0\right).$$

An estimate of $\mathfrak{R}_2$ is required both in the $L_{\mathrm{per}}^q(\mathbb{R} \times \Omega)$ and $L_{\mathrm{per}}^{\frac{3q}{3-q}}(\mathbb{R} \times \Omega)$ norms. Observe that

$$\|\mathcal{P}_\perp u_\infty \, \partial_1 v\|_{L_{\mathrm{per}}^q(\mathbb{R} \times \Omega)} \leq c_{53}\|\mathcal{P}_\perp u_\infty\|_\infty \|\partial_1 v\|_q \leq c_{53}\lambda^{-1}\varepsilon_0\rho. \tag{2.58}$$

With the help of the embedding properties in Theorem 2.1, the other terms in $\mathfrak{R}_2$ can be estimated to obtain

$$\|\mathfrak{R}_2(v, w, \mathcal{V}, \mathcal{W})\|_{L_{\mathrm{per}}^q(\mathbb{R} \times \Omega)} \leq c_{54}\left(\lambda^{-1}\varepsilon_0\rho + \rho^2 + \rho\varepsilon_0 + \varepsilon_0{}^2 + \lambda\varepsilon_0 + \varepsilon_0\right).$$

The embedding properties can also be used to establish an $L_{\mathrm{per},\perp}^{\frac{3q}{3-q}}(\mathbb{R} \times \Omega)$ estimate of $\mathfrak{R}_2$. For example,

$$\|w \cdot \nabla w\|_{L_{\mathrm{per}}^{\frac{3q}{3-q}}(\mathbb{R} \times \Omega)} \leq c_{55}\|w\|_{L_{\mathrm{per}}^{\frac{3q}{3-q}}(\mathbb{R}; L^\infty(\Omega))} \|\nabla w\|_{L_{\mathrm{per}}^\infty(\mathbb{R}; L^{\frac{3q}{3-q}}(\Omega))}$$

$$\leq c_{56}\rho^2,$$

where Theorem 2.1 is utilized with $\alpha = 0$ and $\beta = 1$ in the last inequality. For this particular utilization of Theorem 2.1, it is required that $q \geq \frac{6}{5}$. Further note that

$$\|\mathcal{P}_\perp u_\infty \, \partial_1 v\|_{\frac{3q}{3-q}} \leq \varepsilon_0 \|v\|_{X_\lambda^q} \leq \varepsilon_0\rho,$$

which explains the choice of the exponent $\frac{3q}{3-q}$ in the setting of the mapping $\mathcal{N}$. The rest of the terms in $\mathfrak{R}_2$ can be estimated to conclude

$$\|\mathfrak{R}_2(v, w, \mathcal{V}, \mathcal{W})\|_{L_{\mathrm{per}}^{\frac{3q}{3-q}}(\mathbb{R} \times \Omega)} \leq c_{57}\left(\rho^2 + \rho\varepsilon_0 + \varepsilon_0{}^2 + \lambda\varepsilon_0 + \varepsilon_0\right).$$

We can now conclude from Corollary 2.6 (recalling that $\|A_O^{-1}\|$ is independent of λ) the estimate

$$\|\mathcal{N}(v + w, p + \pi)\|_{\mathcal{K}_\lambda^q} \leq \|A_O^{-1}\| \cdot \left(\|\mathfrak{R}_1\|_{L^q(\Omega)} + \|\mathfrak{R}_2\|_{L_{\mathrm{per}}^{q, \frac{3q}{3-q}}(\mathbb{R} \times \Omega)}\right)$$

$$\leq c_{58}\left(\lambda^{-\frac{3q-3}{q}} \rho^2 + \lambda^{-1}\varepsilon_0\rho + \lambda^{-\frac{1}{2}} \rho\varepsilon_0 + \rho\varepsilon_0 + \varepsilon_0{}^2 + \lambda\varepsilon_0 + \varepsilon_0\right).$$

In particular, $\mathcal{N}$ becomes a self-mapping on B_ρ if

$$c_{58}\left(\lambda^{-\frac{3q-3}{q}}\rho^2 + \lambda^{-1}\varepsilon_0\rho + \lambda^{-\frac{1}{2}}\rho\varepsilon_0 + \rho\varepsilon_0 + \varepsilon_0{}^2 + \lambda\varepsilon_0 + \varepsilon_0\right) \leq \rho.$$

One may choose $\varepsilon_0 := \lambda^2$ and $\rho := \lambda$ to find the above inequality satisfied for sufficiently small λ. For such choice of parameters, one may further verify that $\mathcal{N}$ is also a contraction. By the contraction mapping principle, existence of a fixed point for $\mathcal{N}$ follows. This concludes the proof. $\quad\square$

As for classical anisotropic Sobolev spaces, the trace operator for time-periodic Sobolev spaces is continuous and surjective for any $r \in (1,\infty)$ in the setting

$$\mathrm{Tr}\colon W_{\mathrm{per}}^{1,2,r}(\mathbb{R}\times\Omega) \to W_{\mathrm{per}}^{1-\frac{1}{2r},2-\frac{1}{r},r}(\mathbb{R}\times\partial\Omega),$$

where

$$W_{\mathrm{per}}^{1-\frac{1}{2r},2-\frac{1}{r},r}(\mathbb{R}\times\partial\Omega) := W_{\mathrm{per}}^{1-\frac{1}{2r},r}\left(\mathbb{R};L^r(\Omega)\right) \cap L^r\left(\mathbb{R};W^{2-\frac{1}{r},r}(\Omega)\right).$$

Consequently, for the Sobolev space in (2.47), in which a solution u is established in Theorem 2.7, we find that

$$\mathrm{Tr}\colon W_{\mathrm{per},\perp}^{1,2,q,\frac{3q}{3-q}}(\mathbb{R}\times\Omega) \to W_{\mathrm{per}}^{1-\frac{3-q}{6q},2-\frac{3-q}{3q},\frac{3q}{3-q}}(\mathbb{R}\times\partial\Omega) \qquad (2.59)$$

is continuous and surjective. It therefore seems natural that Theorem 2.7 would hold for arbitrary boundary values u_* in a larger space than (2.45). We leave it as an open question as to whether or not the regularity assumptions on the boundary values in Theorem 2.7 can be weakened.

Acknowledgements

GPG was partially supported by NSF-DMS grant 1614011.

References

Engel, K.-J. & Nagle, R. (2000) *One-parameter semigroups for linear evolution equations.* Graduate Texts in Mathematics, vol. 194. Springer-Verlag, New York.

Galdi, G.P. (2011) *An introduction to the mathematical theory of the Navier–Stokes equations. Steady-state problems.* 2nd ed. Springer, New York.

Galdi, G.P. (2013a) Existence and uniqueness of time-periodic solutions to the Navier–Stokes equations in the whole plane. *Discrete Contin. Dyn. Syst., Ser. S* **6**(5), 1237–1257.

Galdi, G.P. (2013b) On time-periodic flow of a viscous liquid past a moving cylinder. *Arch. Ration. Mech. Anal.* **210**(2), 451–498.

Galdi, G.P. & Silvestre, A.L. (2006) Existence of time-periodic solutions to the Navier–Stokes equations around a moving body. *Pac. J. Math.* **223**(2), 251–267.

Galdi, G.P. & Silvestre, A.L. (2009) On the motion of a rigid body in a Navier–Stokes liquid under the action of a time-periodic force. *Indiana Univ. Math. J.* **58**(6), 2805–2842.

Galdi, G.P. & Sohr, H. (2004) Existence and uniqueness of time-periodic physically reasonable Navier-Stokes flow past a body. *Arch. Ration. Mech. Anal.* **172**(3), 363–406.

Geissert, M., Hieber, M., & Nguyen, T.-H. (2016) A general approach to time periodic incompressible viscous fluid flow problems. *Arch. Ration. Mech. Anal.* **220**(3), 1095–1118.

Giga, Y. (1981) Analyticity of the semigroup generated by the Stokes operator in L_r spaces. *Math. Z.* **178**(3), 297–329.

Giga, Y. & Sohr, H. (1991) Abstract L^p estimates for the Cauchy problem with applications to the Navier–Stokes equations in exterior domains. *J. Funct. Anal.* **102**(1), 72–94.

Grafakos, L. (2008) *Classical Fourier analysis. 2nd ed.* Springer, New York.

Grafakos, L. (2009) *Modern Fourier analysis. 2nd ed.* Springer, New York.

Kaniel, S. & Shinbrot, M. (1967) A reproductive property of the Navier-Stokes equations. *Arch. Rational Mech. Anal.* **24**, 363–369.

Kozono, H. & and Nakao, M. (1996) Periodic solutions of the Navier-Stokes equations in unbounded domains. *Tohoku Math. J. (2)* **48**(1), 33–50.

Kyed, M. (2012) *Time-Periodic Solutions to the Navier–Stokes Equations.* Habilitationsschrift, Technische Universität Darmstadt.

Kyed, M. (2014a) Existence and regularity of time-periodic solutions to the three-dimensional Navier–Stokes equations. *Nonlinearity* **27**(12), 2909–2935.

Kyed, M. (2014b) Maximal regularity of the time-periodic linearized Navier–Stokes system. *J. Math. Fluid Mech.* **16**(3), 523–538.

Kyed, M. (2016) A fundamental solution to the time-periodic Stokes equations. *J. Math. Anal. Appl.* **437**(1), 708719.

Maremonti, P. (1991a) Existence and stability of time-periodic solutions to the Navier-Stokes equations in the whole space. *Nonlinearity* **4**(2), 503–529.

Maremonti, P. (1991b) Some theorems of existence for solutions of the Navier–Stokes equations with slip boundary conditions in half-space. *Ric. Mat.* **40**(1), 81–135.

Maremonti, P. & Padula, M. (1996) Existence, uniqueness and attainability of periodic solutions of the Navier-Stokes equations in exterior domains. *Zap. Nauchn. Sem. S.-Peterburg. Otdel. Mat. Inst. Steklov. (POMI)* **233**(27), 142–182.

Miyakawa, T. & Teramoto, Y. (1982) Existence and periodicity of weak solutions of the Navier-Stokes equations in a time dependent domain. *Hiroshima Math. J.* **12**(3), 513–528.

Morimoto, H. (1972) On existence of periodic weak solutions of the Navier–Stokes equations in regions with periodically moving boundaries. *J. Fac. Sci., Univ. Tokyo, Sect. I A* **18**, 499–524.

Nguyen, T.-H. (2014) Periodic Motions of Stokes and Navier–Stokes Flows Around a Rotating Obstacle. *Arch. Ration. Mech. Anal.* **213**(2), 689–703.

Prodi, G. (1960) Qualche risultato riguardo alle equazioni di Navier-Stokes nel caso bidimensionale. *Rend. Sem. Mat. Univ. Padova* **30**, 1–15.

Prouse, G. (1963) Soluzioni periodiche dell'equazione di Navier-Stokes. *Atti Accad. Naz. Lincei Rend. Cl. Sci. Fis. Mat. Natur. (8)* **35**, 443–447.

Serrin, J. (1959) A note on the existence of periodic solutions of the Navier-Stokes equations. *Arch. Rational Mech. Anal.* **3**, 120–122.

Silvestre, A.L. (2012) Existence and uniqueness of time-periodic solutions with finite kinetic energy for the Navier-Stokes equations in $\mathbb{R}^3$. *Nonlinearity* **25**(1), 37–55.

Simader, C.G. (1990) The weak Dirichlet and Neumann problem for the Laplacian in L^q for bounded and exterior domains. Applications. Pages 180–223 of: *Nonlinear analysis, function spaces and applications, Vol. 4 (Roudnice nad Labem, 1990)*. Teubner-Texte Math., vol. 119. Teubner, Leipzig.

Takeshita, A. (1970) On the reproductive property of the 2-dimensional Navier-Stokes equations. *J. Fac. Sci. Univ. Tokyo Sect. I* **16**, 297–311.

Taniuchi, Y. (2009) On the uniqueness of time-periodic solutions to the Navier–Stokes equations in unbounded domains. *Math. Z.* **261**(3), 597–615.

Teramoto, Y. (1983) On the stability of periodic solutions of the Navier–Stokes equations in a noncylindrical domain. *Hiroshima Math. J.* **13**, 607–625.

Van Baalen, G. & and Wittwer, P. (2011) Time periodic solutions of the Navier–Stokes equations with nonzero constant boundary conditions at infinity. *SIAM J. Math. Anal.* **43**(4), 1787–1809.

Yamazaki, M. (2000) The Navier–Stokes equations in the weak-L^n space with time-dependent external force. *Math. Ann.* **317**(4), 635–675.

Yudovich, V.I. (1960) Periodic motions of a viscous incompressible fluid. *Sov. Math., Dokl.* **1**, 168–172.

3

The Rayleigh–Taylor instability in buoyancy-driven variable density turbulence

John D. Gibbon,
Department of Mathematics, Imperial College London,
London, SW7 2AZ. UK.
`j.d.gibbon@ic.ac.uk`

Pooja Rao,
Department of Aerospace Engineering,
University of Illinois at Urbana-Champaign,
319E Talbot Laboratory MC 236, 104 S. Wright, Urbana, IL 61801. USA.
`poojarao@illinois.edu`

Colm-cille P. Caulfield
BP Institute, University of Cambridge, Madingley Rise,
Madingley Road, Cambridge, CB3 0EZ. UK.
Department of Applied Mathematics & Theoretical Physics,
University of Cambridge, Centre for Mathematical Sciences,
Wilberforce Road, Cambridge, CB3 0WA. UK.
`c.p.caulfield@damtp.cam.ac.uk`

Abstract

This paper reviews and summarizes two recent pieces of work on the Rayleigh–Taylor instability. The first concerns the 3D Cahn–Hilliard–Navier–Stokes (CHNS) equations and the BKM-type theorem proved by Gibbon, Pal, Gupta, & Pandit (2016). The second and more substantial topic concerns the variable density model, which is a buoyancy-driven turbulent flow considered by Cook & Dimotakis (2001) and Livescu & Ristorcelli (2007, 2008). In this model $\rho^*(\boldsymbol{x}, t)$ is the composition density of a mixture of two incompressible miscible fluids with fluid densities $\rho_2^* > \rho_1^*$ and ρ_0^* is a reference normalisation density. Following the work of a previous paper (Rao, Caulfield, & Gibbon, 2017), which used the variable $\theta = \ln \rho^*/\rho_0^*$, data from the publicly available Johns Hopkins Turbulence Database suggests that the L^2-spatial average of the density gradient $\nabla\theta$ can reach extremely large values at intermediate times, even in flows with low Atwood number $At = (\rho_2^* - \rho_1^*)/(\rho_2^* + \rho_1^*) = 0.05$. This implies that very strong mixing of the density field at small scales can potentially arise in buoyancy-driven turbulence thus raising the possibility that the density gradient $\nabla\theta$ might blow up in a finite time.

3.1 Background to the Rayleigh–Taylor instability

The topic of this paper is the Rayleigh–Taylor instability (RTI) which concerns the mixing of two fluids of different densities initially set in varying configurations. RTI flows have been very widely studied (Sharp, 1984; Youngs, 1984, 1989; Glimm et al., 2001; Cook & Dimotakis, 2001; Dimonte et al., 2004; Dimotakis, 2005; Lee et al., 2008; Hyunsun et al., 2008; Andrews & Dalziel, 2010), not only because of their relevance to astrophysics (Cabot & Cook, 2006a), plasma fusion (Petrasso, 1994) and materials science (Hohenberg & Halperin, 1977; Bray, 1994; Chaikin & Lubensky, 2000), but because they also pose several unsolved mathematical questions. Moreover, there are substantial differences in the mixing processes depending on whether the fluids are *miscible* or *immiscible*.

The former case has been widely explored by tank experiments and numerical models in which the initial state is set up such that the heavier fluid sits over the lighter, separated by a removable barrier (Lawrie & Dalziel, 2011; Davies-Wykes & Dalziel, 2014). On removal of the barrier at $t = 0$ a mixing zone between the two fluids develops whose thickness is observed to grow proportionately like αt^2, where the value of the constant α has evoked considerable discussion (Dimonte et al., 2004). In such miscible RTI flows the turbulent mixing zone is not driven by some external forcing mechanism, but is supplied with kinetic energy by the conversion of available potential energy stored in the initial density configuration. This kinetic energy naturally drives a turbulent cascade to small scales, with an increase in the dissipation rate of kinetic energy. Such small scales also lead to substantially enhanced gradients in the density field which, in turn, also leads to irreversible mixing, and hence modification in the density distribution. There is growing evidence that buoyancy-driven turbulence is particularly efficient in driving mixing and certainly more efficient than externally forced turbulent flow (Lawrie & Dalziel, 2011; Davies-Wykes & Dalziel, 2014). This evidence poses the further question of whether there are distinguishing characteristics of the buoyancy-driven turbulent flow that are different from the flow associated with an external forcing, in particular whether these characteristics can be identified as being responsible for the enhanced and efficient mixing (Tailleux, 2013). While much effort has gone into modelling and understanding the growth of the mixing layer thickness, and even into the question of the value of α (Dimonte et al., 2004), our concern here, beginning in Section 3.3, will be with the variable-density model for *incompressible miscible flows* and the growth of the gradient of the composition density (Cook & Dimotakis, 2001; Livescu & Ristorcelli, 2007, 2008).

By contrast, the mixing characteristics of incompressible immiscible binary fluids are different from the corresponding miscible case. In the next section we will briefly summarize some results in this case using the 3D Cahn–Hilliard-Navier–Stokes equations as an example.

3.2 The 3D Cahn–Hilliard-Navier–Stokes equations

Attempts to treat the mixing of immiscible fluids as a two-fluid model not only throw up extreme computational difficulties in the tracking of the interface (see Celani et al., 2009) but also raise serious questions regarding ill-posedness. To circumvent these difficulties condensed-matter physicists have employed what are known as phase-field models (Hohenberg & Halperin, 1977; Bray, 1994; Chaikin & Lubensky, 2000), the most widely used of which is the Cahn–Hilliard equation which displays natural phase separation below a critical temperature (Cahn & Hilliard, 1958). When coupled to the forced Navier–Stokes equations, which govern a turbulent fluid advective velocity field $\boldsymbol{u}$, the Cahn–Hilliard equation is

$$\left(\partial_t + \boldsymbol{u} \cdot \nabla\right) \varphi = \gamma \Delta \mu, \qquad \operatorname{div} \boldsymbol{u} = 0, \tag{3.1}$$

where μ is the chemical potential, given by $\mu = \delta \mathcal{F}/\delta\varphi$, where

$$\mathcal{F} = \int_{\mathcal{V}} \left[\frac{\Lambda}{2} |\nabla\varphi|^2 + \frac{\Lambda}{4\xi^2} \left(\varphi^2 - 1\right)^2 \right] \, \mathrm{d}V \tag{3.2}$$

is the free energy. The chemical potential μ is thus given by

$$\mu = \Lambda \left[-\Delta\varphi + \xi^{-2} \left(\varphi^3 - \varphi\right) \right]. \tag{3.3}$$

The parameter γ in (3.1) is called the mobility (Bray, 1994) and Λ in (3.2) is the energy density. The phase field φ takes the value $\varphi = -1$ for one phase and $\varphi = 1$ for the other. The advantage of such a model is the continuity of the interface, of thickness ξ, that lies between the two fluids; its existence removes the necessity of dealing with the complications of tracking a free boundary. The incompressible 3D Navier–Stokes equations ($\operatorname{div} \boldsymbol{u} = 0$), with a divergence-free, mean-zero forcing function $\boldsymbol{f}(\boldsymbol{x})$, are coupled to (3.1) through a $\varphi\nabla\mu$-term

$$\left(\partial_t + \boldsymbol{u} \cdot \nabla\right) \boldsymbol{u} = \nu\Delta\boldsymbol{u} - \varphi\nabla\mu - \nabla p + \boldsymbol{f}(\boldsymbol{x}). \tag{3.4}$$

The combination of (3.1), (3.3) and (3.4) are known as the Cahn–Hilliard-Navier–Stokes (CHNS) equations. The interfacial dynamics are of particular interest particularly since the immiscible Rayleigh–Taylor instability (RTI) is manifest in this thin mixing layer. While our primary concern is in the three-dimensional case, various results are known in two dimensions, such as the regularity of not only the 2D Navier–Stokes equations (Constantin & Foias, 1988; Foias et al., 2001; Doering & Gibbon, 1995) but also of the stand-alone two-dimensional Cahn–Hilliard equations (Elliott & Songmu, 1986). The regularity problem for the two-dimensional Cahn–Hilliard-Navier–Stokes (CHNS) equations has been solved separately in papers by Abels (2009) and Gal & Grasselli (2010) using different boundary conditions but in three dimensions the issue remains a formidable problem.

The energy $E(t)$ of the CHNS equations (3.1), (3.3) and (3.4) is given by (see Celani et al., 2009)

$$E(t) = \int_{\mathcal{V}} \left\{ \tfrac{1}{2}\Lambda|\nabla\varphi|^2 + \frac{\Lambda}{4\xi^2}(\varphi^2 - 1)^2 + \tfrac{1}{2}|\boldsymbol{u}|^2 \right\} \, dV, \qquad (3.5)$$

whose time derivative obeys (under suitable smoothness conditions and finite initial data $E(0)$)

$$\frac{dE}{dt} = -\int_{\mathcal{V}} \left(\nu|\nabla\boldsymbol{u}|^2 + \gamma|\nabla\mu|^2 \right) \, dV + \int_{\mathcal{V}} \boldsymbol{u}\cdot\boldsymbol{f} \, dV. \qquad (3.6)$$

Thus $dE/dT < 0$ in the absence of any additive forcing, in which case E decays. Thinking of E as an additive set of L^2-norms, the L^∞-equivalent is

$$E_\infty(t) = \tfrac{1}{2}\Lambda\|\nabla\varphi\|_\infty^2 + \frac{\Lambda}{4\xi^2}(\|\varphi\|_\infty^2 - 1)^2 + \tfrac{1}{2}\|\boldsymbol{u}\|_\infty^2. \qquad (3.7)$$

E_∞ plays the role of $\|\boldsymbol{\omega}\|_\infty$ in an equivalent BKM theorem for the 3D CHNS equations (Gibbon et al., 2016, 2017).

The Beale–Kato–Majda (BKM) theorem arises in studies of the 3D incompressible Euler equations. It says the following: suppose that there exists a solution of the 3D Euler equations on $[0, T^*)$ that then loses regularity at T^*, then $\int_0^{T^*} \|\boldsymbol{\omega}\|_\infty \, d\tau = \infty$. Conversely, if for every $T > 0$, $\int_0^{T^*} \|\boldsymbol{\omega}\|_\infty \, d\tau < \infty$ then solutions are regular (Beale, Kato, & Majda, 1984). The criterion in this theorem governs the behaviour of solutions and can be used to check the status of a numerically computed solution.

Likewise, E_∞ plays the same role for the 3D CHNS equations. The theorem of Gibbon et al. (2016) is easily stated. Let

$$H_m(t) = \int_{\mathcal{V}} |\nabla^m \boldsymbol{u}(t)|^2 \, \mathrm{d}V \quad \text{and} \quad P_m(t) = \int_{\mathcal{V}} |\nabla^m \varphi(t)|^2 \, \mathrm{d}V. \quad (3.8)$$

Theorem 3.1 *Consider the CHNS equations posed on a periodic domain $\mathcal{V} = [0, L]^3$ in three spatial dimensions. For $m > 3/2$ suppose that $H_m(0) < \infty$, $P_{m+1}(0) < \infty$, and that there exists a solution on the interval $[0, T^*)$ where T^* is the earliest time that the solution loses regularity, then*

$$\int_0^{T^*} E_\infty(\tau) \, \mathrm{d}\tau = \infty. \quad (3.9)$$

Conversely, there exists a global solution of the 3D CHNS equation if, for every $T > 0$,

$$\int_0^{T} E_\infty(\tau) \, \mathrm{d}\tau < \infty. \quad (3.10)$$

The finiteness, or otherwise, of $E_\infty(t)$ is thus critical to the regularity of solutions. The $\int_0^{T} \|u\|_\infty^2 \, \mathrm{d}\tau$ term is also consistent with one end of the Prodi–Serrin regularity conditions for the Navier–Stokes fluid part of the problem (Serrin, 1962; Robinson, Rodrigo, & Sadowski, 2016). The finiteness in (3.10) needs to be tested numerically from different initial conditions. One way is to plot finite L^m-norms of the energy, namely,

$$E_m(t) = \tfrac{1}{2} \Lambda \|\nabla \varphi\|_m^2 + \frac{\Lambda}{4\xi^2} \left(\|\varphi\|_m^2 - 1 \right)^2 + \tfrac{1}{2} \|u\|_m^2, \quad (3.11)$$

for increasing values of $m \geq 1$. For this see Gibbon et al. (2016, 2017) which showed evidence of convergence of the D_m for increasing values of m in the range $1 \leq m \leq 6$.

3.3 The variable density model for two incompressible miscible fluids

3.3.1 The mathematical model

Consider two incompressible miscible fluids with densities $\rho_2^* > \rho_1^*$. The dimensionless number that expresses the relative difference in these densities is called the Atwood number At, which is defined as

$$At = \frac{\rho_2^* - \rho_1^*}{\rho_2^* + \rho_1^*}. \quad (3.12)$$

Despite the fact that the two fluids are themselves incompressible, molecular mixing generically changes the specific volume of the mixture. This type of flow is called a *variable density* (VD) flow, following the nomenclature suggested by Cook & Dimotakis (2000), and Livescu & Ristorcelli (2007, 2008) and discussed by Rao, Caulfield, & Gibbon (2017) hereafter referred to as RCG2017. In variable density flows, because the specific volume of the mixture is not constant, it is necessary to define a composition density $\rho^*(\boldsymbol{x}, t)$. In dimensionless form, the *composition density* $\rho^*(\boldsymbol{x}, t)$ of a mixture of two constant fluid densities ρ_1^* and ρ_2^* $(\rho_2^* > \rho_1^*)$ is defined as

$$\frac{1}{\rho^*(\boldsymbol{x}, t)} = \frac{Y_1(\boldsymbol{x}, t)}{\rho_1^*} + \frac{Y_2(\boldsymbol{x}, t)}{\rho_2^*}, \tag{3.13}$$

where $Y_i(\boldsymbol{x}, t)$ $(i = 1, 2)$ are the mass fractions of the two fluids and $Y_1 + Y_2 = 1$. Equation (3.13) shows that ρ^* is bounded by

$$\rho_1^* \le \rho^*(\boldsymbol{x}, t) \le \rho_2^*. \tag{3.14}$$

To determine how this couples to a corresponding velocity field $\boldsymbol{u}(\boldsymbol{x}, t)$, we assume that there is Fickian diffusion. Then the mass transport equations for the two species are $(i = 1, 2)$

$$\partial_t\left(\rho^* Y_i\right) + \operatorname{div}\left(\rho^* Y_i \boldsymbol{u}\right) = Pe_0^{-1}\operatorname{div}\left(\rho^* \boldsymbol{\nabla} Y_i\right), \tag{3.15}$$

where Pe_0 is the Péclet number (the dimensionless Reynolds, Schmidt and Péclet numbers are defined in Table 3.1). Since the specific volume $1/\rho^*$ changes due to mixing, a non-zero divergence is induced in the velocity field which affects the conventional continuity equation for mass conservation

$$\partial_t\rho^* + \operatorname{div}\left(\rho^* \boldsymbol{u}\right) = 0, \tag{3.16}$$

which is derived from the sum over the two species in (3.15). Let us now designate $Y_2 = Y$ and $Y_1 = 1 - Y$. Given that the solution of (3.13) shows that $\rho^* Y$ is linear in ρ^* such that

$$\rho^* Y = a\rho^* + b, \qquad a = \frac{\rho_2^*}{\rho_2^* - \rho_1^*}, \qquad b = -\frac{\rho_1^*\rho_2^*}{\rho_2^* - \rho_1^*}, \tag{3.17}$$

equation (3.15) simplifies to

$$b \operatorname{div} \boldsymbol{u} = Pe_0^{-1}\operatorname{div}(\rho^* \boldsymbol{\nabla} Y). \tag{3.18}$$

Noting from (3.17) that $\rho^* \boldsymbol{\nabla} Y = -b\boldsymbol{\nabla}(\ln \rho^*)$ the coefficient b cancels to make (3.18) and (3.16) into

$$\operatorname{div} \boldsymbol{u} = -Pe_0^{-1}\Delta(\ln \rho^*). \tag{3.19}$$

$$(\partial_t + \boldsymbol{u} \cdot \boldsymbol{\nabla}) \rho^* = Pe_0^{-1} \rho^* \Delta(\ln \rho^*), \tag{3.20}$$

coupled to the 3D Navier–Stokes equations (Livescu & Ristorcelli, 2007)

$$\partial_t \left(\rho^* u_i \right) + \left(\rho^* u_i u_j \right)_{,j} = -p_{,i} + \tau_{ij,j} + \frac{1}{Fr^2} \rho^* g_i \tag{3.21}$$

$$\tau_{ij} = \rho^* Re_0^{-1} \left(u_{i,j} + u_{j,i} - \frac{2}{3} \delta_{ij} u_{k,k} \right). \tag{3.22}$$

For single-fluid, Newtonian, incompressible flows, the fact that div $\boldsymbol{u} = 0$ means that the pressure p can be determined from the Poisson equation $\Delta p = -u_{i,j} u_{j,i}$. For general compressible flows, where no separate equation for div $\boldsymbol{u}$ exists, this means that p needs to be a chosen as a function of the density: e.g. $p = c \rho^\gamma$, where γ is chosen differently depending on the molecular nature of the fluid. Here, while div $\boldsymbol{u} \neq 0$, we nevertheless have a separate equation for it, and thus the pressure can be determined from a modified, albeit more complicated, Poisson equation.

3.3.2 The roles played by $\theta = \ln \rho$ and $\boldsymbol{\nabla} \theta$

The central role played by intermittency and anisotropy, as discussed in Livescu & Ristorcelli (2007), suggests we should focus on the time-dependent evolution of nonlinearity within such buoyancy-driven, variable density flows using the full equations (3.19) and (3.20) *without* using the Boussinesq approximation. Using a normalization density $\rho_0^* = \frac{1}{2} (\rho_1^* + \rho_2^*)$, an interesting observation is that with

$$\theta(\boldsymbol{x}, t) = \ln \rho \qquad \rho = \frac{\rho^*}{\rho_0^*}, \tag{3.23}$$

equation (3.19) becomes a diffusion-like equation coupled to a $\boldsymbol{u}$-field whose divergence is linked to two derivatives in θ

$$(\partial_t + \boldsymbol{u} \cdot \boldsymbol{\nabla}) \theta = Pe_0^{-1} \Delta \theta, \qquad \text{with} \qquad \text{div} \boldsymbol{u} = -Pe_0^{-1} \Delta \theta. \tag{3.24}$$

While ρ^* itself is bounded above and below, the two gradients $\boldsymbol{\nabla} \rho$ or $\boldsymbol{\nabla} \theta$ are of much greater interest. Mathematically, understanding the properties of these gradients may reveal something about the regularity or otherwise of solutions of the governing equations. Physically, the magnitudes of these gradients are central to the rate of irreversible mixing within the flow.

Following Livescu & Ristorcelli (2007), there is another way of looking at the growth of these gradients. Consider the equation for θ and introduce a new velocity field

$$v = u + Pe_0^{-1}\nabla\theta. \tag{3.25}$$

The Hopf–Cole-like transformation $\theta = \ln\rho$ in (3.23) then leads to an exact cancellation of the nonlinear terms in (3.24) to give

$$(\partial_t + v \cdot \nabla)\,\rho = Pe_0^{-1}\Delta\rho, \quad \text{with} \quad \text{div}\,v = 0. \tag{3.26}$$

This is the linear advection-diffusion equation driven by a divergence-free velocity field. Note that $\omega = \text{curl}\,u = \text{curl}\,v$. However, the fact that v is actually an (explicit) function of $\nabla\theta$ makes (3.26) less simple than it first appears. Nevertheless, this equation provides a hint as to how we might look at the dynamics in a descriptive way. Consider a one-dimensional horizontal section through a rightward moving wave of ρ at a snapshot in time: in the frame of the advecting velocity u the relevant component of v is greater on the back face of any part of the wave (where $\nabla\rho > 0$) than on the front face (where $\nabla\rho < 0$). Thus in the advecting frame, (3.26) implies that not only is there the usual advection and diffusion but also a natural tendency for the back of a wave to catch up with the front, thus leading to a natural and inevitable steepening of $\nabla\rho$.

3.3.3 Summary of the D_m-method used for the Navier–Stokes equations

The numerically observed steepening process is consistent with experimental observations and raises several theoretical questions. In Rao, Caulfield, & Gibbon (2017) a re-analysis of the dataset of Daniel Livescu was undertaken. This dataset arose from the simulation of a buoyancy-driven flow very similar to that reported in Livescu & Ristorcelli (2007), which is freely available at the Johns Hopkins Turbulence Database (JHTB).

The method of analysis used to address these questions is based on that used to analyze multiple data-sets generated from the incompressible three-dimensional Navier–Stokes equations (Donzis et al., 2013; Gibbon et al., 2014; Gibbon, 2015) on a cubic domain $[0, L]^3$. Here, we review this method for completeness. The method for the 3D Navier–Stokes equations involved taking higher L^p-norms of the vorticity

$$\Omega_m = \left(L^{-3}\int_{\mathcal{V}} |\omega|^{2m}\,dV\right)^{1/2m}, \quad \text{for} \quad m \geq 1, \tag{3.27}$$

which are, in effect, higher global moments of the enstrophy field, each having the dimensions of a frequency. Symmetry considerations in the three-dimensional Navier–Stokes equations suggest that the following scaling is appropriate ($\varpi_0 = \nu L^{-2}$)

$$D_m = \left(\varpi_0^{-1}\Omega_m\right)^{\alpha_m}, \tag{3.28}$$

$$\alpha_m = \frac{2m}{4m - 3}. \tag{3.29}$$

In most theoretical analyses it is impossible to avoid gradients of $\boldsymbol{\omega}$ in expressions for the higher moments Ω_m thus causing great difficulties with closure. It has been shown in Gibbon (2015) that the sequence of D_m can be connected with D_1 in the following way ($m \geq 2$)

$$D_m = C_m D_1^{A_{m,\lambda}}, \tag{3.30}$$

and

$$A_{m,\lambda} = \frac{\lambda_m(t)(m - 1) + 1}{4m - 3}, \tag{3.31}$$

where the set of exponents $\{\lambda_m(t)\}$, subject to $1 \leq \lambda_m \leq 4$, are time dependent, and the C_m a set of positive constants. One way of explaining the use of the relation in (3.30) is this: instead of using higher L^{2m}-norms in the sequence $\{D_1 \,;\, D_m(t)\}$, one examines the sequence $\{D_1 \,;\, \lambda_m(t) \,;\, C_m\}$. Thus, the enstrophy D_1 is taken as the main variable and the exponents $\{\lambda_m(t)\}$ are then monitored numerically. Following how these exponents vary in time thus captures how the scaled moments D_m vary in time. The lower bound $\lambda_m \geq 1$ comes from the fact that $\Omega_1 \leq \Omega_m$ which can be shown to be equivalent to $D_1^{\alpha_m/2} \leq D_m$. The case $\lambda_m = 1$ lies at the lower bound. At $\lambda_m = 4$, $A_m = 1$, and so $D_m = C_m D_1$ in (3.30). In this situation, it is apparent that the Navier–Stokes nonlinearity is fully saturated. Computations in Donzis et al. (2013) and Gibbon et al. (2014) have shown that the λ-parameter used there lies in the range $1.15 \leq \lambda \leq 1.5$. It is this that we refer to as 'nonlinear depletion', because the higher order, appropriately scaled moments D_m do not 'saturate' nonlinearly.

As shown in Section 3.4.1, in the buoyancy-driven flow considered here, it is possible to define not only higher L^p-norms of the vorticity Ω_m (and hence D_m) as in (3.27) and (3.28), (which we relabel now as $\Omega_{m,\lambda}$ and $D_{m,\omega}$ to make the dependence on vorticity explicit) but also equivalent norms of $\nabla\theta$, which we refer to as $\Omega_{m,\theta}$ and $D_{m,\theta}$. We can then investigate the equivalent nonlinear depletion in the density gradients through considering the rate of growth of $D_{1,\theta}$ and then monitoring the evolution of the exponents $\lambda_{m,\theta}(t)$ in the nonlinear growth terms, while also taking into account $D_{1,\omega}$. While it is not currently possible to answer definitively the question posed about blow-up of the density gradients at intermediate times, before diffusive mixing might be presumed to smooth out the density distribution, this paper is meant to set the scene as a way of analyzing and interpreting future computations.

3.4 Some L^{2m}-estimates on $\nabla\theta$ and ω

3.4.1 Definitions

It is clear from (3.13) that the composition density ρ^* is bounded above and below: $\rho_1^* \leq \rho^* \leq \rho_2^*$. However, our interest lies more in $\nabla\rho^*$, both mathematically through its analogy with the vorticity field ω, and physically through its central role in irreversible mixing; but it is difficult to work with this quantity alone. To circumvent this problem, the variable θ introduced in (3.23) and satisfying (3.24) is an easier variable with which to work. The idea is to consider both $\nabla\theta$ and $\omega = \operatorname{curl} u$ in the higher (unscaled) norms $L^{2m}(\mathcal{V})$, defined for $(1 \leq m < \infty)$ by

$$\Omega_{m,\theta} = \left(\int_{\mathcal{V}} |\nabla\theta|^{2m} \, dV \right)^{1/2m}, \tag{3.32}$$

$$\Omega_{m,\omega} = \left(\int_{\mathcal{V}} |\omega|^{2m} \, dV \right)^{1/2m}, \tag{3.33}$$

where L_0 is the non-dimensionalization length in the the database which rescales the volume of integration here to be $\mathcal{V} = [0, L/L_0]^3$ (note that these are analogous to, but slightly different from, those defined in (3.27) and (3.28)). The natural sequence of Hölder inequalities

$$\Omega_{m,\theta} \leq \left(\int_{\mathcal{V}} |\nabla\theta|^{2(m+1)} \, dV \right)^{\frac{1}{2(m+1)}} \left(\int_{\mathcal{V}} 1 \, dV \right)^{\frac{1}{2m(m+1)}}$$

$$= \left(\frac{L}{L_0} \right)^{\frac{3}{2m(m+1)}} \Omega_{m+1,\theta}, \tag{3.34}$$

has a multiplicative factor which is only unity when $L = L_0$. Remembering the definition of α_m from (3.28),

$$\alpha_m = \frac{2m}{4m - 3}, \tag{3.35}$$

the exponent on L/L_0 in (3.34) is related to α_m and α_{m+1} by

$$\frac{3}{2m(m+1)} = \frac{1}{\alpha_{m+1}} - \frac{1}{\alpha_m}. \tag{3.36}$$

In turn, this leads us to define a natural dimensionless length

$$\ell_m = (L/L_0)^{1/\alpha_m}, \tag{3.37}$$

which turns (3.34) into $\ell_m\Omega_{m,\theta} \leq \ell_{m+1}\Omega_{m+1,\theta}$. Use of these length scales allows us to investigate the relationship between the various moments of the density gradients in the most natural fashion, to allow us to determine whether they exhibit nonlinear depletion (or not). The aim is to assume there exists a solution of (3.24) in tandem with the vorticity field $\boldsymbol{\omega}$. Motivated by the depletion properties studied in Donzis et al. (2013) and Gibbon et al. (2014) for the Navier–Stokes equations, the following definitions are made

$$D_{m,\theta} = (\ell_m\Omega_{m,\theta})^{\alpha_m}, \tag{3.38}$$

with a similar definition using $\boldsymbol{\omega}$

$$D_{m,\omega} = (\ell_m\Omega_{m,\omega})^{\alpha_m}. \tag{3.39}$$

In the JHT database the dimensionless domain size is 2π thus indicating that $L/L_0 = 2\pi$. The α_m-scaling in (3.38) has its origins in scaling properties of the three-dimensional Navier–Stokes equations (see Gibbon *et al* 2014). Then, as in (3.28) above, we consider

$$D_{m,\theta}(t) = D_{1,\theta}^{A_{m,\theta}(t)}, \tag{3.40}$$

where the multiplicative set of constants C_m have here been taken as unity. Following this, the JHT database shows that the relation between $D_{m,\theta}$ and $D_{1,\theta}$ takes the form of (3.40). The data are consistent with $A_{m,\theta}(t)$ being expressed as

$$A_{m,\theta}(t) = \frac{\lambda_{m,\theta}(t)(m-1)+1}{4m-3}, \tag{3.41}$$

with λ_m lying in the range

$$1 \leq \lambda_{m,\theta} \leq 4. \tag{3.42}$$

The $D_{m,\theta}(t)$ are thus defined in terms of the set $\{D_{1,\theta}(t), \lambda_{m,\theta}(t)\}$.

There is a lower bound $\lambda_{m,\theta} \geq 1$ for the equivalent quantities D_m and D_1, as defined in (3.28). As explained in Section 3.5, numerically the $\lambda_{m,\theta}(t)$ will be calculated from the JHT database by considering

$\ln D_{m,\theta}/\ln D_{1,\theta}$. Note that the ordering observed in (3.34) does not necessarily hold for the $D_{m,\theta}$ or the $D_{m,\omega}$ because α_m *decreases* with m.

3.4.2 The evolution of $D_{1,\theta}$

Formally consider the time evolution of $\int_{\mathcal{V}} |\boldsymbol{\nabla}\theta|^2\,dV$ using (3.24) and the relation for $\boldsymbol{\nabla}\cdot\boldsymbol{u}$ in (3.24)

$$
\begin{aligned}
\frac{1}{2}\frac{d}{dt}\int_{\mathcal{V}}|\boldsymbol{\nabla}\theta|^2\,dV &= \int_{\mathcal{V}}\boldsymbol{\nabla}\theta\cdot\left(Pe_0^{-1}\Delta - \boldsymbol{\nabla}\boldsymbol{u}\right)\cdot\boldsymbol{\nabla}\theta\,dV \\
&\quad + \frac{1}{2}\int_{\mathcal{V}}|\boldsymbol{\nabla}\theta|^2(\boldsymbol{\nabla}\cdot\boldsymbol{u})\,dV \\
&\leq -Pe_0^{-1}\int_{\mathcal{V}}|\Delta\theta|^2\,dV + \int_{\mathcal{V}}|\boldsymbol{\nabla}\theta|^2|\boldsymbol{\nabla}\boldsymbol{u}|\,dV \\
&\quad + \frac{1}{2}Pe_0^{-1}\int_{\mathcal{V}}|\boldsymbol{\nabla}\theta|^2|\Delta\theta|\,dV.
\end{aligned}
\tag{3.43}
$$

Before going further it should be noted that a standard result connects $\boldsymbol{\nabla}\boldsymbol{u}$, $\boldsymbol{\omega}$ and div $\boldsymbol{u}$

$$
\begin{aligned}
\int_{\mathcal{V}}|\boldsymbol{\nabla}\boldsymbol{u}|^2\,dV &= \int_{\mathcal{V}}|\boldsymbol{\omega}|^2\,dV + \int_{\mathcal{V}}|\text{div }\boldsymbol{u}|^2\,dV \\
&= \int_{\mathcal{V}}|\boldsymbol{\omega}|^2\,dV + Pe_0^{-2}\int_{\mathcal{V}}|\Delta\theta|^2\,dV
\end{aligned}
\tag{3.44}
$$

For $m \geq 2$, and noting that $\frac{m-2}{2(m-1)} + \frac{m}{2(m-1)} = 1$, consider the term

$$
\int_{\mathcal{V}}|\boldsymbol{\nabla}\theta|^2|\boldsymbol{\nabla}\boldsymbol{u}|\,dV \leq c_{1,m}\Omega_{1,\omega}^{\frac{m-2}{m-1}}\Omega_{m,\theta}^{\frac{m}{m-1}}\left[\Omega_{1,\omega} + Pe_0^{-1}\|\Delta\theta\|_2\right]
\tag{3.45}
$$

$$
\leq c_{2,m}\,D_{1,\theta}^{\frac{m-2}{2(m-1)}}\,D_{m,\theta}^{\frac{m}{\alpha_m(m-1)}}\left[\frac{1}{\sqrt{2\pi}}D_{1,\omega}^{1/2} + Pe_0^{-1}\|\Delta\theta\|_2\right],
$$

where the factors of ℓ_m have been absorbed into the dimensionless constants $c_{1,m}$ and $c_{2,m}$. Now we turn to the idea introduced in Section 3.4.1 that connects $D_{m,\theta}$ with $D_{1,\theta}$ by using the formula (3.40) and (3.41) for $A_{m,\theta}$ and the set $\{\lambda_{m,\theta}(t)\}$. Inserting (3.40) into the right hand side of (3.45) gives (factors of 2π have also been absorbed into the constants)

$$
\begin{aligned}
\int_{\mathcal{V}}|\boldsymbol{\nabla}\theta|^2|\boldsymbol{\nabla}\boldsymbol{u}|\,dV &\leq c_{2,m}D_{1,\theta}^{(1+\lambda_{m,\theta})/2}\left[(2\pi)^{-1/2}D_{1,\omega}^{1/2} + Pe_0^{-1}\|\Delta\theta\|_2\right] \\
&\leq \frac{1}{2}Pe_0 D_{1,\omega} + c_{3,m}Pe_0^{-1}D_{1,\theta}^{1+\lambda_{m,\theta}} \\
&\quad + \frac{1}{2}Pe_0^{-1}\|\Delta\theta\|_2^2,
\end{aligned}
\tag{3.46}
$$

where the use of a Hölder inequality has split up the terms of the right hand side of (3.45). The same idea is used to estimate the last term in (3.43) with $|\nabla u|$ replaced by $|\Delta\theta|$:

$$Pe_0^{-1}\int_{\mathcal{V}}|\nabla\theta|^2|\Delta\theta|\,\mathrm{d}V \leq \left(Pe_0^{-1}\|\Delta\theta\|_2^2\right)^{1/2}\left(c_{3,m}Pe_0^{-1}D_{1,\theta}^{1+\lambda_{m,\theta}}\right)^{1/2}$$

$$\leq \frac{1}{2}Pe_0^{-1}\|\Delta\theta\|_2^2 + c_{4,m}Pe_0^{-1}D_{1,\theta}^{1+\lambda_{m,\theta}}. \quad (3.47)$$

Altogether, (3.43) becomes

$$\frac{1}{4\pi}\dot{D}_{1,\theta} \leq -\frac{1}{4}Pe_0^{-1}\|\Delta\theta\|_2^2 + c_{5,m}Pe_0^{-1}D_{1,\theta}^{1+\lambda_{m,\theta}} + \frac{1}{2}Pe_0 D_{1,\omega}. \quad (3.48)$$

A simple integration by parts shows that

$$\|\nabla\theta\|_2^2 \leq \|\Delta\theta\|_2\|\theta\|_2, \quad (3.49)$$

which leads to

$$\frac{\mathrm{d}}{\mathrm{d}t}D_{1,\theta} \leq -\frac{D_{1,\theta}^2}{4\pi Pe_0\|\theta\|_2^2} + 4\pi c_m Pe_0^{-1}D_{1,\theta}^{1+\lambda_{m,\theta}(t)} + 2\pi Pe_0 D_{1,\omega}, \quad (3.50)$$

in which the constant $c_{5,m}$ has been re-labelled as c_m.

Given that $\lambda_{m,\theta} \leq 4$ it is easy to show by direct integration that there is a short-time existence result

$$D_1(t) \leq \frac{\text{const}}{(t_0 - t)^{1/4}}, \quad (3.51)$$

where t_0 is dependent on initial data $D_1(0)$ but, other than this, there are no weak solution results available as in the 3D Navier–Stokes equations. In the next section we therefore turn to numerical results based on the JHT Database.

3.5 Some results from the JHT Database

The $D_{m,\theta}(t)$ are defined in terms of the set $\{D_{1,\theta}(t), \lambda_{m,\theta}(t)\}$ using equations (3.40), (3.41) and (3.42). There is also an equivalent formula for $D_{m,\omega}$ in terms of the set $\{D_{1,\omega}(t), \lambda_{m,\omega}(t)\}$

$$D_{m,\omega}(t) = D_{1,\omega}^{A_{m,\omega}(t)}, \quad (3.52)$$

where the multiplicative set of constants C_m have here been taken as unity. The data are consistent with $A_{m,\omega}(t)$ being expressed as

$$A_{m,\omega}(t) = \frac{\lambda_{m,\omega}(t)(m-1)+1}{4m-3}, \quad (3.53)$$

Reynolds number	$Re_0 = \rho_0^* L_0 U_0 / \mu_0$	12500
Froude number	$Fr = U_0 / \sqrt{gL_0}$	1
Schmidt number	$Sc = \mu_0 / D\rho_0^*$	1
Péclet number	$Pe_0 = Re_0 Sc$	12500
Atwood number	$At = (\rho_2^* - \rho_1^*)/(\rho_2^* + \rho_1^*)$	0.05
Domain		$(2\pi)^3$
Domain Length	L	2π
Non-dimensionalization Length	L_0	1

Table 3.1 *Parameters in the simulations of Livescu on the JHT Database.*

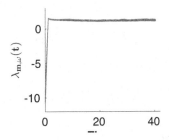

Figure 3.1 Time variation of $\lambda_{m,\omega}(t)$, calculated using the relations (3.52) and (3.53).

with $\lambda_{m,\omega}$ lying in the range

$$1 \leq \lambda_{m,\omega} \leq 4. \tag{3.54}$$

In Figure 3.1, the JHT database shows that the relation between $D_{m,\omega}$ and $D_{1,\omega}$ takes the form of (3.52). While the scale is large – shown this way to contrast with panel 1 of Figure 3.2 – it can be seen that $\lambda_{m,\omega}$ lies only just greater than unity with no evidence of strong growth. This is consistent with results for plain Navier–Stokes turbulence in Gibbon et al. (2014) and thus implies that the Navier–Stokes component remains unchanged.

However, the left-hand panel of Figure 3.2 shows that the $\lambda_{m,\theta}$ for several values of m undergo heavy growth. We have been unable to establish an analytical relation between $D_{1,\theta}$ and $D_{1,\omega}$ but the JHT Database provides us with the relation with a further exponent which we call $\beta(t)$

$$D_{1,\omega} = D_{1,\theta}^{\beta(t)} \tag{3.55}$$

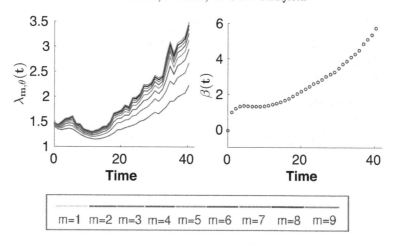

Figure 3.2 Time variation of: (a) $\lambda_{m,\theta}(t)$, as defined in (3.41), which fan out and grow with time; (b) $\beta(t)$ as defined in (3.55).

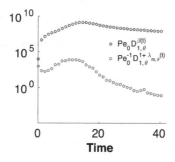

Figure 3.3 Time variation of $Pe_0 D_{1,\theta}^{\beta(t)}$ (plotted with squares) and $Pe_0^{-1} D_{1,\theta}^{1+\lambda_{m,\theta}(t)}$-term (plotted with circles) where $\lambda_{m,\theta}(t)$ is chosen to be the maximum value over m at each time step with $Pe_0 = 12,500$.

where the growth in the exponent $\beta(t)$ is shown in the right-hand panel of Figure 3.2. This too is evidence of very strong growth.

3.6 Conclusion

In Section 3.2 we summarized the results of Gibbon et al. (2016, 2017) on the CHNS equations and saw that the equivalent BKM-criterion, namely

$\int_0^t E_\infty \, d\tau$, is bounded in simulations performed in that paper. According to Theorem 3.1 the finiteness of this quantity guarantees the smoothness of solutions. However, results from the JHT database with respect to the variable-density model, combined with the analysis of Section 3.4.2, raise several unanswered questions.

Firstly, can the growth of the density field gradient $\nabla\theta$ be shown to be bounded? Such bounds could yield insights into the regularity of the density field and the uniform validity of the Boussinesq approximation for flows with $At \ll 1$, which may explain the mixing efficiency associated with buoyancy-driven turbulence.

Secondly, if the growth of density gradients fails to saturate, as in Figure 3.2, and instead accelerates by continuing to draw on available kinetic energy, is this process fast enough for a singularity to form in a finite time? The rapid growth observed by Luo & Hou (2014a, 2014b) in the vorticity field of a symmetric incompressible Euler flow in a cylinder (with $u \cdot \hat{n} = 0$ on the boundary) provides strong evidence of a singularity in that system. It is possible that buoyancy-driven variable density turbulence is another example, despite the difference in boundary conditions and the presence of diffusion. Neither the data nor the methods used here are sufficient to answer this question definitively.

Acknowledgments: We acknowledge, with thanks, the staff of IPAM UCLA where this collaboration began in the Autumn of 2014 on the programme "Mathematics of Turbulence". We would also like to thank Charles Doering and Daniel Livescu for useful discussions. The research activity of CPC is supported by EPSRC Programme Grant on "Mathematical Underpinnings of Stratified Turbulence" (EP/K034529/1). All the numerical data used is from the Johns Hopkins Turbulence Database (Livescu et al., 2014), a publicly available direct numerical simulation (DNS) database. For more information, see

http://turbulence.pha.jhu.edu/.

References

Abels, H. (2009) Longtime behavior of solutions of a Navier–Stokes/Cahn–Hilliard system, Pages 9–19 *Nonlocal and Abstract Parabolic Equations and Their Applications*, Banach Center Publ., Polish Acad. Sci., **86**.

Andrews, M.J. & Dalziel, S.B. (2010) Small Atwood number Rayleigh–Taylor experiments. *Phil. Trans. R. Soc. Ser. A* **368**, 1663–79.

Beale, J.T., Kato, T., & Majda, A.J. (1984) Remarks on the breakdown of smooth solutions for the 3D Euler equations. *Commun. Math. Phys.* **94** 61–66.

Bray, A.J. (1994) Theory of Phase Ordering Kinetics. *Adv. Phys.* **43**, 357–459.

Cabot, W.H. & Cook, A.W. (2006) Reynolds number effects on Rayleigh–Taylor instability with possible implications for type Ia supernovae. *Nat. Phys.* **2** (8), 562–568.

Cahn, J.W. & Hilliard, J.E. (1958) Free energy of a non-uniform system. I. Interfacial free energy. *J. Chem. Phys.* **28**, 258–267.

Celani, A., Mazzino, A., Muratore–Ginanneschi, P., & Vozella, L. (2009) Phase-field model for the RayleighTaylor instability of immiscible fluids. *J. Fluid Mech.* **622**, 115–134.

Chaikin, P.M. & Lubensky, T.C. (2000) *Principles of Condensed Matter Physics.* Cambridge University Press, Cambridge, UK, reprint edition.

Constantin, P. & Foias, C. (1988) *Navier–Stokes Equations.* Chicago University Press.

Cook, A.W. & Dimotakis, P.E. (2001) Transition stages of Rayleigh–Taylor instability between miscible fluids. *J. Fluid Mech.* **443**, 69–99.

Davies-Wykes, M.S. & Dalziel, S.B. (2014) Efficient mixing in stratified flows: experimental study of a Rayleigh–Taylor unstable interface within an otherwise stable stratification. *J. Fluid Mech.* **756**, 1027–1057.

Dimonte, G., Youngs, D.L., Dimits, A., Weber, S., Marinak, M., Wunsch, S., Garasi, C., Robinson, A., Andrews, M.J., Ramaprabhu, P., Calder, A. C., Fryxell, B., Biello, J., Dursi, L., MacNeice, P., Olson, K., Ricker, P., Rosner, R., Timmes, F., Tufo, H., Young, Y.-N., & Zingale, M. (2004) A comparative study of the turbulent Rayleigh–Taylor instability using high-resolution three-dimensional numerical simulations: The Alpha-Group collaboration. *Phys. Fluids* **16**, 1668–1693.

Dimotakis, P.E. (2005) Turbulent mixing. *Annu. Rev. Fluid Mech.* **37**, 329–356.

Doering, C.R. & Gibbon, J.D. (1995) *Applied Analysis of the Navier–Stokes Equations.* Cambridge University Press, Cambridge, UK.

Donzis, D.A., Gibbon, J.D., Kerr, R.M., Gupta, A., Pandit, R., & Vincenzi, D. (2013) Vorticity moments in four numerical simulations of the 3d Navier–Stokes equations. *J. Fluid Mech.* **732**, 316–331.

Elliott, C.M. & Songmu, Z. (1986) On the Cahn–Hilliard equation. *Arch. Rat. Mech. Anal.* **96** (4), 339–357.

Foias, C., Manley, O., Rosa, R., & Temam, R. (2001) *Navier–Stokes equations and Turbulence.* Cambridge University Press, Cambridge, UK.

Gal, C.G. & Graselli, M. (2010) Asymptotic behavior of a Cahn–Hilliard-Navier–Stokes system in 2D. *Ann. Inst. H. Poincaré* **27**, 401–436.

Gibbon, J.D., Donzis, D.A., Kerr, R.M., Gupta, A., Pandit, R., & Vincenzi, D. (2014) Regimes of nonlinear depletion and regularity in the 3D Navier–Stokes equations. *Nonlinearity* **27**, 1–19.

Gibbon, J.D. (2016) High-low frequency slaving and regularity issues in the 3D Navier–Stokes equations. *IMA J. Appl. Math.* 81(2), 308–320.

Gibbon, J.D., Pal, N., Gupta, A., & Pandit, R. (2016) Regularity criterion for solutions of the three-dimensional Cahn–Hilliard-Navier–Stokes equations and associated computations. *Phys. Rev. E* **94**, 063103.

Gibbon, J.D., Gupta, A., Pal, N., & Pandit, R. (2017) The role of the BKM theorem in 3D Euler, Navier–Stokes and Cahn–Hilliard-Navier–Stokes analysis. *Physica D*, to appear. DOI:10.1016/j.physd.2017.11.007.

Glimm, J., Grove, J.W., Li, X.L., Oh, W., & Sharp, D.H. (2001) A critical analysis of Rayleigh–Taylor growth rates. *J. Comp. Phys.* **169** (2), 652–677.

Hohenberg P.C. & Halperin, B.I. (1977) Theory of dynamic critical phenomena. *Rev. Mod. Phys.* **49** 435–480.

Hyunsun, L., Hyeonseong, J., Yan, Y., & Glimm, J. (2008) On validation of turbulent mixing simulations for Rayleigh–Taylor instability. *Phys. Fluids* **20**, 012102.

Lawrie, A.G.W. & Dalziel, S.B. (2011) Rayleigh–Taylor mixing in an otherwise stable stratification. *J. Fluid Mech.* **688**, 507–527.

Lee, H., Jin, H., Yu, Y., & Glimm, J. (2008) On validation of turbulent mixing simulations for Rayleigh–Taylor instability. *Phys. Fluids* **20**, 1–8.

Livescu, D. & Ristorcelli, J.R. (2007) Buoyancy-driven variable-density turbulence. *J. Fluid Mech.* **591**, 43–71.

Livescu, D. & Ristorcelli, J.R. (2008) Variable-density mixing in buoyancy-driven turbulence. *J. Fluid Mech.* **605**, 145–180.

Luo, G. & Hou, T. (2014a) Potentially singular solutions of the 3D axisymmetric Euler equations. *Proc. Nat. Acad. Sci.* **111**, 12968–12973.

Luo, G. & Hou, T. (2014b) Toward the finite time blow-up of the 3D incompressible Euler equations: a numerical investigation. *Multiscale Model. Simul.* **12**, 1722–1776.

Petrasso, R.D. (1994) Rayleigh's challenge endures. *Nat. Phys.* **367**, 217–218.

Rao, P., Caulfield, C.P., & Gibbon, J.D. (2017) Nonlinear effects in buoyancy-driven variable density turbulence. *J. Fluid Mech.* **810**, 362–377.

Robinson, J.C., Rodrigo, J.L., & Sadowski, W. (2016) *The three-dimensional Navier–Stokes equations.* Cambridge Studies in Advanced Mathematics 157. Cambridge University Press, Cambridge, UK.

Serrin, J. (1962) On the interior regularity of weak solutions of the Navier–Stokes equations. *Arch. Rat. Mech. Anal.* **9**, 187–191.

Sharp, D.H. (1984) An overview of Rayleigh–Taylor Instability. *Physica D* **12D**, 3–18.

Tailleux, R. (2013) Available potential energy and exergy in stratified fluids. *Annu. Rev. Fluid Mech.* **45**, 35–58.

Youngs, D.L. (1984) Numerical simulation of turbulent mixing by Rayleigh–Taylor instability. *Physica D: Nonlinear Phenomena* **12**, (1-3), 3244.

Youngs, D.L. (1989) Modelling turbulent mixing by Rayleigh–Taylor instability. *Physica D: Nonlinear Phenomena* **37**, 270–287.

4

On localization and quantitative uniqueness for elliptic partial differential equations

Guher Camliyurt

Department of Mathematics
University of Southern California
Los Angeles, CA 90089
camliyur@usc.edu

Igor Kukavica

Department of Mathematics
University of Southern California
Los Angeles, CA 90089
kukavica@usc.edu

Fei Wang

Department of Mathematics
University of Maryland
College Park, MD 20742
fwang256@umd.edu

Abstract

We address the decay and the quantitative uniqueness properties for solutions of the elliptic equation with a gradient term, $\Delta u = W \cdot \nabla u$. We prove that there exists a solution in a complement of the unit ball which satisfies $|u(x)| \leq C \exp(-C^{-1}|x|^2)$ where W is a certain function bounded by a constant. Next, we revisit the quantitative uniqueness for the equation $-\Delta u = W \cdot \nabla u$ and provide an example of a solution vanishing at a point with the rate $\mathrm{const}\|W\|_{L^\infty}^2$. We also review decay and vanishing results for the equation $\Delta u = V u$.

4.1 Introduction

In this paper, we address the spatial decay and the quantitative uniqueness properties for the elliptic equations

$$\Delta u = W \cdot \nabla u \tag{4.1}$$

and

$$\Delta u = V u. \tag{4.2}$$

Published as part of *Partial Differential Equations in Fluid Mechanics*, edited by C.L. Fefferman, J.C. Robinson, & J.L. Rodrigo. © Cambridge University Press 2018.

Meshkov (1991) constructed a complex-valued solution of (4.2) for which $|V| \leq C$ in the exterior domain $B_1^c = \{x : |x| \geq 1\}$, such that u satisfies

$$|u(x)| \leq C \exp(-|x|^{4/3}/C),$$

where C is a sufficiently large constant. This was a counterexample to Landis's conjecture (Kondrat & Landis, 1988), which predicted the largest allowable exponent to be 1 rather than 4/3. There are other important aspects of this example: it first demonstrates the optimality of Hörmander's Carleman inequality (cf. (4.10) below). Also, Kenig (2007, p. 34) showed a connection with the qualitative behavior at a point by constructing a sequence of radii $r_j \rightarrow 0$, potentials V_j, with $\|V_j\|_{L^\infty} = M_j$, and solutions u_j to $\Delta u_j + V_j u_j = 0$, with $\|u_j\| \leq C_0$ and $\|u_j\|_{L^\infty(B_1)} \geq 1$ such that

$$\max_{|x|<r_j} |u_j(x)| \leq C r_j^{C M_j^{2/3}}.$$

This is obtained by a translation and a dilation of the Meshkov solution. Bakri & Casteras (2014) constructed a sequence of solutions u_j of

$$\Delta u_j = V_j u_j, \tag{4.3}$$

with $\|V_j\|_{L^\infty} \rightarrow \infty$ such that u_j vanishes *at the point* ∞ of order $\|V_j\|_{L^\infty}^{2/3}$, up to a constant. Furthermore, Camliyurt, Kukavica, & Wang (2018) constructed solutions of (4.3), satisfying Dirichlet or periodic boundary conditions, that vanish at *a finite point* of order $\|V_j\|_{L^\infty}^{2/3}$ (this is higher than $\|V\|_{L^\infty}^{1/2}$ conjectured in Kukavica, 1998). We point out that all the examples, starting with Meshkov's, are complex valued; thus the question of optimality of such upper bound remains open for real-valued solutions.

Camliyurt & Kukavica (2017) used the frequency method (see for example Almgren, 1979; Garofalo & Lin, 1986; Kukavica, 1998, 1999; Kurata, 1994; Poon, 1996) to estimate from above the order of vanishing for solutions of the parabolic equation

$$\partial_t u - \Delta u = W \nabla u + V u \tag{4.4}$$

with an upper bound

$$C \|V\|_{L^\infty}^{2/3} + C \|W\|_{L^\infty}^2, \tag{4.5}$$

where C depends on the Dirichlet quotient of u. The primary motivation is to be able to address the two-dimensional Navier–Stokes equations, which in vorticity formulation read

$$\partial_t \omega + u \cdot \nabla \omega = 0$$

or the three-dimensional Navier–Stokes equation

$$\partial_t \omega + u \cdot \nabla \omega - \omega \cdot \nabla u = 0,$$

where u is connected to ω through the Biot–Savart law.

Based on Camliyurt et al. (2018), the exponent $2/3$ for V in (4.5) is optimal, but the question remains if the exponent 2 for W in (4.5) is the best possible as well. It is easy to check that for the elliptic type equation (4.1), Hörmander's Carleman inequality leads to the same upper bound (4.5) for the order of vanishing of solutions of

$$\Delta u = W \cdot \nabla u + V u, \tag{4.6}$$

with the Dirichlet boundary condition (for example), where C depends only on the domain. A similar question may be asked for the equivalent of the Landis–Meshkov question for the equation (4.1). In this paper, we solve both questions by proving the following assertions.

Regarding the decay at infinity for solutions of (4.1), we prove that there exists a solution u in B_1^c, such that $\|W\|_{L^\infty} \leq C$ and

$$|u(x)| \leq C \exp(-|x|^2/C).$$

The construction is provided in Section 4.4 below. Also, in Section 4.3, we prove that this type of localization cannot be improved upon. Regarding the degree of vanishing, the construction is similar to the one with V provided in Camliyurt et al. (2018). For the equation (4.1), subject to either periodic or Dirichlet boundary conditions, we construct a solution that achieves the upper bound $C\|W\|_{L^\infty}^2$ as in Camliyurt & Kukavica (2017). Finally in the last section, we recall, without proofs, some recent results on the decay and the order of vanishing for the equation (4.2).

The field of unique continuation has a very rich history, and we shall not survey it here (cf. instead Kenig, 1989, 2007, 2008; Vessella, 2003); however, we point out some key results for the equations (4.1), (4.2), and (4.4). The theory was initiated by Carleman (1939), who used a weighted-type inequality to obtain a unique continuation result for equation (4.2) in two dimensions. An extension to a general spatial dimension for regular elliptic equations was provided by Aronszajn (1957) and Cordes (1956). Hörmander (1983) provided a Carleman estimate (cf. (4.10) below), that gives a strong unique continuation property for equation (4.6) with V and W in certain L^p spaces. (Recall that an equation satisfies the strong unique continuation property if every solution that vanishes of infinite order at a point has to vanish identically.) A sharp L^p condition on V that guarantees a strong unique continuation property was obtained by Jerison & Kenig (1985), Koch & Tataru (2001), and Wolff (1995). For equations of the type (4.4), sharp L^p conditions were obtained in Escauriaza (2000), Escauriaza & Vega (2001), and Koch & Tataru (2009). For some other results on unique continuation, see Agmon (1965), Agmon & Nirenberg (1967), Alessandrini & Escauriaza (2008), Alessandrini et al. (2009), Alessandrini & Vessella (1988), Almgren (1979), Bakri (2013), Camliyurt & Kukavica (2016), Escauriaza & Fernández (2003), Escauriaza, Fernández, & Vesella (2006), Garofalo & Lin (1986), Kurata (1994), Lin (1991), and Poon (1996) while for quantitative uniqueness, see Donnelly & Fefferman (1988), Donelly & Fefferman (1990), Escauriaza & Vessella (2003), Kenig, Silvestre, & Wang (2015), Kukavica (2000), Kukavica (1999), and Kukavica (2003).

The paper is organized as follows. In Section 4.2, we provide a lower bound for the decay of the equation (4.4). In Section 4.3, we construct a complex-valued example showing that the rate provided in Section 4.2 for (4.1) is optimal. In Section 4.4, we construct a solution of (4.1) vanishing of the optimal order (4.5). Finally, in the last section, we summarize some known results for (4.2).

4.2 A lower bound for the decay of $\Delta u = W\nabla u + Vu$

In this section, we provide a lower bound on the decay of a solution of the equation

$$\Delta u = W \cdot \nabla u + Vu \quad \text{in } \mathbb{R}^N, \tag{4.7}$$

where $N \geq 1$.

Proposition 4.1 *Let u be a solution to the equation (4.7) satisfying $u(0) = 1$. Assume that*

$$|W|, |V|, |u| \leq \widetilde{C}. \tag{4.8}$$

Then for any $x_0 \in \mathbb{R}^N$ with $|x_0| = R \geq 1$, we have the lower bound

$$\max_{|x - x_0| \leq 1} |u(x)| \geq \frac{1}{C_0} \exp\left(-C_0 R^2 \log R\right)$$

for some positive constant C_0 depending on $\widetilde{C}$ in (4.8).

Note that all constants are allowed to depend on the dimension N. The proof follows the approach in Bourgain & Kenig (2005).

First, recall Hörmander's Carleman type inequality from Bourgain & Kenig (2005) (see also Hörmander, 1983). There exists a (radial) increasing function $\psi(r)$ on $[0,1]$ for which

$$\frac{r}{C} \le \psi(r) \le Cr, \qquad r \in [0,1] \tag{4.9}$$

and such that

$$\lambda^3 \int u^2 \psi^{-1-2\lambda} + \lambda \int |\nabla u|^2 \psi^{1-2\lambda} \le C \int (\Delta u)^2 \psi^{2-2\lambda}, \qquad \lambda \ge C \tag{4.10}$$

for all $u \in C_0^\infty(B_1 \backslash \{0\})$.

Proof of Proposition 4.1 Since (4.7) is translation invariant, we may assume $x_0 = 0$. Define

$$\overline{u}(x) = u(KRx)$$

where $K > 0$ is a large constant to be specified below. Then $\overline{u}$ satisfies

$$\Delta \overline{u} = KR\overline{W} \cdot \nabla \overline{u} + (KR)^2 \overline{V}\overline{u}$$

where we set $\overline{W}(x) = W(KRx)$ and $\overline{V}(x) = V(KRx)$. From (4.8), we obtain $|\overline{W}|, |\overline{V}|, |\overline{u}| \le C$. Next, by the Sobolev embedding we may estimate

$$\|\nabla u\|_{L^\infty(B_r)} \le C\|u\|_{L^p(B_r)}^{1-N/2p}\|D^2 u\|_{L^p(B_r)}^{N/2p} + \frac{C}{r^{1+N/p}}\|u\|_{L^p(B_r)} \tag{4.11}$$

where $r > 0$ and $p > N/2$ is sufficiently large. The interior regularity for elliptic equations (Gilbarg & Trudinger, 2001) gives

$$\|D^2 u\|_{L^p(B_r)} \le \frac{C}{r^2}\|u\|_{L^p(B_{2r})} \le \frac{C}{r^{2-N/p}}\|u\|_{L^\infty(B_{2r})}.$$

Using the above estimate in (4.11), we arrive at

$$\|\nabla u\|_{L^\infty(B_r)} \le \frac{C}{r}\|u\|_{L^\infty(B_{2r})}$$

which implies, by rescaling,

$$\|\nabla \overline{u}\|_{L^\infty(B_r)} \le \frac{C}{r}\|\overline{u}\|_{L^\infty(B_{2r})}, \qquad r > 0. \tag{4.12}$$

Choose a cut-off function ζ with the properties

$$0 \leq \zeta \leq 1,$$

$$\operatorname{supp}\zeta \subseteq \left\{ \frac{1}{4KR} \leq |x| \leq 4 \right\},$$

$$\zeta = 1 \quad \text{in} \quad \left\{ \frac{1}{3KR} \leq |x| \leq 3 \right\}, \tag{4.13}$$

$$|\partial_\beta \zeta| \leq C_\beta R^{|\beta|} \quad \text{if } |x| \leq \frac{1}{3R},$$

$$|\partial_\beta \zeta| \leq C_\beta \quad \text{if } |x| \geq 3,$$

with the last two inequalities holding for every multi-index $\beta \in \mathbb{N}_0^N$. Applying the Carleman estimate (4.10) with $\zeta \overline{u}$, for $\lambda \geq C$ we obtain

$$\lambda^3 \int \psi^{-1-2\lambda}(\zeta \overline{u})^2 + \lambda \int \psi^{1-2\lambda}|\nabla(\zeta \overline{u})|^2 \leq C \int \psi^{2-2\lambda}(\Delta(\zeta \overline{u}))^2$$

$$\leq C \int \psi^{2-2\lambda}(\zeta \Delta \overline{u}))^2 + C \int \psi^{2-2\lambda}(\nabla \zeta \cdot \nabla \overline{u}))^2 + C \int \psi^{2-2\lambda}(\overline{u}\Delta \zeta)^2, \tag{4.14}$$

for some increasing function ψ satisfying

$$\frac{r}{C} \leq \psi(r) \leq Cr \quad \text{for } 0 \leq r \leq 4.$$

For the first term on the right-hand side of (4.14), we have

$$\int \psi^{2-2\lambda}(\zeta \Delta \overline{u})^2$$

$$\leq C(KR)^2 \int \psi^{2-2\lambda}\zeta^2|\overline{W}|^2|\nabla \overline{u}|^2 + C(KR)^4 \int \psi^{2-2\lambda}\zeta^2|\overline{V}|^2\overline{u}^2$$

$$\leq C(KR)^2 \int \psi^{2-2\lambda}|\nabla(\zeta \overline{u}) - \overline{u}\nabla \zeta|^2 + C(KR)^4 \int \psi^{2-2\lambda}\zeta^2|\overline{V}|^2\overline{u}^2$$

$$\leq C(KR)^2 \int \psi^{1-2\lambda}|\nabla(\zeta \overline{u})|^2 + C(KR)^4 \int \psi^{-1-2\lambda}\zeta^2|\overline{V}|^2\overline{u}^2$$

$$+ C(KR)^2 \int \psi^{2-2\lambda}\overline{u}|\nabla \zeta|^2, \tag{4.15}$$

where we have used $\psi \leq C$ for $|x| \leq 4$. Note that

$$\operatorname{supp}\nabla\zeta, \ \operatorname{supp}\Delta\zeta \subseteq \left\{ \frac{1}{4KR} \leq |x| \leq \frac{1}{3KR} \right\} \cup \{3 \leq |x| \leq 4\}.$$

Therefore, the second and the third terms on the right-hand side of (4.14) as well as the last term on the right-hand side of (4.15) may be decomposed into integrals over the annuli $\{1/4KR \leq |x| \leq 1/3KR\}$ and $\{3 \leq |x| \leq 4\}$.

We first consider the integrals over the inner annulus

$$\{1/4KR \leq |x| \leq 1/3KR\}.$$

We have

$$
\int_{\{1/4KR \leq |x| \leq 1/3KR\}} \psi^{2-2\lambda} \left(|\nabla \zeta \overline{u}|^2 + (\nabla \zeta \cdot \nabla \overline{u})^2 + |\overline{u} \Delta \zeta|^2 \right)
$$
$$
\leq CR^{2\lambda-2-N} R^2 \left(\sup |\overline{u}|^2 + \sup |\nabla \overline{u}|^2 + R^2 \sup |\overline{u}|^2 \right), \tag{4.16}
$$

where the suprema are all taken over $|x| \leq 1/3KR$ and we have used (4.9). By (4.12), we get

$$
\sup_{|x| \leq 1/3KR} |\nabla \overline{u}|^2 \leq CR^2 \sup_{|x| \leq 2/3KR} |\overline{u}|^2.
$$

Therefore, we have

$$
\int_{\{1/4KR \leq |x| \leq 1/3KR\}} \psi^{2-2\lambda} \left(|\nabla \zeta \overline{u}|^2 + (\nabla \zeta \cdot \nabla \overline{u})^2 + |\overline{u} \Delta \zeta|^2 \right)
$$
$$
\leq CR^{2\lambda-N+2} \sup_{|x| \leq 2/3KR} |\overline{u}|^2. \tag{4.17}
$$

For the outer annulus $\{3 \leq |x| \leq 4\}$, we use

$$\psi(x) \geq \psi(3), \qquad |x| \geq 3,$$

to deduce that

$$
\int_{|x| \geq 3} \psi^{2-2\lambda} \left(|\overline{u} \nabla \zeta|^2 + (\nabla \zeta \cdot \nabla \overline{u})^2 + |\overline{u} \Delta \zeta|^2 \right)
$$
$$
\leq C\psi^{2-2\lambda}(3)(C + CR^{2+2N/p} + C) \leq C\psi^{2-2\lambda}(3)R^3 \tag{4.18}
$$

if we choose $p \geq 2N$, where we also used $2 - 2\lambda < 0$. Combining (4.14), (4.15), (4.17), and (4.18) gives

$$
\lambda^3 \int \psi^{-1-2\lambda} (\zeta \overline{u})^2 + \lambda \int \psi^{1-2\lambda} |\nabla(\zeta \overline{u})|^2
$$
$$
\leq C(KR)^2 \int \psi^{1-2\lambda} |\nabla(\zeta \overline{u})|^2 + C(KR)^4 \int \psi^{-1-2\lambda} \zeta^2 \overline{u}^2 \tag{4.19}
$$
$$
+ CR^{2\lambda-N+2} \sup_{|x| \leq 2/3KR} |\overline{u}|^2 + C\psi^{2-2\lambda}(3)R^3.
$$

Setting

$$\lambda = CR^2, \tag{4.20}$$

with C sufficiently large, the first and the second terms on the right-hand side are absorbed by the left-hand side. By the assumptions of the lemma, there exists $a \in \mathbb{R}^N$ with $|a| = 1/K$ such that

$$\overline{u}(a) = 1.$$

In view of the gradient estimate (4.12), we get

$$|\overline{u}(x) - \overline{u}(a)| \leq C|x - a| \leq \frac{1}{2}$$

if $|x - a| \leq 1/2C$ and C is sufficiently large. Then we obtain a lower bound on the first term on the left side of (4.19) which reads

$$\lambda^3 \int \psi^{-1-2\lambda}(\zeta \overline{u})^2 \geq \frac{1}{C}\lambda^3 \psi^{-1-2\lambda}\left(\frac{2}{K}\right).$$

Let $C_0 > 0$ be a constant. Using (4.2), we may take K sufficiently large and ensure that $\psi(2/K) \leq \psi(3)/C_0$. Using this inequality and (4.20), we get

$$\frac{1}{C}\lambda^3 \left(\psi\left(\frac{2}{K}\right)\right)^{-1-2\lambda} \geq \frac{1}{C}R^3 \left(\frac{C_0}{\psi(3)}\right)^{1+2\lambda},$$

where we used $R^6 \geq R^3$. If C_0 is sufficiently large, the last term of the right side of (4.19) is absorbed by the first term on the left side, and thus only the third term on the right side of (4.19) remains. We obtain

$$\frac{1}{C}\lambda^3 \psi^{-1-2\lambda}\left(\frac{2}{K}\right) \leq CR^{2\lambda-N+2} \sup_{|x|\leq 2/3KR} |\overline{u}|^2,$$

from where, using (4.9) and (4.20),

$$\sup_{|x|\leq 2/3KR} |\overline{u}| \geq \frac{1}{C}\exp\left(-CR^2 \log R\right),$$

concluding the proof. $\qquad\square$

4.3 A construction of a localized solution

In this section, we construct a solution of equation (4.1) with the optimal decay provided in the previous section.

Theorem 4.2 *There exists a complex-valued solution u of equation (4.1) in $\mathbb{R}^2 \setminus \overline{B}_1$, with W complex valued satisfying*

$$|W| \leq C$$

such that

$$|u(x)| \leq C \exp\left(-\frac{1}{C}|x|^2\right),$$

where $C > 0$.

From here on, denote

$$A_{r_1,r_2} = \left\{x \in \mathbb{R}^2 : r_1 \leq |x| \leq r_2\right\}, \qquad x \in \mathbb{R}^2$$

the annulus with inner radius r_1 and outer r_2.

Proof of Theorem 4.2   Let $m \in \mathbb{N}$ be sufficiently large, i.e., $m \geq C$. Set

$$n = m^2$$

and

$$k = 12m + 16. \qquad (4.21)$$

(so that we have (4.37) below). Throughout the proof, we work with complex-valued functions and use polar coordinates (r, θ). First, we choose smooth, radially-symmetric cut off functions ψ_1 and ψ_2 such that

$$\psi_1 = \begin{cases} 1, & r \leq m + 1.5 \\ 0, & r \geq m + 1.9 \end{cases}$$

and

$$\psi_2 = \begin{cases} 0, & r \leq m + 0.1 \\ 1, & r \geq m + 0.5. \end{cases}$$

In addition, we require

$$|\psi_i^{(j)}| \leq C, \qquad i = 1, 2, \qquad j = 0, 1, 2.$$

Note that the function

$$u_1 = \frac{1}{r^n} e^{-in\theta}$$

is harmonic. Also, set

$$u_2 = -b r^{-n+2k} \exp\left(i(n + 2k)\theta + i\Phi(\theta)\right) \qquad (4.22)$$

where $b > 0$ and the real valued function Φ is determined below. In order to define it, first note that the solutions of

$$e^{-in\theta} - e^{i(n+2k)\theta} = 0, \qquad \theta \in [0, 2\pi).$$

are given by

$$\theta_l = \frac{\pi l}{n+k}, \qquad l = 0, 1, \ldots, 2n + 2k - 1.$$

Let $T = \pi/(n+k)$. Then Φ is defined as a T-periodic smooth function such that

$$\Phi(\theta) = -4k(\theta - \theta_l), \qquad |\theta - \theta_l| \le \frac{T}{5}.$$

It is clear that

$$\Phi(\theta_l) = \Phi(lT) = 0, \qquad l = 0, 1, \ldots, 2n + 2k - 1.$$

Furthermore, we may assume that for all $\theta \in [0, 2\pi)$ we have

(i) $|\Phi(\theta)| \le 5kT$,
(ii) $|\Phi'(\theta)| \le 5k$, and
(iii) $|\Phi''(\theta)| \le Ckn$.

Due to this choice of Φ, the function u_2 is harmonic in the sectors

$$|\theta - \theta_l| \le T/5.$$

Outside of these sectors, we have

$$|\Delta u_2| = \left| \left(\frac{(n-2k)^2}{r^2} - \frac{(n+2k+\Phi')^2}{r^2} + i\frac{\Phi''}{r^2} \right) u_2 \right| \le \frac{Ckn}{r^2} |u_2|. \quad (4.23)$$

The factor b in (4.22) is chosen so that

$$|\partial_r u_1| = |\partial_r u_2|, \qquad r = m + 1$$

for all θ; this gives

$$b = \frac{n}{n-2k} \frac{1}{(m+1)^{2k}}.$$

Now, we define

$$u = \psi_1 u_1 + \psi_2 u_2,$$

and we will prove that the ratio $|\Delta u|/|\nabla u|$ stays bounded on the annulus $A_{m,m+2}$ whenever $|\nabla u| > 0$. This immediately implies that W is bounded in these regions.

For this purpose, we subdivide $A_{m,m+2}$ into

$$A_{m,m+0.5} \cup A_{m+0.5,m+1.5} \cup A_{m+1.5,m+2}$$

and consider each one of them in turn.

(i) On the first annulus $A_{m,m+0.5}$, we have $\psi_1 = 1$ and $0 \le \psi_2 \le 1$. Therefore, we have

$$\nabla u = \begin{pmatrix} \frac{u_\theta}{r} \\ u_r \end{pmatrix} = \begin{pmatrix} -i\frac{n}{r}u_1 + i\psi_2\frac{n+2k+\Phi'}{r}u_2 \\ -\frac{n}{r}u_1 + \frac{-n+2k}{r}\psi_2 u_2 + \psi_2' u_2 \end{pmatrix}.$$

Using the bounds on ψ_2' and Φ', we obtain

$$|\partial_r u| \ge \frac{n}{r}|u_1| - \frac{n-2k}{r}|u_2| - 4|u_2|. \tag{4.24}$$

Note that

$$\left|\frac{u_1}{u_2}\right| = \frac{r^{-n}}{|b|r^{-n+2k}} = \frac{1}{|b|r^{2k}} \ge \frac{n-2k}{n}\left(\frac{m+1}{m+1/2}\right)^{2k}.$$

By (4.21), we arrive at

$$\left|\frac{u_1}{u_2}\right| \ge \frac{1}{2}\frac{(1+1/m)^{24m-2C_0}}{(1+1/2m)^{24m+2C_0}} \ge \frac{1}{2}\frac{1-\varepsilon}{1+\varepsilon}\frac{\mathrm{e}^{24}}{\mathrm{e}^{12}} \ge \frac{\mathrm{e}^{10}}{2} \tag{4.25}$$

for $\varepsilon > 0$ sufficiently small. Therefore,

$$|\partial_r u| \ge \frac{n}{r}\frac{\mathrm{e}^{10}}{2}|u_2| - \frac{2(n-2k)}{r}|u_2| \ge \frac{1}{C}\frac{n}{r}|u_2|.$$

Moreover,

$$\Delta u = \Delta(\psi_2 u_2) = \psi_2\Delta u_2 + 2\psi_2'\frac{\partial u_2}{\partial r} + \left(\frac{1}{r}\psi_2' + \psi_2''\right)u_2. \tag{4.26}$$

By the bounds on $|\psi_2'|$ and $|\psi_2''|$, we obtain

$$\left|2\psi_2'\frac{\partial u_2}{\partial r}\right| = 2|\psi_2'|\frac{n-2k}{r}|u_2| \le \frac{Cn}{r}|u_2| \tag{4.27}$$

and

$$\left(\left|\frac{1}{r}\psi_2'\right| + |\psi_2''|\right)|u_2| \le C|u_2| \le \frac{Cn}{r}|u_2|, \tag{4.28}$$

where we also used $k/r = \mathcal{O}(1)$ and $n/r = \mathcal{O}(m)$. Combining (4.23), (4.26), (4.27), and (4.28), we get

$$|\Delta u| \le \frac{Cnk}{r^2}|u_2|.$$

Therefore,

$$\frac{|\Delta u|}{|\nabla u|} \le \frac{Cnk}{r^2}|u_2|\frac{r}{n|u_2|} \le \frac{Ck}{r} = \mathcal{O}(1),$$

which completes the proof in the annulus $A_{m,m+0.5}$.

(ii) On $A_{m+1.5,m+2}$, we have $\psi_2 = 1$ and $0 \leq \psi_1 \leq 1$. Then

$$\Delta u = \Delta u_2 + 2\psi_1' \frac{\partial u_1}{\partial r} + \left(\frac{\psi_1'}{r} + \psi_1'' \right) u_1.$$

A similar calculation as (4.25) gives

$$\left| \frac{u_1}{u_2} \right| \leq \frac{n - 2k}{n} \frac{(m+1)^{2k}}{(m+1.5)^{2k}} \leq \frac{e^{-10}}{2}.$$

Therefore,

$$|\Delta u| \leq \frac{Ckn}{r^2} |u_2| + \frac{Cn}{r} |u_1| + C|u_1| \leq \frac{Ckn}{r^2} |u_2|.$$

On the other hand

$$|\partial_r u| \geq \frac{n - 2k}{r} |u_2| - |\psi_1'||u_1| \geq \frac{n}{Cr} |u_2|.$$

Hence,

$$\frac{|\Delta u|}{|\partial_r u|} \leq \frac{Ck}{r} = \mathcal{O}(1)$$

on $A_{m+1.5,m+2}$.

(iii) On $A_{m+0.5,m+1.5}$, we have $\psi_1 = \psi_2 = 1$. Note that on this annulus $|\nabla u_1|$ and $|\nabla u_2|$ are comparable, and thus $|\nabla u|$ may vanish. Define,

$$A_l = \left\{ (r, \theta) : m + 0.5 \leq r \leq m + 1.5, |\theta - \theta_l| \leq \frac{T}{5} \right\}$$

for $l = 0, 1, \ldots, 2n + 2k - 1$. We claim that the zeros of $|\nabla u|$ can only be in the sectors A_l. In fact,

$$|\partial_r u| = \left| \frac{n}{r} u_1 + \frac{n - 2k}{r} u_2 \right| = \frac{n}{r} \frac{r^{-n+2k}}{(m+1)^{2k}} \left| \exp(iS(\theta)) - \frac{(m+1)^{2k}}{r^{2k}} \right|,$$

where $S(\theta) = (2n + 2k)\theta + \Phi(\theta)$. Since $S'(\theta) = 2n + 2k + \Phi' \geq n > 0$, the function $S(\theta)$ is monotone on the sector $[\theta_l, \theta_{l+1}]$ taking the values $S(\theta_l) = 2\pi l$ and $S(\theta_{l+1}) = 2\pi(l + 1)$ at the endpoints.

If $\theta \notin A_l$ for all $l \in \{0, 1, \ldots, 2n + 2k - 1\}$, there exists l such that

$$\theta_l + \frac{T}{5} \leq \theta \leq \theta_l + \frac{4T}{5}. \tag{4.29}$$

Using the monotonicity of S and evaluating it at the endpoints in (4.29), we deduce that

$$2m\pi + \frac{nT}{5} \leq S(\theta) \leq 2\pi(m + 1) - \frac{nT}{5}.$$

Note that $nT/5 = n\pi/5(n+k)$ and $k = \mathcal{O}(n^{1/2})$, so we may assume

$$\frac{nT}{5} \geq \frac{\pi}{7},$$

which implies

$$2\pi m + \frac{\pi}{7} \leq S(\theta) \leq 2\pi(m+1) - \frac{\pi}{7}.$$

Hence,

$$\left| \exp(iS(\theta)) - \frac{(m+1)^{2k}}{r^{2k}} \right| \geq \sin\frac{\pi}{7} > 0,$$

from where we arrive at

$$|\partial_r u| \geq \frac{n}{r} \frac{r^{-n+2k}}{(m+1)^{2k}} \sin\frac{\pi}{7} > 0. \qquad (4.30)$$

Thus we have proven that the zeros of $|\nabla u|$ can only be located in some A_l. Since in A_l the function u_2 is harmonic, we get

$$\Delta u = \Delta u_1 + \Delta u_2 = 0.$$

In these sectors, we simply set $W = 0$. Outside of these sectors, we have

$$|\Delta u| = |\Delta u_2| \leq C\frac{kn}{r}|u_2|.$$

Combining this with (4.30), we get

$$\frac{|\Delta u|}{|\partial_r u|} \leq \frac{Ckn}{r^2}|u_2|\frac{r}{n}\frac{\sin(\pi/7)}{|u_2|} = \frac{Ck}{r} = \mathcal{O}(1)$$

on the annulus $A_{m+0.5,m+1.5}$.

In the next step, we provide u on the annulus $A_{m+2,m+3}$. We choose a C^∞ cut off function ψ such that

$$\psi = \begin{cases} 1, & r \leq m + 2.5 \\ 0, & r \geq m + 2.75 \end{cases}$$

with

$$|\psi^{(j)}| \leq C, \qquad j = 0, 1, 2.$$

Let

$$u = -br^{-n+2k} \exp\Big(i(\psi(r)\Phi(\theta) + (n+2k)\theta)\Big).$$

Note that $u(r, \theta) = u_2(r, \theta)$ for $r = m + 2$. On $A_{m+2,m+3}$ we have

$$\Delta u = \left(\left(\frac{-n + 2k}{r} + i\psi' \Phi \right)^2 + i\Phi \frac{\psi'}{r + \psi''} - \frac{(n + 2k + 4\Phi')^2}{r^2} + \frac{2\psi\Phi''}{r^2} \right) u$$

$$\leq \frac{8nk}{r^2} |u| + |\psi'|^2 |\Phi|^2 |u| + \frac{n}{r} |\psi'||\Phi||u|$$

$$+ |\Phi||\psi''||u| + \frac{n}{r^2} |\psi||\Phi'||u| + \frac{|\psi|^2|\Phi'|^2|u|}{r^2} + \frac{|\psi||\Phi''||u|}{r^2}.$$

By the assumptions on Φ, we have

$$|\Delta u| \leq \frac{Cnk}{r^2} |u|.$$

On the other hand

$$\partial_r u = \frac{-n + 2k}{r} u + i\psi'(r)\Phi(r)u,$$

which implies

$$|\partial_r u| \geq \frac{n - 2k}{r} |u| - \frac{Ck}{n} |u| \geq \frac{n}{Cr} |u|$$

provided n is larger than a constant. Hence, on the annulus $A_{m+2,m+3}$, we obtain

$$\frac{|\Delta u|}{|\partial_r u|} = \mathcal{O}(1).$$

In the next step, we consider the annulus $A_{m+3,m+4}$, where we set

$$u = -br^{-n+2k} e^{i(n+2k)\theta} h(r),$$

where $h(r) = \psi(r) + (1 - \psi(r))g(r)$ with

$$\psi = \begin{cases} 1, & r \leq m + 3.3, \\ 0, & r \geq m + 3.7. \end{cases} \tag{4.31}$$

Then define

$$g(r) = dr^{-4k},$$

where $d = (m+3)^{4k}$ so that $g(m+3) = 1$. A straightforward computation shows that $g(r) \geq e^{-48}$ on $A_{m+3,m+4}$, which implies

$$e^{-48} \leq h(r) \leq 1.$$

Note that

$$\partial_r u = \begin{cases} \frac{-n+2k}{r} u, & r \in A_{m+3,m+3.3} \\ \frac{-n-2k}{r} u_3 g, & r \in A_{m+3.7,m+4} \end{cases} \tag{4.32}$$

where

$$u_3 = -br^{-n+2k}e^{i(n+2k)\theta}.$$

Then we have

$$|\partial_r u| \geq \frac{n}{Cr}|u_3| \tag{4.33}$$

on the above two annuli. On $A_{m+3.3,m+3.7}$, we have $|h'| \leq C$, which implies

$$|\partial_r u| \geq \frac{n-2k}{r}|u_3|e^{-48} - C|u_3| \geq \frac{n}{Cr}|u_3| > 0.$$

One may also check that

$$|\Delta u| \leq \frac{Cnk}{r^2}|u_3| \quad \text{on } A_{m+3,m+4}.$$

Therefore,

$$\frac{|\Delta u|}{|\nabla u|} \leq C \quad \text{on } A_{m+3,m+4}.$$

By a similar construction as in the interval $A_{m,m+2}$, we may rearrange the function $u_4 = -bdr^{-(n+2k)}e^{i(n+2k)\theta}$ to $u_5 = a_1 r^{-(n+2k)}e^{-i(n+k)\theta}$ on $A_{m+4,m+6}$, where

$$a_1 = bd\frac{n+2k}{(n+k)(m+5)^k}.$$

This completes the first step of the construction, with $u = r^{-m^2}e^{-im^2\theta}$ in a neighborhood of $\{r = m\}$ and $u = a_1 r^{-(m+6)^2}e^{-i(m+6)^2\theta}$ in a neighborhood of $\{r = m+6\}$.

Before we start with the main construction on the plane, we estimate $|u|$. Define $v(r) = \max\{|u(r,\theta)| : 0 \leq \theta \leq 2\pi\}$. Then we consider the moduli of the functions u_i for $i = 1,\ldots,5$ which gives an upper bound for $v(r)$. Note that

$$\begin{aligned}
U(r) &= r^{-n}, & r &\in A_{m,m+1} \\
U(r) &= br^{-n+2k}, & r &\in A_{m+1,m+3} \\
U(r) &= br^{-n+2k}h(r), & r &\in A_{m+3,m+4} \\
U(r) &= bdr^{-n-2k}, & r &\in A_{m+4,m+5} \\
U(r) &= a_1 r^{-n-k}, & r &\in A_{m+5,m+6}
\end{aligned} \tag{4.34}$$

yields a piecewise smooth function which satisfies $v(r) \leq 2U(r)$ for every $r \in [m, m+6]$. Also, $v(m) = U(m)$. Furthermore,

$$\frac{d}{dr}\ln U(r) = \frac{-n + O(k)}{r} \leq -\frac{m^2}{4(m+6)} \leq -\frac{r}{C}.$$

Therefore,

$$\ln v(r) - \ln v(m) \leq \ln 2 + \int_m^r \frac{d}{dt} U(t)\, dt \qquad (4.35)$$

$$\leq \ln 2 - \frac{1}{C} \int_m^r s\, ds. \qquad (4.36)$$

In order to extend the construction to the plane we repeat the same procedure on further annuli of the same length, and set $m_1 = m$, where m was our initial radius choice. Then, we define $m_{j+1} = m_j + 6$ and set $n_j = m_j^2$. Thus

$$k_j = n_{j+1} - n_j = 12 m_j + 36. \qquad (4.37)$$

Let u_j denote the solution of (4.1) constructed on $A_{m_j, m_{j+1}}$. Then for $r \geq m$, there exists $j \in \mathbb{N}$ such that $m_j \leq r \leq m_{j+1}$, and a solution of (4.1) on $\{r \geq m\}$ is given by

$$u = \Pi_{i=1}^{j-1} a_i u_j(r, \theta).$$

For the ball $A_{0,m}$, we define

$$h(r) = \begin{cases} r^{m^2+1}, & r \leq \frac{1}{2} \\ 1, & r \geq 1 \end{cases}$$

such that $h \geq 0$ is smooth. Then we consider $u = h(r) r^{-m^2} e^{-im^2 \theta}$. Since $|\partial_r u| \geq 1$ for $r \leq 1/2$ and $|\partial_\theta u| \geq 1/C$ for $r \geq 1/2$, we have

$$\frac{|\Delta u|}{|\nabla u|} \leq C,$$

since $|\Delta u| \leq C$ on $A_{0,m}$. This finishes the construction of the example on the plane.

Moreover, notice that given a radius $r > 0$, there exists $l \in \mathbb{N}$ such that $m + 6l \leq r \leq m + 6(l+1)$. Then,

$$\ln v(r) = \ln v(r) - \ln v(m + 6l)$$

$$+ \sum_{j=0}^{l-1} (v(m + 6(j+1)) - v(m + 6j)) + \ln v(m).$$

Using the estimate in (4.36) we deduce that

$$\ln v(r) \leq l \ln 2 - \frac{1}{C} \int_m^r s\, ds + v(m) \leq C - \frac{1}{C} r^2.$$

As a result,

$$v(r) \leq C \exp(-r^2/C).$$

The proof of the theorem is thus complete.   □

4.4 A construction of a solution vanishing of high order

The following is the main result on construction to a solution to the problem

$$\Delta u = W \cdot \nabla u. \tag{4.38}$$

Theorem 4.3 *Let $n \in \mathbb{N}$. There exists a smooth complex-valued functions W and u such that (4.33) holds with*

$$|W(x)| \leq Cn, \tag{4.39}$$

the function u vanishes at 0 of order n^2, and $|u|$ is constant in $B_1 \backslash B_{1/2}$.

The initial step of the construction relies on the following lemma.

Lemma 4.4 *Let $n \in \mathbb{N}$ be sufficiently large and let $k = 2n - 1$. Then there is a C^∞ function u defined on the annulus $A_{1,1+8/k}$ such that*

$$u = r^{n^2} e^{-in^2\theta} \quad \text{on } A_{1,1+0.25/k}$$

and

$$u = -br^{(n-1)^2} e^{i(n-1)^2\theta} \quad \text{on } A_{1+7.75/k,1+8/k},$$

where

$$b = \frac{1}{(1+7/k)^k (1+2/k)^{2k}}. \tag{4.40}$$

Moreover, the function u satisfies (4.38) on $A_{1,1+8/k}$ with

$$|W| \leq Cn,$$

where C is a sufficiently large constant.

On the other hand, the next lemma provides the mth step of the iteration, addressing the construction in the annulus A_{ρ_{m-1},ρ_m} where the radii are defined iteratively by

$$\rho_m = \rho_{m-1} + \frac{8}{K_m}, \qquad m \in \mathbb{N},$$

$$\rho_0 = 1,$$

where $n_m = n - m + 1$ with $k_m = 2n_m - 1$ and

$$K_m = \frac{1}{C} \left(\frac{1}{n_m^{1/2} n^{1/2}} + \frac{1}{n^{1/3} n_m} + \frac{1}{n_m^{5/4}} \right)^{-1}, \qquad m \in \mathbb{N}.$$

Note that the expression for K_m is comparable with

$$\frac{1}{C} \min \left\{ n_m^{1/2} n^{1/2}, n^{1/3} n_m, n_m^{5/4} \right\}.$$

Lemma 4.5 *Let $n > m$ be positive integers such that $n - m \geq C$, with C a sufficiently large constant. Then there exists a C^∞ function u defined on the annulus A_{ρ_{m-1}, ρ_m} such that*

$$u = r^{(n-m+1)^2} e^{-i(n-m+1)^2 \theta} \quad on \quad A_{\rho_{m-1}, \rho_{m-1}+0.25/K_m}$$

and

$$u = -b r^{(n-m)^2} e^{i(n-m)^2 \theta} \quad on \quad A_{\rho_{m-1}+7.75/K_m, \rho_{m-1}+8/K_m},$$

where

$$b = \frac{n_m^2}{n_m^2 + 2k_m} \frac{1}{(\rho_{m-1} + 7/K_m)^{k_m} (\rho_{m-1} + 2/K_m)^{2k_m}}. \tag{4.41}$$

Moreover, u satisfies (4.38) and (4.39) on $A_{\rho_{m-1}, \rho_{m-1}+8/K_m}$.

The next lemma then covers the case when n is smaller than a constant and connects the solution to $u = e^{in\theta}$.

Lemma 4.6 *Assume that $n \leq C$. Then there exists a C^∞ function u defined on the annulus $A_{2,3}$ such that*

$$u = r^n e^{in\theta} \quad on \quad A_{2,2.3}$$

and

$$u = e^{in\theta} \quad on \quad A_{2.6,3}.$$

Moreover, the function u obeys (4.38) and (4.39) on $A_{2,3}$.

Proof of Lemma 4.4 Since the proof uses ideas similar to those in Camliyurt et al. (2018), we only give a sketch the proof. Assume that n is sufficiently large.

1. Construction of u on $A_{0,1+4/k}$: in the first step, we connect the harmonic functions $u_1 = r^{n^2} e^{-in^2 \theta}$ and $u_2 = -a r^{n^2 + 2k} e^{i((n^2 - 2k)\theta + \Phi(\theta))}$ using

$$u = \psi_1 u_1 + \psi_2 u_2,$$

where ψ_1 and ψ_2 are C^∞ smooth cut-off functions between 0 and 1 such that

$$\psi_1 = \begin{cases} 1, & r \leq 1 + \frac{3}{k} \\ 0, & r \geq 1 + \frac{3.75}{k} \end{cases} \tag{4.42}$$

and

$$\psi_2 = \begin{cases} 0, & r \leq 1 + \frac{0.25}{k} \\ 1, & r \geq 1 + \frac{1}{k}. \end{cases} \tag{4.43}$$

Above, the value

$$a = \left(1 + \frac{2}{k}\right)^{-2k} \in (0,1)$$

is chosen so that the leading terms of $|\nabla u_1|$ and $|\nabla u_2|$ agree at $r = 1 + 2/k$. Also, $\Phi(\theta)$ is a function to be determined below. We shall tacitly assume that all cut off functions are chosen in an optimal way in terms of bounds on derivatives. Here, we assume $\psi_i^{(j)} \leq Ck^j$ for $j = 0, 1, 2$ and $i = 1, 2$. In order to specify Φ, we point out that the zeros of the equation

$$e^{-in^2\theta} - e^{i(n^2 - 2k)\theta} = 0$$

in $[0, 2\pi)$ are

$$\theta_l = \frac{2\pi l}{2n^2 - 2k}, \qquad l = 0, 1, \dots, 2n^2 - 2k - 1.$$

Now, we define

$$\Phi(\theta) = 4k(\theta - \theta_l)$$

in the neighborhood $|\theta - \theta_l| \leq d/5$ of the point θ_l, where

$$d = \frac{\pi}{n^2 - k}.$$

Additionally, we may also assume the optimality on the bounds for the derivatives, i.e.,

(i) $|\Phi(\theta)| \leq 5kd,$

(ii) $|\Phi'(\theta)| \leq 5k,$ \hfill (4.44)

(iii) $|\Phi''(\theta)| \leq Ckn.$

In the sectors $|\theta - \theta_l| \leq d/5$, where $l = 0, 1, \dots, 2n^2 - 2k - 1$, we have $\Delta u_2 = 0$, while on the complement

$$|\Delta u_2| \leq \frac{Cn^2 k}{r^2}|u_2| + \frac{Cn^2|\Phi'||u_2|}{r^2} + \frac{|\Phi'|^2|u_2|}{r^2} + \frac{|\Phi''||u_2|}{r^2} \leq \frac{Cn^2 k}{r^2}|u_2|$$

using

$$\Delta u = \frac{\partial^2 u}{\partial r^2} + \frac{1}{r}\frac{\partial u}{\partial r} + \frac{1}{r^2}\frac{\partial^2 u}{\partial^2 \theta}.$$

We need to check that $|\Delta u|/|\nabla u|$ is bounded by a constant in $A_{1,1+4/k}$ on the complement of the union of all the sectors. To establish this, we divide $A_{1,1+4/k}$ into three regions as in (4.42) and (4.43), but we only provide details for the inner annulus $A_{1,1+1/k}$. There, we have have $\psi_1 = 1$ and $0 \le \psi_2 \le 1$. Since u_1 is harmonic, we have

$$
\begin{aligned}
|\Delta u| &= |\Delta(u_1 + \psi_2 u_2)| = |\Delta(\psi_2 u_2)| \\
&= \left| \psi_2 \Delta u_2 + 2\psi_2' \frac{\partial u_2}{\partial r} + \left(\frac{1}{r}\psi_2' + \psi_2'' \right) u_2 \right| \\
&\le |\Delta u_2| + Ck(n^2 + 2k)|u_2| + C(k + k^2)|u_2| \le Cn^2 k|u_2|.
\end{aligned}
\tag{4.45}
$$

On the other hand, we have

$$
\nabla u = \begin{bmatrix} \partial_r \psi_2 u_2 + \psi_2 \partial_r u_2 + \partial_r u_1 \\ \psi_2 r^{-1} \partial_\theta u_2 + r^{-1} \partial_\theta u_1 \end{bmatrix}.
$$

Next we consider the ratio

$$\frac{|\partial_\theta u_1|}{|\partial_\theta u_2|} = \frac{1}{ar^{2k}} \frac{n^2}{n^2 - 2k + \Phi'} \tag{4.46}$$

from where using (4.44) we further get

$$\frac{|\partial_\theta u_1|}{|\partial_\theta u_2|} \ge \frac{1}{a(1+1/k)^{2k}} \frac{1}{1+1/C} = \frac{(1+2/k)^{2k}}{(1+1/k)^{2k}} \frac{1}{1+\varepsilon} \ge 1 + \frac{1}{C}$$

for n sufficiently large. We conclude that we may bound $|\nabla u|$ from below as

$$|\nabla u| \ge |\partial_\theta u_1| - |\partial_\theta u_2| \ge \left(1 + \frac{1}{C} \right)|\partial_\theta u_2| - |\partial_\theta u_2| \ge \frac{1}{C}|\partial_\theta u_2| \ge \frac{n^2}{C}|u_2|. \tag{4.47}$$

Using (4.45), we get $|\Delta u|/|\nabla u| \le Ck$ from where

$$\frac{|\Delta u|}{|\nabla u|} \le Cn. \tag{4.48}$$

For the proof in the intermediate and outer annuli, we proceed similarly (cf. Camliyurt et al. (2018) with necessary modifications).

2. Construction of u on $A_{1+4/k,1+5/k}$: here we connect the function u_2 from the first step with $u_3 = -ar^{n^2+2k}e^{i(n^2-2k)\theta}$. Consider a smooth

cut off function such that

$$\psi(r) = \begin{cases} 1, & r \le 1 + \frac{4.5}{k} \\ 0, & r \ge 1 + \frac{4.75}{k}. \end{cases}$$

The function

$$u = -ar^{n^2+2k} \exp\left(i\psi(r)\Phi(\theta) + i(n^2 - 2k)\theta\right)$$

defined in $A_{1+4/k,1+5/k}$, satisfies

$$|\Delta u|$$

$$= \left| \frac{(n^2+2k)^2}{r^2} u + \frac{2(n^2+2k)+1}{r} i\psi'\Phi u + \left(i\psi'\Phi - (\psi'\Phi)^2 + \frac{i\psi\Phi''}{r^2} \right) u \right.$$

$$\left. - \frac{(n^2-2k)^2}{r^2} u - \frac{(\psi\Phi')^2}{r^2} u - \frac{2\psi\Phi'(n^2-2k)}{r^2} u \right| \le \frac{Cn^2k}{r^2} |u|$$

and

$$|\partial_r u| \ge \left| \frac{n^2+2k}{r} \right| |u| - |\psi'||u| \ge \frac{n^2}{C} |u|.$$

Therefore, we may divide and obtain (4.24).

3. Construction of u on $r \in A_{1+5/k,1+6/k}$: here, we let

$$u = -ah(r)r^{n^2+2k} e^{i(n^2-2k)\theta} = h(r)u_3$$

where $h(r) = \psi(r) + (1 - \psi(r))g(r)$ and $g(r) = 1/r^{4k}$ with the cutoff function ψ that satisfies

$$\psi = \begin{cases} 1, & r \le 1 + \frac{5}{k} + \frac{1}{3k} \\ 0, & r \ge 1 + \frac{5}{k} + \frac{2}{3k}. \end{cases}$$

Note that

$$|g(r)| \ge \left(1 + \frac{6}{k}\right)^{-4k} \ge \frac{1}{C}$$

and thus $|h| \ge 1/C$. On the annulus $A_{1+5/k,1+5/k+1/3k}$, we have $h = 1$ and $u = u_3 = -r^{n^2+2k} e^{i(n^2-2k)}\theta$ from where

$$|\Delta u| = \left| \frac{(n^2+2k)(n^2+2k-1)}{r^2} u_3 + \frac{n^2+2k}{r^2} u_3 - \frac{(n^2-2k)^2}{r^2} u_3 \right|$$

$$= \left| \frac{(n^2+2k)^2 - (n^2-2k)^2}{r^2} \right| |u_3| \le Cn^2k|u_3|.$$

Since

$$|\nabla u_3| \ge \frac{n^2}{C} |u_3|,$$

we get (4.48).

In the annulus $A_{1+5/k+1/3k,1+5/k+2/3k}$, the proof is carried on in a similar manner. Finally, on $A_{1+5/k+2/3k,1+6/k}$, we have $\Delta u = 0$, and thus we may set $W = 0$.

4. Construction of u on $A_{1+6/k,1+8/k}$: here we connect the harmonic functions $u_4 = -ar^{n^2-2k}e^{i(n^2-2k)\theta}$ and $u_5 = br^{n^2-k}e^{-i(n^2-k)\theta}$, where $b = (1 + 7/k)^{-k}a$. The proof is similar to the arguments as in the construction on $A_{0,1+4/k}$ above and it is thus omitted. $\square$

Proof of Lemma 4.5 Construction of u on A_{ρ_{m-1},ρ_m}: let $u = \psi_1 u_1 + \psi_2 u_2$, where ψ_1 and ψ_2 are cut-offs which connect $u_1 = r^{n_m^2}e^{-in_m^2\theta}$ and $u_2 = -a_m r^{n_m^2+2k_m}e^{i((n_m^2-2k_m)\theta+\Phi(\theta))}$ with $\Phi(\theta)$ specified below and

$$a_m = \frac{n_m^2}{n_m^2 + 2k_m}\left(\rho_{m-1} + \frac{2}{K_m}\right)^{-2k_m}. \qquad (4.49)$$

The cut-offs ψ_1 and ψ_2 are monotone and satisfy

$$\psi_1 = \begin{cases} 1, & r \leq 1 + \frac{3}{K_m} \\ 0, & r \geq 1 + \frac{3.75}{K_m} \end{cases}$$

and

$$\psi_2 = \begin{cases} 0, & r \leq 1 + \frac{0.25}{K_m} \\ 1, & r \geq 1 + \frac{1}{K_m}. \end{cases}$$

Recall that the zeros of the equation

$$e^{-in_m^2\theta} - e^{i(n_m^2-2k_m)\theta} = 0$$

on $[0, 2\pi)$ are

$$\theta_l = \frac{2\pi l}{2n_m^2 - 2k_m}, \qquad l = 0, 1, \ldots, 2n_m^2 - 2k_m - 1. \qquad (4.50)$$

Now define $\Phi(\theta) = 4k_m(\theta - \theta_l)$ in the neighborhood $|\theta - \theta_l| \leq d/5$ where $d = \pi/(n_m^2 - k_m)$. As above, we assume that $|\Phi(\theta)| \leq 5k_m d$ with $|\Phi'(\theta)| \leq 5k_m$ and $|\Phi''(\theta)| \leq Ck_m n_m$. In the sectors $|\theta - \theta_l| \leq d/5$, for $l = 0, 1, \ldots, 2n^2 - 2k - 1$, u_2 is harmonic, while outside of the sectors

$$|\Delta u_2| =$$
$$\left|\left(\frac{(n_m^2 + 2k_m)^2}{r^2} - \frac{(n_m^2 - 2k_m)^2}{r^2} - \frac{2(n_m^2 - 2k_m)\Phi'}{r^2} - \frac{|\Phi'|^2}{r^2} + i\frac{\Phi''}{r^2}\right)u_2\right|$$
$$\leq \frac{Cn_m^2 k_m}{r^2}|u_2|. \qquad (4.51)$$

We still need to check that $|\Delta u|/|\nabla u|$ stays bounded on the annulus $A_{1,1+4/K_m}$ where $\Delta u \neq 0$.

For this purpose, we separate the annulus into three regions, inner, outer, and intermediate as in the proof above. We only show details for the inner annulus $A_{\rho_{m-1},\rho_{m-1}+1/K_m}$.

In the inner annulus, we have $u = u_1 + \psi_2 u_2$, as $\psi_1 = 1$ and $0 \leq \psi_2 \leq 1$, whence

$$\Delta u = \Delta(u_1 + \psi_2 u_2) = \psi_2 \Delta u_2 + 2\psi_2' \frac{\partial u_2}{\partial r} + \left(\frac{1}{r}\psi_2' + \psi_2''\right) u_2.$$

Using (4.51), we get

$$
\begin{aligned}
|\Delta u| &\leq |\Delta u_2| + CK_m(n_m^2 + 2k_m)|u_2| + C(K_m + K_m^2)|u_2| \\
&\leq C\left(n_m^2 k_m + K_m(n_m^2 + 2k_m) + K_m + K_m^2\right)|u_2| \qquad (4.52) \\
&\leq C(K_m n_m^2 + K_m^2)|u_2|.
\end{aligned}
$$

Next, we have

$$
\begin{aligned}
\frac{|\partial_r u_1|}{|\partial_r u_2|} &= \frac{1}{a_m r^{2k_m}} \frac{n_m^2}{n_m^2 + 2k_m} \geq \frac{1}{a_m(1 + 1/K_m)^{2k_m}} \frac{n_m^2}{n_m^2 + 2k_m} \\
&= \frac{(1 + 2/K_m)^{2k_m}}{(1 + 1/K_m)^{2k_m}} = \left(1 + \frac{1}{1 + K_m}\right)^{k_m} \geq \left(1 + \frac{1}{2K_m}\right)^{k_m} \\
&\geq 1 + \frac{k_m}{4K_m} \geq 1 + \frac{n_m}{4K_m},
\end{aligned}
$$
$$(4.53)$$

where we used $K_m \geq C$. Therefore

$$
\begin{aligned}
|\nabla u| &\geq |\partial_r u_1| - |\partial_r u_2| - |\partial_r \psi_2||u_2| \\
&\geq \left(1 + \frac{1}{C}\frac{n_m}{K_m}\right) \frac{n_m^2 + 2k_m}{r}|u_2| - \frac{n_m^2 + 2k_m}{r}|u_2| - CK_m|u_2| \\
&= \frac{n_m}{K_m}\left(\frac{1}{C}\frac{n_m^2 + 2k_m}{r} - C\frac{K_m^2}{n_m}\right)|u_2| \geq \frac{n_m^3}{CK_m}|u_2|,
\end{aligned}
$$

assuming r is bounded and $n - m$ is sufficiently large, since

$$\frac{(n_m^2 + 2k_m/r)}{K_m^2/n_m} \geq C$$

by the definition of K_m.

By (4.47) and (4.52) we get

$$\frac{|\Delta u|}{|\nabla u|} \leq C\left(\frac{K_m^2}{n_m} + \frac{K_m^3}{n_m^3}\right) \leq C\frac{nn_m}{n_m} + C\frac{nn_m^3}{n_m^3} \leq Cn. \qquad (4.54)$$

The treatment in the other annuli is the same as in the previous proof with the changes shown above. Thus we omit further details.   □

Proof of Lemma 4.6   Assume that $n \le C$, and let

$$u(x) = h(r)r^n e^{in\theta}$$

where h is a C^∞ function such that

$$h = \begin{cases} 1, & r \le 2.3\,, \\ r^{-n}, & r \ge 2.6\,. \end{cases}$$

Now, we claim that $\Delta u = W \cdot \nabla u$ on $A_{2,3}$, for some W with $|W| \le C$. Since u is harmonic on $A_{2,2.3}$, we only need to consider the annulus $A_{2.3,2.6}$, where

$$\Delta u = \Delta r^n e^{in\theta} h + 2\partial_r (r^n e^{in\theta}) \partial_r h$$

which gives $|\Delta u| \le C$. On the other hand, we get

$$\frac{\partial_\theta u}{r} = \frac{in}{r} h r^n e^{in\theta},$$

from where $|\nabla u| \ge 1/C$ and thus $|\Delta u|/|\nabla u| \le C$.   □

Theorem 4.3 is based on the steps provided in Lemmas 4.4–4.6.

Proof of Theorem 4.3   Due to Lemma 4.6, we may assume that $n \ge C$. First, let

$$u = r^{n^2} e^{-in^2\theta}, \qquad |x| \le 1 + \frac{8}{k}.$$

Denote

$$b_1 = \frac{1}{(1 + 7/k)^k (1 + 2/k)^{2k}}$$

(the right-hand side of (4.40)) and

$$b_m = \frac{n_m^2}{n_m^2 + 2k_m} \frac{1}{(\rho_{m-1} + 7/K_m)^{k_m} (\rho_{m-1} + 2/K_m)^{2k_m}}$$

(the right-hand side of (4.41)).

Applying Lemma 4.4 and then Lemma 4.5 successively, we may extend u to the annulus $r \le \rho_{\overline{m}}$, where $\overline{m}$ is the largest integers such that $n - \overline{m} \ge C$ with C as in Lemma 4.5. Note that we have

$$u = (-1)^{m+1} b_1 \cdots b_{m-1} r^{(n-m+1)^2} \exp\left((-1)^m i(n-m+1)^2 \theta\right)$$

on $A_{\rho_{m-1},\rho_{m-1}+0.25/K_m}$ for $m = 1, 2, \ldots, \overline{m}$. After the final application of Lemma 4.5, we get

$$u = (-1)^{\overline{m}} b_1 \cdots b_{\overline{m}} r^{(n-\overline{m})^2} \exp\left((-1)^{\overline{m}+1} i(n - \overline{m})^2 \theta\right) \qquad (4.55)$$

on $A_{\rho_{\overline{m}-1}+7.75/K_m, \rho_{\overline{m}-1}+8/K_{\overline{m}}}$. But then Lemma 4.6 allows us to connect the function in (4.55) to $e^{in\eta}$, where $n \le C$; the last step can be performed only if

$$\sum_{m=1}^{n} \frac{1}{K_m} \le C,$$

which holds since

$$\sum_{m=1}^{n} \frac{1}{K_m} = C \sum_{m=1}^{n} \left(\frac{1}{n_m^{1/2} n^{1/2}} + \frac{1}{n^{1/3} n_m} + \frac{1}{n_m^{5/4}}\right) \le C \qquad (4.56)$$

and the theorem is established. $\qquad\qquad\qquad\qquad\qquad\qquad\qquad\qquad\qquad\square$

4.5 The equation $\Delta u = V u$

In this section, we state, without proofs, the vanishing and localization results for the equation

$$\Delta u = V u.$$

Regarding the localization, we have the important result of Meshkov (1991).

Theorem 4.7 (Meshkov, 1991) *There exist smooth u and V on $\mathbb{R}^2 \backslash B_1$ such that (4.5) holds, and we have $\|V\|_{L^\infty(B_1^c)}$ and*

$$|u(x)| \le C \exp(-|x|^{4/3}/C) \qquad (4.57)$$

for all $x \in B_1^c$.

Camliyurt et al. (2018) constructed a solution of (4.5) in B_1 vanishing of high order.

Theorem 4.8 (Camliyurt et al., 2018) *For every $n \in \mathbb{N}$, there exists a smooth complex-valued function u on B_1 that satisfies (4.5) with the Dirichlet boundary conditions such that*

$$|V(x)| \le C n^3, \qquad x \in B_1,$$

and the function u vanishes at 0 of order n^2.

Since the construction in Camliyurt et al. (2018) is carried out so that the solution is constant on $B_{3/4}\backslash B_{1/2}$, it is easy to adjust the example to be periodic, satisfy Neumann boundary conditions etc. It is also possible to use the solution constructed for Theorem 4.7 to obtain a solution of a parabolic equation

$$\partial_t u - \Delta u = V u$$

on $\mathbb{R}^2 \times (-1, 1)$, with 1-periodic boundary condition on the sides with the following property.

Theorem 4.9 (Camliyurt et al., 2018) *For every $n \in \mathbb{N}$, there exists a smooth complex-valued solution u of the problem above, that satisfies*

$$\sup_{t \in (-1,0)} q(t) \le n^3$$

where

$$q(t) = \frac{\int_{(0,1)^2} |\nabla u(\cdot, t)|^2}{\int_{(0,1)^2} |u(\cdot, t)|^2}$$

and

$$|V(x, t)| \le C n^3$$

for $\mathbb{R}^2 \times (-1, 0)$. Moreover, the function u vanishes of order n^2 at $(0, 0)$.

In particular, this theorem shows the optimality of the estimate (4.5) (when $W = 0$) in Camliyurt & Kukavica (2017).

Acknowledgments

The authors were supported in part by the NSF grant DMS-1615239.

References

Agmon, S. (1966) *Unicité et convexité dans les problèmes différentiels.* Séminaire de Mathématiques Supérieures, No. 13 (Été, 1965). Les Presses de l'Université de Montréal, Montreal, Que.

Agmon, S. & Nirenberg, L. (1967) Lower bounds and uniqueness theorems for solutions of differential equations in a Hilbert space. *Comm. Pure Appl. Math.* **20**, 207–229.

Alessandrini, G. & Escauriaza, L. (2008) Null-controllability of one dimensional parabolic equations. *ESAIM Control Optim. Calc. Var.* **14**, 284–293.

Alessandrini, G., Morassi, A., Rosset, E., & Vessella, S. (2009) On doubling inequalities for elliptic systems. *J. Math. Anal. Appl.* **357**, 349–355.

Alessandrini, G. & Vessella, S. (1988) Local behaviour of solutions to parabolic equations. *Comm. Partial Differential Equations* **13**, 1041–1058.

Almgren, Jr., F.J. (1979) *Dirichlet's problem for multiple valued functions and the regularity of mass minimizing integral currents.* Minimal submanifolds and geodesics (Proc. Japan-United States Sem., Tokyo, 1977). North-Holland, Amsterdam-New York, 1–6.

Aronszajn, N. (1957) A unique continuation theorem for solutions of elliptic partial differential equations or inequalities of second order. *J. Math. Pures Appl.* **36**, 235–249.

Bakri, L. (2012) Quantitative uniqueness for Schrödinger operator. *Indiana Univ. Math. J.* **61**, 1565–1580.

Bakri, L. (2013) Carleman estimates for the Schrödinger operator. Applications to quantitative uniqueness. *Comm. Partial Differential Equations* **38**, 69–91.

Bakri, L. & Casteras, J.-B. (2014) Quantitative uniqueness for Schrödinger operator with regular potentials. *Math. Methods Appl. Sci.* **37**, 1992–2008.

Bourgain, J. & Kenig, C.E. (2005) On localization in the continuous Anderson-Bernoulli model in higher dimension. *Invent. Math.* **161**, 389–426.

Camliyurt, G. & Kukavica, I. (2016) A local asymptotic expansion for a solution of the Stokes system. *Evol. Equ. Control Theory* **5**, 647–659.

Camliyurt, G. & Kukavica, I. (2017) Quantitative unique continuation for a parabolic equation. *Indiana Univ. Math. J.* **67**, no. 2, 657–678.

Camliyurt, G., Kukavica, I., & Wang, F. (2018) On quantitative uniqueness for elliptic equations. *Math Z.*, to appear.

Carleman, T. (1939) Sur un problème d'unicité pur les systèmes d'équations aux dérivées partielles à deux variables indépendantes. *Ark. Mat., Astr. Fys.* **26**, 9.

Cordes, H.O. (1956) Über die eindeutige Bestimmtheit der Lösungen elliptischer Differentialgleichungen durch Anfangsvorgaben. *Nachr. Akad. Wiss. Göttingen. Math.-Phys. Kl. IIa.*, 239–258.

Donnelly, H. & Fefferman, C. (1988) Nodal sets of eigenfunctions on Riemannian manifolds. *Invent. Math.* **93**, 161–183.

Donnelly, H. & Fefferman, C. (1990) Nodal sets for eigenfunctions of the Laplacian on surfaces. *J. Amer. Math. Soc.* **3**, 333–353.

Escauriaza, L. (2000) Carleman inequalities and the heat operator. *Duke Math. J.* **104**, 113–127.

Escauriaza, L. & Fernández, F.J. (2003) Unique continuation for parabolic operators. *Ark. Mat.* **41**, 35–60.

Escauriaza, L., Fernández, F.J., & Vessella, S. (2006) Doubling properties of caloric functions. *Appl. Anal.* **85**, 205–223.

Escauriaza, L. & Vega, L. (2001) Carleman inequalities and the heat operator. II. *Indiana Univ. Math. J.* **50**, 1149–1169.

Escauriaza, L. & Vessella, S. (2003) Optimal three cylinder inequalities for solutions to parabolic equations with Lipschitz leading coefficients. Inverse problems: theory and applications (Cortona/Pisa, 2002). *Contemp. Math.* **333**, Amer. Math. Soc., Providence, RI, 79–87.

Garofalo, N. & Lin, F.H. (1986) Monotonicity properties of variational integrals, A_p weights and unique continuation. *Indiana Univ. Math. J.* **35**, 245–268.

Gilbarg, D. & Trudinger, N.S. (2001) *Elliptic partial differential equations of second order.* Reprint of the 1998 edition. Classics in Mathematics, Springer-Verlag, Berlin.

Hörmander, L. (1983) Uniqueness theorems for second order elliptic differential equations. *Comm. Partial Differential Equations* **8**, 21–64.

Jerison, D. & Kenig, C.E. (1985) Unique continuation and absence of positive eigenvalues for Schrödinger operators, with an appendix by E.M. Stein. *Ann. of Math.* **121**, 463–494.

Kenig, C. (1989) Restriction theorems, Carleman estimates, uniform Sobolev inequalities and unique continuation, in *Harmonic analysis and partial differential equations (El Escorial, 1987).* Lecture Notes in Math. **1384**, Springer, Berlin, 69–90.

Kenig, C. (2007) *Some recent applications of unique continuation*, Recent developments in nonlinear partial differential equations, Contemp. Math., **439**, Amer. Math. Soc., Providence, RI, 25–56.

Kenig, C. (2008) Quantitative unique continuation, logarithmic convexity of Gaussian means and Hardy's uncertainty principle. *Perspectives in partial differential equations, harmonic analysis and applications.* Proc. Sympos. Pure Math. **79**, Amer. Math. Soc., Providence, RI, 207–227.

Kenig, C., Silvestre, L., & Wang, J.-N. (2015) On Landis' conjecture in the plane. *Comm. Partial Differential Equations* **40**, 766–789.

Koch, H. & Tataru, D. (2001) Carleman estimates and unique continuation for second-order elliptic equations with nonsmooth coefficients. *Comm. Pure Appl. Math.* **54**, 339–360.

Koch, H. & Tataru, D. (2009) Carleman estimates and unique continuation for second order parabolic equations with nonsmooth coefficients. *Comm. Partial Differential Equations* **34**, 305–366.

Kondratév, V.A. & Landis, E.M. (1988) Qualitative theory of second-order linear partial differential equations. *Partial differential equations* **3** (Russian), Itogi Nauki i Tekhniki, Akad. Nauk SSSR, Vsesoyuz. Inst. Nauchn. i Tekhn. Inform., Moscow, 99–215.

Kukavica, I. (1998) Quantitative uniqueness for second-order elliptic operators. *Duke Math. J.* **91**, 225–240.

Kukavica, I. (1999) Self-similar variables and the complex Ginzburg–Landau equation. *Comm. Partial Differential Equations* **24**, 545–562.

Kukavica, I. (2000) Quantitative uniqueness and vortex degree estimates for solutions of the Ginzburg–Landau equation. *Electron. J. Differential Equations* **61**, 15 pp. (electronic).

Kukavica, I. (2003) Length of vorticity nodal sets for solutions of the 2D Navier–Stokes equations. *Comm. Partial Differential Equations* **28**, 771–793.

Kurata, K. (1994) On a backward estimate for solutions of parabolic differential equations and its application to unique continuation. *Spectral and scattering theory and applications*, Adv. Stud. Pure Math. **23**, Math. Soc. Japan, Tokyo, 247–257.

Lin, F.H. (1991) Nodal sets of solutions of elliptic and parabolic equations. *Comm. Pure Appl. Math.* **44**, 287–308.

Meshkov, V.Z. (1991) On the possible rate of decrease at infinity of the solutions of second-order partial differential equations. *Mat. Sb.* **182**, 364–383.

Poon, C.C. (1996) Unique continuation for parabolic equations. *Comm. Partial Differential Equations* **21**, 521–539.

Vessella, S. (2003) Carleman estimates, optimal three cylinder inequalities and unique continuation properties for parabolic operators. *Progress in analysis* Vol. I, II (Berlin, 2001), World Sci. Publ., River Edge, NJ, 485–492.

Wolff, T.H. (1995) Recent work on sharp estimates in second order elliptic unique continuation problems. *Fourier analysis and partial differential equations (Miraflores de la Sierra, 1992)*, Stud. Adv. Math., CRC, Boca Raton, FL, 99–128.

5

Quasi-invariance for the Navier–Stokes equations

Koji Ohkitani

School of Mathematics and Statistics,
The University of Sheffield,
Hicks Building, Hounsfield Road, Sheffield, S3 7RH. UK.
`k.ohkitani@sheffield.ac.uk`

Abstract In this contribution we focus on a few results regarding the study of the three-dimensional Navier–Stokes equations with the use of vector potentials. These dependent variables are critical in the sense that they are scale invariant. By surveying recent results utilising criticality of various norms, we emphasise the advantages of working with scale-invariant variables.

The Navier–Stokes equations, which are invariant under static scaling transforms, are not invariant under dynamic scaling transforms. Using the vector potential, we introduce scale invariance in a weaker form, that is, invariance under dynamic scaling *modulo* a martingale (Maruyama–Girsanov density) when the equations are cast into Wiener path-integrals. We discuss the implications of this *quasi-invariance* for the basic issues of the Navier–Stokes equations.

5.1 Introduction

Many of the results in Navier–Stokes theory have been obtained by paying attention to scale-invariant properties of norms, stemming from the work of Kato & Fujita (1962), for example. Recent work in this direction includes the exclusion of self-similar blowup in Nečas, Růžička, & Šverák (1996) [also Chae (2007) and Hou & Li (2007)], the regularity criterion by the L^3-norm of velocity in Escauriaza, Seregin, & Sverak (2003) and the global regularity result for small initial data in BMO^{-1} in Koch & Tataru (2001). These results are motivated, at least partly, by static (i.e. frozen time) scale-invariant considerations of the Navier–Stokes equations under the usual parabolic transformations.

There is a more general kind of transformation, which is dynamical in

Published as part of *Partial Differential Equations in Fluid Mechanics*, edited by C.L. Fefferman, J.C. Robinson, & J.L. Rodrigo © Cambridge University Press 2018.

nature. In view of these successful results with static scaling transformations, it seems promising to pursue further development with dynamic
scaling transformations. Through such efforts, it has been recognised recently that with the scale-invariant dependent variables we would benefit
from some advantages in the analysis. In particular, we can characterise
the concept of scale invariance in its most generalised sense. The purpose
of this contribution is to survey a number of results on the basic problems of the Navier–Stokes equations obtained in this spirit. This survey
is not intended to be an exhaustive list of literature, but it is rather an
idiosyncratic review.

The rest of this paper is constructed as follows. In Section 5.2, we
describe the usual reformulation of the Navier–Stokes equations as integral equations using the velocity variable. In Section 5.3, we recall how
we may solve the forced Burgers equations by linearisation and a path
integral. In Section 5.4, we recall the Navier–Stokes equations written
using vector potentials and some regularity conditions using critical and
subcritical norms. In Section 5.5, we apply an analogue of the Cole–
Hopf transform and the Feynman–Kac formula just as we did for the
Burgers equations. In Section 5.6, using a dynamic scaling transform
we derive the Leray equations. Using probabilistic tools we compare the
Navier–Stokes and Leray equations in detail, thereby recognising their
quasi-invariance. Section 5.7 is devoted to a summary.

5.2 Navier–Stokes equations

We are interested in the Navier–Stokes equations in $\mathbb{R}^3$:

$$\frac{\partial u}{\partial t} + u \cdot \nabla u = -\nabla p + \frac{1}{2}\triangle u, \tag{5.1}$$

$$\nabla \cdot u = 0,$$

$$u(x,0) = u_0(x).$$

Starting from Leray (1934), there has been much work on the analysis
of the Navier–Stokes equations. General references include Constantin
& Foias (1988); Doering (2009); Doering & Gibbon (1995); Robinson,
Rodrigo, & Sadowski (2016).

It is useful to recall how we can convert the above equations to the
conventional integral equations. Using the heat kernel

$$g_t = \frac{1}{(2\pi t)^{3/2}} \exp\left(-\frac{|x|^2}{2t}\right),$$

we apply the Duhamel principle to the Navier–Stokes equations to obtain

$$\boldsymbol{u}(t) = g_t * \boldsymbol{u}_0 - \int_0^t g_{t-s} * \mathbb{P} \nabla \cdot (\boldsymbol{u} \otimes \boldsymbol{u})(\cdot, s) \, \mathrm{d}s, \tag{5.2}$$

$$= \mathbb{E}[\boldsymbol{u}_0(\boldsymbol{W}_t)] - \int_0^t \mathbb{E}[\mathbb{P} \nabla \cdot (\boldsymbol{u} \otimes \boldsymbol{u})(\boldsymbol{W}_s, t - s)] \, \mathrm{d}s,$$

where $\mathbb{P} = \boldsymbol{I} - \nabla \triangle^{-1} \nabla \cdot$ denotes solenoidal projection and $*$ convolution. Here $\boldsymbol{W}_t$ denotes standard three-dimensional Brownian motion starting from $\boldsymbol{x}$ at $t = 0$ $\boldsymbol{W}_0 = \boldsymbol{x}$ and $\mathbb{E}[\cdot]$ an average with respect to a probability measure associated with $\boldsymbol{W}_t$ (see Appendix A).

The following condition due to Serrin (1963)

$$\int_0^T \|\boldsymbol{u}\|_{L^p}^{\frac{2p}{p-3}} \, \mathrm{d}t < \infty, \qquad 3 < p \leq \infty, \tag{5.3}$$

guarantees uniqueness and smoothness of the solution on $[0, T)$. In particular, we have in the limit $p \to \infty$

$$\int_0^T \|\boldsymbol{u}\|_{L^\infty}^2 \, \mathrm{d}t < \infty, \tag{5.4}$$

which is probably the best-known criterion for regularity.

Different arguments - based on different ways to estimate the right-hand side of (5.2) - are required to obtain the two regularity criteria in (5.3) and (5.4) (see Section 6.3.3 in the contribution by Ożański & Pooley (2018) in this volume). The condition (5.4) can be compared with (5.17) in the equation (5.12) below, which is obtained as a boundedness condition for the exponential process to be a martingale. This illustrates an advantage of working with critical dependent variables.

5.3 Burgers equation

To illustrate the basic ideas, we consider the Burgers equations in $\mathbb{R}^3$ subject to an external forcing of the form $-\nabla V(\boldsymbol{x}, t)$

$$\frac{\partial \boldsymbol{v}}{\partial t} + \boldsymbol{v} \cdot \nabla \boldsymbol{v} = -\nabla V + \frac{1}{2} \triangle \boldsymbol{v}, \tag{5.5}$$

$$\boldsymbol{v}(\boldsymbol{x}, 0) = \boldsymbol{v}_0(\boldsymbol{x}).$$

(More generally, the following argument holds in any $\mathbb{R}^n$, $n \geq 1$.) Here we restrict ourselves to the special class of potential flows $\boldsymbol{v} = \nabla \varphi$. The variable $\boldsymbol{v}$ satisfies the following well-known scale-invariance:

if $v(x, t)$ is a solution then so is $\lambda v(\lambda x, \lambda^2 t)$ for any $\lambda > 0$.

Integrating the equation (5.5) and taking the constant of integration to be zero, we find the following Hamilton–Jacobi equation

$$\frac{\partial \varphi}{\partial t} + \frac{1}{2} |\nabla \varphi|^2 + V = \frac{1}{2} \triangle \varphi, \tag{5.6}$$

$$\varphi(x, 0) = \varphi_0(x).$$

In terms of the variable φ, scale invariance now reads

if $\varphi(x, t)$ is a solution then so is $\varphi(\lambda x, \lambda^2 t)$ for any $\lambda > 0$.

An observation made by Cole (1949, 1951) is that φ lacks a prefactor after the transformation. This is because φ has the same physical dimension as kinematic viscosity $\nu (= 1/2)$. Applying a transform $\varphi = k \log \theta$, with a constant k of the same dimension as ν, we rewrite (5.6) as

$$\frac{\partial \theta}{\partial t} = \frac{1}{2} \triangle \theta - \left(\frac{k+1}{2} \frac{|\nabla \theta|^2}{\theta^2} + \frac{V}{k} \right) \theta,$$

$$\theta(x, 0) = \theta_0(x).$$

Choosing $k = -1$ and following Cole (1951) and Hopf (1950), we can linearise the Burgers equation to a heat equation with a potential term, i.e. the Schrödinger equation at imaginary times

$$\frac{\partial \theta}{\partial t} = \frac{1}{2} \triangle \theta + V\theta. \tag{5.7}$$

If the potential term is bounded in the sense that

$$\int_0^t \sup_x |V(x, s)| \, \mathrm{d}s < \infty,$$

the equation (5.7) is soluble by the Feynman–Kac formula as

$$\theta(x, t) = E \left[\theta_0(W_t) \exp \left(\int_0^t V(W_s, s) \, \mathrm{d}s \right) \right]. \tag{5.8}$$

This representation can be obtained by applying a time-dependent Trotter formula, see e.g. Taylor (1996). See Appendix B for alternative forms of functional integrals. Note that solutions θ_k for $k \neq -1$ can be obtained as $\theta_k = \theta^{-1/k}$.

We refer to (5.8) as the Cole–Hopf-Feynman–Kac formula for the Burgers equations. We will take a brute-force approach to obtain a similar expression for the Navier–Stokes equations.

	BMO^{-1}	L^3	$\dot{H}^{1/2}$	L^p $(p > 3)$	$\dot{H}^1$
small data→ global regularity	Yes KT('01)	Yes	Yes KF('62)	NA	NA
time of local existence	NA	NA	NA	$\|u_0\|^{-\frac{2p}{p-3}}$	$\nu^3 / \|u_0\|^4_{\dot{H}^1}$
blowup criterion	?	Yes	Yes	Yes	Yes

Table 5.1

Here u_0 denotes the initial velocity and $\|u\|_{BMO^{-1}} \approx \|\psi\|_{BMO}$. Note that NA's appear in a staggered manner; on the left for the three critical norms and on the right for the two subcritical norms. KT('01) refers to Koch & Tataru (2001) and KF('62) to Kato & Fujita (1962).

5.4 Use of critical dependent variables

We introduce the vector potential ψ defined such that $u = \nabla \times \psi$ and $\nabla \cdot \psi = 0$. The Navier–Stokes equations have been written as a nonlocal version of the Hamilton–Jacobi equations in Ohkitani (2015)

$$\frac{\partial \psi}{\partial t} - \frac{1}{2}\triangle \psi = T[\nabla \psi], \qquad (5.9)$$

where

$$T[\nabla \psi] \equiv \frac{3}{4\pi} \fint_{\mathbb{R}^3} \frac{r \times (\nabla \times \psi(y))\, r \cdot (\nabla \times \psi(y))}{|r|^5} \, dy, \qquad (5.10)$$

with $r = x - y$ and where $\fint$ denotes a principal-value integral. We assume that $|\psi(x, t)| \to 0$ as $|x| \to \infty$ for all $t \geq 0$. It can be checked that $\nabla \cdot T[\nabla \psi] = 0$ is satisfied.

In Table 5.4 we compare a number of known results concerning Navier–Stokes regularity. One kind of theorem claims global regularity for small initial data, while the other kind guarantees local existence for general initial data. We list results obtained with critical BMO^{-1}, L^3 and $\dot{H}^{1/2}$-norms and those with subcritical L^p $(p > 3)$ and $\dot{H}^1$-norms. If any one of NA's *were* available, that would imply global regularity immediately. See also Ohkitani (2016) for an asymptotic analysis related to $\dot{H}^{1/2}$-norm.

Experience shows that these two kinds of theorems go together; in view of the embedding

$$\|\psi\|_{BMO} \lesssim \|u\|_{L^3},$$

we may ask whether

$$\|\psi\|_{BMO} \to \infty \text{ as } t \to t_*$$

for a possible blowup at t_*. Apparently, this remains an open question (hence a question mark in Table 5.4) and we will briefly remark on it, in connection with the Cole–Hopf transform at the end of Section 5.5.

At the moment, it is known from Ohkitani (2017a) that the condition

$$\int_0^{t_*} \|\boldsymbol{T}[\nabla\boldsymbol{\psi}]\|_{L^\infty} \, dt = \infty \qquad (5.11)$$

holds for blowup at $t = t_*$.

A possibility of an even weaker norm $\|\boldsymbol{u}\|_{B^{-1}_{\infty,\infty}}$ serving as a blowup criterion has been explored in Cheskidov & Shvydkoy (2010). Because

$$\|\boldsymbol{u}\|_{B^{-1}_{\infty,\infty}} \lesssim \|\boldsymbol{u}\|_{\mathrm{BMO}^{-1}} \simeq \|\boldsymbol{\psi}\|_{\mathrm{BMO}},$$

the motivation is more ambitious than ours. Cheskidov & Shvydkoy obtained a dichotomy-type result: upon a possible singularity, either i) $\|\boldsymbol{u}\|_{B^{-1}_{\infty,\infty}}$ becomes unbounded, or ii) it is bounded but there is a jump of $O(\nu)$ in the norm (as a function of time) near the critical time. It is not known whether $\|\boldsymbol{u}\|_{B^{-1}_{\infty,\infty}}$ becomes unbounded or not.

In a corresponding analysis in two dimensions, the Navier–Stokes equation in the stream function formulation was derived in Ohkitani (2008) and its applications are described in Ohkitani (2017a,c).

5.5 Cole–Hopf transform and Feynman–Kac formula

We consider an analogue of the Cole–Hopf transform for the Navier–Stokes equations, introduced component-wise in Ohkitani (2017c), by

$$\psi_j = k \log \theta_j, \quad (j = 1, 2, 3), \qquad (5.12)$$

with a constant $k(\neq 0)$ and derive equations for $\boldsymbol{\theta}$ (see also Vanon & Ohkitani, 2018).

The derivation of the equations for $\boldsymbol{\theta}$ is straightforward, but best stated here for completeness

$$\frac{\partial \psi_j}{\partial t} - T_j[\nabla \psi_j] - \frac{1}{2}\triangle \psi_j$$

$$= \frac{k}{\theta_j}\frac{\partial \theta_j}{\partial t} - k^2 T_j\left[\frac{\nabla\theta_1}{\theta_1}, \frac{\nabla\theta_2}{\theta_2}, \frac{\nabla\theta_3}{\theta_3}\right] - \frac{1}{2}k\left(\frac{\triangle\theta_j}{\theta_j} - \frac{|\nabla\theta_j|^2}{\theta_j^2}\right)$$

$$= k\left\{\frac{1}{\theta_j}\left(\frac{\partial\theta_j}{\partial t} - \frac{1}{2}\triangle\theta_j\right) - \left(k\,T_j\left[\frac{\nabla\theta_1}{\theta_1}, \frac{\nabla\theta_2}{\theta_2}, \frac{\nabla\theta_3}{\theta_3}\right] - \frac{1}{2}\frac{|\nabla\theta_j|^2}{\theta_j^2}\right)\right\},$$

where no summation over j is implied. Setting the right-hand side equal to zero, we obtain a system of heat equations with a potential term

$$\frac{\partial \theta_j}{\partial t} = \frac{1}{2} \triangle \theta_j + f_j[\boldsymbol{\theta}](\boldsymbol{x}, t)\theta_j, \tag{5.13}$$

where

$$f_j[\boldsymbol{\theta}](\boldsymbol{x}, t) \equiv kT_j \left[\frac{\nabla \theta_1}{\theta_1}, \frac{\nabla \theta_2}{\theta_2}, \frac{\nabla \theta_3}{\theta_3} \right] - \frac{1}{2} \frac{|\nabla \theta_j|^2}{\theta_j^2}, \quad (j = 1, 2, 3.) \tag{5.14}$$

Hereafter no summation is implicit with respect to j in f_j.

Regarding the nonlinear term as forcing in the spirit of Duhamel principle, we convert (5.13) into path-integral equations by the Feynman–Kac formula

$$\theta_j(\boldsymbol{x}, t) = \mathbb{E} \left[\theta_j(\boldsymbol{W}_t, 0) \exp \left(\int_0^t f_j[\boldsymbol{\theta}](\boldsymbol{W}_s, s) \, ds \right) \right]. \tag{5.15}$$

For proof, see Ohkitani (2017c). The path-integral representation (5.15) is just another way of writing down the Navier–Stokes equations. While the formula contains complicated contents, we note that it is fully explicit, with $\boldsymbol{f}[\boldsymbol{\theta}]$ defined by (5.14) and $\boldsymbol{T}[\nabla \boldsymbol{\psi}]$ by (5.10).

For convenience, we use the following notation hereafter

$$F_j[\boldsymbol{\theta}](\boldsymbol{W}_t) \equiv \theta_j(\boldsymbol{W}_t, 0) \exp \left(\int_0^t f_j[\boldsymbol{\theta}](\boldsymbol{W}_s, s) \, ds \right). \tag{5.16}$$

The exponential term f, which corresponds to the potential V in the forced Burgers equations, controls the regularity of the Navier–Stokes equations. We emphasise that a regularity condition readily follows from (5.15). Namely, we have

$$\int_0^t \|\boldsymbol{f}[\boldsymbol{\theta}]\|_{L^\infty} \, ds < \infty, \quad \text{for } \textit{some } k(\neq 0) \quad \Rightarrow \quad \text{smooth up to time } t, \tag{5.17}$$

or, equivalently

$$\text{blowup at time } t \quad \Rightarrow \quad \int_0^t \|\boldsymbol{f}[\boldsymbol{\theta}]\|_{L^\infty} \, ds = \infty, \quad \text{for } \textit{all } k(\neq 0). \tag{5.18}$$

These conditions are similar to Serrin's, but slightly different because of the first term in $\boldsymbol{f}$. It was noted in Ohkitani (2017a) that if blowup takes place, it is impossible to cancel out the two unbounded integrals $\int_0^{t*} \|\boldsymbol{u}\|_{L^\infty}^2 \, dt = \infty$ and $\int_0^{t*} \|\boldsymbol{T}[\nabla \boldsymbol{\psi}]\|_{L^\infty} \, dt = \infty$ so as to make $\boldsymbol{f}$ remain bounded, no matter how carefully k is chosen.

Before closing this section, a brief remark on the blowup criterion is in order. We distinguish two possible scenarios regarding (5.12):

(1) $\theta_j \to 0$ and therefore $\|\psi\|_{L^\infty} \to \infty$. (Note that if only countably many zeros appear in the flow field then we still have $\|\psi\|_{\mathrm{BMO}} < \infty$.)
(2) $\theta_j > 0$, but becomes non-differentiable whilst $\|\psi\|_{L^\infty} < \infty$.

In connection with the above open problem, if $\|\psi\|_{\mathrm{BMO}} \to \infty$ upon singularity, uncountably many zeros in θ_j must appear at the time of breakdown.

5.6 Dynamic scaling transform

We will apply the Cole–Hopf–Feynman–Kac formula to the dynamically-scaled version of the Navier–Stokes equations.

Invariance under dynamic scaling implies that

$$\psi(\boldsymbol{x}, t) = \boldsymbol{\Psi}(\boldsymbol{\xi}, \tau) \tag{5.19}$$

and $\boldsymbol{\Psi}$ satisfies the Leray equations of the form

$$\frac{\partial \boldsymbol{\Psi}}{\partial \tau} - \frac{1}{2}\triangle_{\boldsymbol{\xi}}\boldsymbol{\Psi} + a\boldsymbol{\xi}\cdot\nabla_{\boldsymbol{\xi}}\boldsymbol{\Psi} = \frac{3}{4\pi}\int_{\mathbb{R}^3}\frac{\boldsymbol{\rho}\times(\nabla\times\boldsymbol{\Psi}(\boldsymbol{\xi}'))\,\boldsymbol{\rho}\cdot(\nabla\times\boldsymbol{\Psi}(\boldsymbol{\xi}'))}{|\boldsymbol{\rho}|^5}\,\mathrm{d}\boldsymbol{\xi}', \tag{5.20}$$

where $\boldsymbol{\rho} = \boldsymbol{\xi} - \boldsymbol{\xi}'$ and $\psi(\cdot, 0) = \boldsymbol{\Psi}(\cdot, 0)$. The difference between (5.9) and (5.20) is just one drift term, which is minimal due to the critical nature of ψ. Setting $\Psi_j = k\log\Theta_j$, $(j = 1, 2, 3)$, we obtain as above

$$\frac{\partial\Theta_j}{\partial\tau} = \frac{1}{2}\triangle_{\boldsymbol{\xi}}\Theta_j - a\boldsymbol{\xi}\cdot\nabla_{\boldsymbol{\xi}}\Theta_j + f_j[\Theta](\boldsymbol{\xi}, \tau)\Theta_j. \tag{5.21}$$

These can also be converted into a path-integral form

$$\Theta_j(\boldsymbol{x}, t) = \mathbb{E}\left[\Theta_j(\boldsymbol{X}_t, 0)\exp\left(\int_0^t f_j[\Theta](\boldsymbol{X}_s, s)\,\mathrm{d}s\right)\right], \tag{5.22}$$

where $\boldsymbol{X}_t$ denotes the Ornstein–Uhlenbeck process, generated by the modified dissipative operator $\frac{1}{2}\triangle_{\boldsymbol{\xi}} - a\boldsymbol{\xi}\cdot\nabla_{\boldsymbol{\xi}}$, i.e. the Laplace operator with a drift term.

5.6.1 Change of probability measures

We are in a position to make a detailed comparison between the Navier–Stokes equations and their dynamically-scaled counterparts (the Leray equations), using path-integral representations. Such a comparison without the Feynman–Kac formula has been carried out in Ohkitani (2017b), and with the Feynman–Kac formula in Ohkitani (2017d).

Scale invariance

$\theta(x,t)$ $=\!=\!=$ $\Theta(\xi,\tau)$

N-S $\big\|$ Leray $\big\|$

$\mathbb{E}\,[\boldsymbol{F}[\theta](\boldsymbol{W}_t)]$ $\mathbb{E}\,[\boldsymbol{F}[\Theta](\boldsymbol{W}_\tau + a\boldsymbol{h}(\tau))]$

M-G $\big\|$ C-M $\big\|$

$\mathbb{E}\left[\boldsymbol{F}[\theta](\boldsymbol{W}_t + a\boldsymbol{h}(t))\widehat{G}_a(t)\right]$ $\mathbb{E}\,[\boldsymbol{F}[\Theta](\boldsymbol{W}_\tau)G_a(\tau)]$

Figure 5.1
Scale-invariance, the dynamical equations and the transformation of probability measures; N-S stands for the Navier–Stokes equations, M-G for the Maruyama–Girsanov Theorem and C-M for the Cameron–Martin Theorem.

Necessary tools are taken from stochastic analysis, whose general references include Bell (2006); Malliavin & Thalmaier (2006); Nunno, Oksendal, & Proske (2009); Sanz-Solé (2005); Shigekawa (2004); Steele (2001); Üstünel & Zakai (2010); Üstünel (2015). See Bru & Yor (2002) for historical remarks including the measure-changing theorems.

5.6.2 Leray equations

We consider the Leray equations first, because they have a global smooth solution by assumption (i.e. by construction). Defining Θ by $\Psi_j = k\log\Theta_j$, $(j = 1, 2, 3)$ the scale-invariance becomes

$$\theta(x,t) = \Theta(\xi,\tau).$$

Let us take the drift term as $\boldsymbol{b}(x) = -x$ and $\boldsymbol{h}(t) = \int_0^t \boldsymbol{b}(\boldsymbol{W}_s)\,\mathrm{d}s$. For a simpler comparison, we write (x,t) for (ξ,τ). (See Figure 5.6.1 for a list of relationships with independent variables distinguished.)

The transformed variable Θ satisfies the following equations

$$\Theta = \mathbb{E}\big[\boldsymbol{F}[\Theta](\boldsymbol{W}_t + a\boldsymbol{h}(t))\big], \quad \text{all } t \ge 0 \qquad (5.23)$$

$$= \mathbb{E}\big[\boldsymbol{F}[\Theta](\boldsymbol{W}_t)G_a(t)\big], \quad 0 \le t < \frac{\sqrt{2}}{a} \qquad (5.24)$$

where $G_a(t)$ denotes the Maruyama–Girsanov density

$$G_a(t) = \exp\left(a\int_0^t \boldsymbol{b}(\boldsymbol{W}_s)\cdot\mathrm{d}\boldsymbol{W}_s - \frac{a^2}{2}\int_0^t |\boldsymbol{b}(\boldsymbol{W}_s)|^2\,\mathrm{d}s\right).$$

Here use has been made of the Cameron–Martin Theorem

$$\mathbb{E}\left[F(\boldsymbol{W}_t + \boldsymbol{h})\right] = \mathbb{E}\left[F(\boldsymbol{W}_t) \exp\left(\int_0^t \dot{\boldsymbol{h}}(s) \cdot \mathrm{d}\boldsymbol{W}_s - \frac{1}{2}\int_0^t |\dot{\boldsymbol{h}}(s)|^2\,\mathrm{d}s\right)\right],$$
$$(5.25)$$

where F denotes an arbitrary functional. The time scale $\sqrt{2}/a$ has been determined by the Novikov condition for $G_a(t)$ to be a martingale. It is important that this time scale is larger than $1/2a$, because the following comparison cannot be made otherwise.

In (5.23), "all $t \geq 0$" means that it has a smooth solution in $t \geq 0$.

We next consider the case with finite a and characterise the difference in an additive manner. Subtracting $\mathbb{E}\big[F[\Theta](\boldsymbol{W}_t)\big]$, we have

$$\Theta - \mathbb{E}\big[F[\Theta](\boldsymbol{W}_t)\big] = \mathbb{E}\big[F[\Theta](\boldsymbol{W}_t + ah(t)) - F[\Theta](\boldsymbol{W}_t)\big] \quad (5.26)$$

$$= \mathbb{E}\big[F[\Theta](\boldsymbol{W}_t)(G_a(t) - 1)\big], \quad\quad\quad (5.27)$$

$$\equiv \mathbb{E}\big[\langle D\boldsymbol{F}[\Theta](\boldsymbol{W}_t + \mu h(t)), ah\rangle\big], \quad\quad (5.28)$$

which is valid for $t < \sqrt{2}/a$. Applying the usual mean-value theorem to $G_a(t)$ for fixed t, we find

$$\frac{G_a - 1}{a} = \left.\frac{\partial G_a}{\partial a}\right|_{a=\mu}, \qquad \text{for some } \mu \in (0, a),$$

where

$$\left.\frac{\partial G_a}{\partial a}\right|_{a=\mu} = \left(\int_0^t \boldsymbol{b}(\boldsymbol{W}_s) \cdot \mathrm{d}\boldsymbol{W}_s - \mu \int_0^t |\boldsymbol{b}(\boldsymbol{W}_s)|^2\,\mathrm{d}s\right) G_\mu.$$

The equation (5.28) can be regarded as a result of an application of "the mean-value theorem"[1] to (5.26), whose precise meaning is given by (5.27). The equation (5.28) shows that the Leray equations have an extra additive term in the form of the Malliavin H-derivative, on top of those of the Navier–Stokes equations.

For finite a, we have

$$\Theta - \mathbb{E}\big[F[\Theta](\boldsymbol{W}_t)\big] = \mathbb{E}\big[\langle D\boldsymbol{F}[\Theta](\boldsymbol{W}_t + \mu h(t)), ah\rangle\big] \quad (5.29)$$

$$= a\mathbb{E}\left[F[\Theta](\boldsymbol{W}_t)\left.\frac{\partial G_a}{\partial a}\right|_{a=\mu}\right]. \quad\quad (5.30)$$

We stress that the left-hand side alone defines the Navier–Stokes equations.

---

[1] This is reminiscent of an application of the elementary mean-value theorem $f(x + a) = f(x) + af'(x + \mu)$ for some $\mu \in (0, a)$.

In passing, we note that as $a \to 0$

$$\lim_{a \to 0} \left(\Theta - \mathbb{E}\left[F[\Theta](W_t) \right] \right) = 0,$$

but that

$$\lim_{a \to 0} \frac{1}{a} \left(\Theta - \mathbb{E}\left[F[\Theta](W_t) \right] \right) = \mathbb{E}\left[\langle DF[\Theta](W_t), h \rangle \right]. \tag{5.31}$$

This limit, however is not very useful as we are assuming that no finite-time blowup takes place for the Navier–Stokes equations ($t_* = 1/2a \to \infty$).

5.6.3 Navier–Stokes equations

We now turn our attention to the Navier–Stokes equations in the form (5.15) and carry out an analysis in a parallel fashion. By assumption, it has a short-lived solution θ for $t < 1/2a (= t_*)$, which satisfies

$$\theta = \mathbb{E}\left[F[\theta](W_t) \right], \quad 0 \le t < \frac{1}{2a} \tag{5.32}$$

$$= \mathbb{E}\left[F[\theta](W_t + ah(t))\widehat{G}_a(t) \right]. \tag{5.33}$$

Here $\widehat{G}_a(t)$ denotes the Maruyama–Girsanov density

$$\widehat{G}_a(t) = \exp\left(-a \int_0^t b(W_s) \cdot dW_s - \frac{a^2}{2} \int_0^t |b(W_s)|^2 \, ds \right),$$

from the Maruyama–Girsanov Theorem

$$\mathbb{E}\left[F(W_t) \right] = \mathbb{E}\left[F(W_t + h) \exp\left(-\int_0^t \dot{h}(s) \cdot dW_s - \frac{1}{2} \int_0^t |\dot{h}(s)|^2 \, ds \right) \right]. \tag{5.34}$$

As above, we have

$$\theta - \mathbb{E}\left[F[\theta](W_t + ah(t)) \right] = \mathbb{E}\left[F[\theta](W_t) - F[\theta](W_t + ah(t)) \right]$$

$$= \mathbb{E}\left[F[\theta](W_t + ah(t))(\widehat{G}_a - 1) \right]$$

$$= -\mathbb{E}\left[\langle DF[\theta](W_t + \mu'h(t)), ah \rangle \right],$$

where

$$\frac{\widehat{G}_a - 1}{a} = \left. \frac{\partial \widehat{G}_a}{\partial a} \right|_{a = \mu'} \qquad \text{for some } \mu' \in (0, a)$$

and

$$\left. \frac{\partial \widehat{G}_a}{\partial a} \right|_{a=\mu'} = - \left(\int_0^t b(W_s) \cdot \mathrm{d}W_s + \mu' \int_0^t |b(W_s)|^2 \, \mathrm{d}s \right) \widehat{G}_{\mu'}.$$

For finite a, we find

$$\theta - \mathbb{E}\big[F[\theta](W_t + ah(t))\big] = -\mathbb{E}\big[(DF[\theta](W_t + \mu'h(t)), ah)\big] \quad (5.35)$$

$$= a\mathbb{E}\left[F[\theta](W_t) \left. \frac{\partial \widehat{G}_a}{\partial a} \right|_{a=\mu'} \right]. \quad (5.36)$$

Again, note that the left-hand side alone defines the Leray equations. The Navier–Stokes equations can be regarded as a perturbed version of the Leray equations.

5.7 Summary

In this paper we have surveyed some results on the basic issues of the Navier–Stokes equations, which have been obtained by paying attention to the scale-invariant nature of the equations.

We then show how we can generalise the concept of invariance under dynamic scaling transforms. The key step is to write down the basic equations in the dependent variables, which themselves are already scale invariant. In three dimensions, they are simply the vector potentials. Using dynamic scaling (as a push-forward), we obtain the Leray equations, where the dissipative operator changes from the Laplacian to the Ornstein–Uhlenbeck operator. If we move onto path-integral representations, the probability measures can be made explicit. By using the Cameron–Martin-Maruyama–Girsanov transforms (as a pull-back), we retrieve the Navier–Stokes equations *modulo* a Maruyama–Girsanov density G.

Hence it seems natural to define quasi-invariance by equivalence *modulo G* in path integral representations. Under dynamic scaling, Navier–Stokes equations change their forms only slightly when written in the vector potentials. We have also seen that the difference can be interpreted in terms of the H-derivative.

It is of interest to study the implications of quasi-invariance on the basic issues. Now that the two equations have been shown to be very close, while the behaviour of their solutions are totally different, such a close similarity can impose constraints on the possibility of blowup.

Particularly, it is of interest to study which specific properties of $T[\nabla\psi]$, if any, can make the solution to the pulled-back Navier–Stokes equations outlive the original one so that we would possibly get a contradiction.

Finally, we note that all these arguments hold in $\mathbb{R}^n$ for any $n \geq 2$.

Acknowledgement This work has been supported by an EPSRC grant: EP/N022548/1.

Appendix A Wiener process

In the triplet $(\Omega, F, \boldsymbol{P})$, Ω is a set of $\mathbb{R}^d$-valued continuous functions defined for $t \in [0, \infty)$, F the σ-algebra on Ω and $\boldsymbol{P}$ the probability measure on it.

The solution to the heat equation

$$\frac{\partial u}{\partial t} = \frac{1}{2}\triangle u$$

can be written as

$$u(\boldsymbol{x}, t) = \mathbb{E}[f(\boldsymbol{x} + \boldsymbol{W}_t)|\boldsymbol{W}_0 = 0],$$

or, equivalently

$$u(\boldsymbol{x}, t) = \mathbb{E}[f(\boldsymbol{W}_t)|\boldsymbol{W}_0 = \boldsymbol{x}].$$

In the text, an abridged notation $u(\boldsymbol{x}, t) = \mathbb{E}[f(\boldsymbol{W}_t)]$ has been used throughout.

A link to the Gaussian probability measure $p(\boldsymbol{x}, \boldsymbol{y}, t)$ can be made explicit (Ikeda & Watanabe, 1988) by noting that

$$\mathbb{E}[f(\boldsymbol{x} + \boldsymbol{W}_t)] = \int f(\boldsymbol{y})\mathbb{E}[\delta_{\boldsymbol{y}}(\boldsymbol{x} + \boldsymbol{W}_t)]\,d\boldsymbol{y},$$

that is,

$$p(\boldsymbol{x}, \boldsymbol{y}, t) = \mathbb{E}[\delta_{\boldsymbol{y}}(\boldsymbol{x} + \boldsymbol{W}_t)] = \frac{1}{(2\pi t)^{d/2}} \exp\left(-\frac{|\boldsymbol{x} - \boldsymbol{y}|^2}{2t}\right),$$

where $\delta_{\boldsymbol{y}}(\boldsymbol{x})$ denotes the Dirac mass supported at $\boldsymbol{y}$.

Appendix B Feynman–Kac formula for time-dependent potential

For given $f_j(\boldsymbol{x}, t)$, $j = 1, 2, 3$ a number of different representations are available for the (unique) solution to (5.13). To distinguish them properly, we assume here that Brownian motion starts from the origin $W_0 = 0$, as opposed to the assumption $W_0 = x$ in the main text. (Here no summation is implied on j.)

The expression (5.15)

$$\theta_j(x, t) = \mathbb{E}\left[\theta_j(x + W_t, 0) \exp\left(\int_0^t f_j(x + W_s, s)\,ds\right)\right] \qquad \text{(B.37)}$$

can be obtained by applying the time-dependent Trotter formula, see Section 11.2 of Taylor (1996).

Another form

$$\theta_j(x, t) = \mathbb{E}\left[\theta_j(x + W_t, 0) \exp\left(\int_0^t f_j(x + W_s, t - s)\,ds\right)\right] \qquad \text{(B.38)}$$

can be found in Freidlin (1985).

Yet another form

$$\theta_j(x, t) = \mathbb{E}\left[\theta_j(x + W_t, 0) \exp\left(\int_0^t f_j(x + W_t - W_s, s)\,ds\right)\right] \qquad \text{(B.39)}$$

can be found in Friedrichs et al. (1957). The expression (B.39) can be extended to the case where the potential term f_j itself is stochastic (Chow, 2014).

We can make use of the alternative forms of functional integrals by changing the all arguments in $f_j(\cdot, \cdot)$ accordingly.

References

Bell, D.R. (2006) *The Malliavin Calculus*. Dover, New York.

Bru, B. & Yor, M. (2002) Comments on the life and mathematical legacy of Wolfgang Doeblin. *Finance and Stochastics* **6**, 3-47.

Chae, D. (2007) Nonexistence of asymptotically self-similar singularities in the Euler and the Navier–Stokes equations. *Math. Ann.* **338**, 435–449.

Cheskidov, A. & Shvydkoy, R. (2010) The Regularity of Weak Solutions of the 3D Navier–Stokes Equations in $B_{\infty,\infty}^{-1}$. *Arch. Rat. Mech. Anal.* **195**, 159–169.

Chow, P.L. (2014) *Stochastic Partial Differential Equations*. CRC Press, Boca Raton.

Cole, J.D. (1951) On a linear quasilinear parabolic equation in aerodynamics. *Q. Appl. Math.* **9**, 225–236.

Cole, J.D. (1949) *Problems in transonic flow*. PhD Thesis, California Institute of Technology (pages 56 and onward).

Constantin, P. & Foias, C. (1988) *Navier–Stokes Equations*. University of Chicago Press, Chicago.

Doering, C.R. (2009) The 3D Navier–Stokes problem. *Annu. Rev. Fluid Mech.* **41**, 109–128.

Doering, C.R. & Gibbon, J.D. (1995) *Applied Analysis of the Navier–Stokes Equations.* Cambridge University Press, Cambridge.

Escauriaza, L., Seregin, G., & Sverak, V. (2003) $L_{3,\infty}$-solutions of the Navier–Stokes equations and backward uniqueness. *Russ. Math. Surv.* **58**, 211–250.

Freidlin, M.I. (1985) *Functional Integration and Partial Differential Equations.* Annals of Mathematics Studies 109, Princeton University Press, Princeton.

Friedrichs, K.O., Seidman, T., Wendroff, B., Shapiro, H.N., & Schwartz, J. (1957) *Integration of functionals.* New York University, Institute of Mathematical Sciences.

Hopf, E. (1950) The partial differential equation $u_t + uu_x = \mu u_{xx}$. *Commun. Pure Appl. Math.* **3**, 201–230.

Hou, T. & Li, R. (2007) Nonexistence of locally self-similar blow-up for the 3D incompressible Navier–Stokes equations. *Dis. Cont. Dyn. Sys.* **18**, 637–642.

Ikeda, I. & Watanabe, S. (1988) *Stochastic Differential Equations and Diffusion Processes.* North Holland, Amsterdam.

Kato, T. & Fujita, H. (1962) On the nonstationary Navier–Stokes system. *Rendiconti del Seminario Matematico della Universitá di Padova* **32**, 243–260.

Koch, K. & Tataru, D. (2001) Well-posedness for the Navier–Stokes equations. *Adv. Math.* **157**, 22–35.

Leray, J. (1934) Essai sur le mouvement d'un liquide visqueux emplissant l'espace. *Acta Math.* **63**, 193–248.

Malliavin, P. & Thalmaier, A. (2006) *Stochastic Calculus of Variations in Mathematical Finance.* Springer, Berlin, Heidelberg.

Nečas, J., Růžička, M., & Šverák, V. (1996) On Leray's self-similar solutions of the Navier–Stokes equations. *Acta Math.* **176**, 283–294.

Nunno, G.D., Oksendal, B., & Proske, F. (2009) *Malliavin Calculus for Lévy Processes with Applications to Finance.* Berlin, Springer.

Ohkitani, K. (2008) A miscellany of basic issues on incompressible fluid equations. *Nonlinearity* **21**, 255–271.

Ohkitani, K. (2015) Dynamical equations for the vector potential and the velocity potential in incompressible irrottational Euler flows: a refined Bernoulli theorem. *Phys. Rev. E.* **92**, 033010.

Ohkitani, K. (2016) Study of the Navier–Stokes regularity problem with critical norms. *Fluid. Dyn. Res.* **48**, 021401.

Ohkitani, K. (2017a) Characterization of blowup for the Navier–Stokes equations using vector potentials. *AIP Advances* **7**, 015211

Ohkitani, K. (2017b) Near-invariance under dynamic scaling for the Navier–Stokes equations in critical spaces: a probabilistic approach to regularity problems. *J. Phys. A: Math. Theor.* **50**, 045501.

Ohkitani, K. (2017c) Analogue of the Cole–Hopf transform for the incompressible Navier–Stokes equations and its application. *Journal of Turbulence* **18**, 465–479.

Ohkitani, K. (2017d) Cole–Hopf–Feynman–Kac formula and quasi-invariance for Navier–Stokes equations. *J. Phys. A: Math. Theor.* **50**, 405501.

Ożański, W. & Pooley, B. (2018) Leray's fundamental work on the Navier–Stokes equations: a modern review of *"Sur le mouvement d'un liquide visqueux emplissant l'espace"*. In Fefferman, C.L., Robinson, J.C., & Rodrigo, J.L. (Eds.) *Partial Differential Equations in Fluid Mechanics*. LMS Lecture Notes in Mathematics. Cambridge University Press, Cambridge, UK.

Robinson, J.C., Rodrigo, J.L., & Sadowski, W. (2016) *The three-dimensional Navier–Stokes equations. Classical Theory*. Cambridge University Press, Cambridge, UK.

Sanz-Solé, M. (2005) *Malliavin Calculus with Applications to Stochastic Partial Differential Equations*. EPFL Press, Lausanne.

Serrin, J. (1963) The initial value problem for the Navier–Stokes equations. In Langer, R.E. (Ed.) *Nonlinear Problems*, 69–98, University of Wisconsin Press, Madison.

Shigekawa, I. (2004) *Stochastic Analysis*. American Mathematical Society, Providence, RI.

Steele, J.M. (2001) *Stochastic Calculus and Financial Applications*. Springer, New York.

Taylor, M. (1996) *Partial Differential Equations II: Qualitative Studies of Linear Equations*. Springer, New York.

Üstünel, A.S. (2015) *Analysis on Wiener Space and Applications*. arXiv preprint arXiv:1003.1649.

Üstünel, A.S. & Zakai, M. (2010) *Transformation of Measure on Wiener Space*. Springer, Berlin Heidelberg.

Vanon, R. & Ohkitani, K. (2018) Applications of a Cole–Hopf transform to the 3D Navier–Stokes equations. *Journal of Turbulence* **19**, 322–333.

6

Leray's fundamental work on the Navier–Stokes equations: a modern review of *"Sur le mouvement d'un liquide visqueux emplissant l'espace"*

Wojciech S. Ożański

Mathematics Institute
University of Warwick
Coventry, CV4 7AL. UK
w.s.ozanski@warwick.ac.uk

Benjamin C. Pooley

Mathematics Institute
University of Warwick
Coventry, CV4 7AL. UK
b.pooley@warwick.ac.uk

Abstract

This article offers a modern perspective which reveals the many contributions of Leray in his celebrated work on the three-dimensional incompressible Navier–Stokes equations from 1934. Although the importance of his work is widely acknowledged, the precise contents of his paper are perhaps less well known.

The purpose of this article is to rectify this. We follow Leray's results in detail: we prove local existence of strong solutions starting from divergence-free initial data that is either smooth, or belongs to H^1 or $L^2 \cap L^p$ (with $p \in (3, \infty]$), as well as lower bounds on the norms $\|\nabla u(t)\|_2$ and $\|u(t)\|_p$ ($p \in (3, \infty]$) as t approaches a putative blow-up time. We show global existence of a weak solution and weak-strong uniqueness. We present Leray's characterisation of the set of singular times for the weak solution, from which we deduce that its upper box-counting dimension is at most $\frac{1}{2}$.

Throughout the text we provide additional details and clarifications for the modern reader and we expand on all ideas left implicit in the original work, some of which we have not found in the literature. We

Published as part of *Partial Differential Equations in Fluid Mechanics*, edited by C.L. Fefferman, J.C. Robinson, & J.L. Rodrigo. © Cambridge University Press 2018.

use some modern mathematical tools to bypass some technical details in Leray's work, and thus expose the elegance of his approach.

6.1 Introduction

The Navier–Stokes equations,

$$\partial_t u + (u \cdot \nabla)u - \nu \Delta u + \nabla p = 0,$$

$$\nabla \cdot u = 0,$$

where u denotes the velocity of a fluid, p denotes the scalar pressure and $\nu > 0$ denotes the viscosity of the fluid, comprise the fundamental model for the flow of an incompressible viscous fluid. They are named in recognition of Claude-Louis Navier (1822) and George Stokes (1845) who first formulated them, and they form the basis for many models in engineering and mathematical fluid mechanics. The equations have been studied extensively and a number of excellent textbooks on the subject are available, see for instance Constantin (2008), Constantin & Foias (1988), Lemarié-Rieusset (2002), Robinson, Rodrigo, & Sadowski (2016), Sohr (2001) and Temam (2001). However, the fundamental issue of the well-posedness of the equations in three dimensions remains unsolved.

In this article we focus solely on the work of Jean Leray (1934b), which to this day remains of fundamental importance in the study of the Navier–Stokes equations. Leray was the first to study the Navier–Stokes equations in the context of *weak solutions*. It is remarkable that such solutions are defined using a distributional form of the equations while distribution theory was only formally introduced later by Schwartz (1950).

Many of the ideas in the modern treatment of these equations, and in a number of other systems, originate from Leray (1934b) (which we shall often refer to simply as "Leray's work" or "Leray's paper"). The importance of this work is evidenced by the fact that it is one of the most cited works in mathematical fluid mechanics.

Leray studied the Navier–Stokes equations on the whole space ($\mathbb{R}^3$). Unlike later authors who have largely adopted Faedo-Galerkin techniques (see Hopf (1951) and Kiselev & Ladyzhenskaya (1957) for early examples), a characteristic of earlier works, including Leray's, is the use of explicit kernels. In this instance, Leray applied the Oseen kernel (herein denoted by $\mathcal{T}$), as derived by Oseen (1911), to obtain solutions of the *nonhomogeneous Stokes equations*:

$$\partial_t u - \nu \Delta u + \nabla p = F, \qquad \nabla \cdot u = 0,$$

for a given forcing F. At the time these were also known as the equations for *infinitely slow motion*.

Oseen had previously applied this kernel iteratively to prove local well-posedness of the Navier–Stokes equations in $C^2(\mathbb{R}^3)$ (with bounded velocity, decay conditions on $\omega = \nabla \times u$, and polynomial growth estimates on $\nabla \omega$ and ∇u; see Section 3.8 of Lemarié-Rieusset (2016) for a more complete discussion of Oseen's contributions). Leray applied a more elegant iteration scheme (a Picard iteration) to prove existence, uniqueness, and regularity results for local-in-time strong solutions for initial data $u_0 \in H^1 \cap L^\infty \cap C^1$ (Leray used the term *regular solutions*). We will see that in fact, $u_0 \in L^2 \cap L^\infty$ suffices to construct strong solutions when his arguments are rewritten using a distributional form of the equations.

Leray then derives lower bounds on various norms of the strong solution $u(t)$ as t approaches the maximal time of existence T_0 if T_0 is finite, which indicate the rate of blow-up of a strong solution if such a blow-up occurs. He leaves open the issue of the existence of blow-ups.

Next, Leray considers a generalised notion of solution of the Navier–Stokes equations, *weak solutions*. For this he studies the so-called *regularised equations*, which are obtained by replacing the nonlinear term $(u \cdot \nabla) u$ by $(J_\varepsilon u \cdot \nabla) u$ (where J_ε is the standard mollification operator). He shows that the regularised equations admit local well-posedness results similar to those for strong solutions of the Navier–Stokes equations, with an additional global-in-time estimate on the L^∞ norm of the velocity. This extra property results in global-in-time existence and uniqueness of the solution u_ε for each $\varepsilon > 0$. By a careful compactness argument, he constructs a sequence of solutions $\{u_{\varepsilon_n}\}$ converging to a global-in-time weak solution to the Navier–Stokes equations. Such a solution, which he termed a *turbulent solution*, can be thought of as a weak continuation of the strong solution beyond the blow-up time, a revolutionary idea at the time. He then shows that any such weak solution is strong locally-in-time except on a certain compact set of singular times of Lebesgue measure zero. To this end he uses a certain *weak-strong uniqueness* property.

We now briefly highlight a few important developments that proceeded from Leray's work. Eberhard Hopf (1951) performed a study of the Navier–Stokes equations on a bounded domain $\Omega \subset \mathbb{R}^3$, and proved global-in-time existence of weak solutions. Ladyženskaja (1959) proved the existence and uniqueness of global-in-time strong solutions to the Navier–Stokes equations in bounded two-dimensional domains (Leray, 1933, dealt with well-posedness in $\mathbb{R}^2$ in his thesis but was less successful in studying the case of bounded domains, see Leray, 1934a). Fujita & Kato (1964) used fractional powers of operators and the theory of semigroups to construct local-in-time strong solutions of the three-dimensional Navier–Stokes equations on bounded domains (Kato, 1984 used similar methods in the case of unbounded domains). As for the smoothness of weak solutions, it follows from the works of Serrin (1962), Prodi (1959), and Ladyženskaja (1967), that if a weak solution u belongs to $L^r((a,b); L^s)$ with $2/r + 3/s \leq 1$, $s > 3$ then u is a strong solution on the time interval (a,b); the critical case $r = \infty$, $s = 3$ was settled later by Escauriaza, Seregin, & Šverák (2003). In addition, Beale, Kato, & Majda (1984) showed that if curl $u \in L^1((a,b); L^\infty)$ then u is a strong solution on the time interval $(a,b]$.

Scheffer (1977) was the first to study the size of the singular set in both time and in space. Subsequently, Caffarelli, Kohn, & Nirenberg (1982) proved that the one-dimensional Hausdorff measure of the singular set is zero. We refer the reader to the textbooks above for a wider description of the contributions to the theory of the Navier–Stokes equations in the last 80 years.

It is remarkable that despite many significant contributions, it is still not known whether the (unique) local-in-time strong solutions to the Navier–Stokes equations develop blow-ups or whether the global-in-time weak solutions are unique. This remains one of the most important open problems in mathematics and, at the turn of the millennium, Clay Mathematics Institute announced it as one of seven Millennium Problems, see Fefferman (2006).

A number of concepts and methods that found early use in Leray's work are now ubiquitous in the analysis of PDEs. These include: weak compactness of bounded sequences in L^2, the concept of weak derivatives (called *quasi-derivatives* by Leray), the mollification operation and the fact that a weakly convergent sequence in L^2 converges strongly if and only if the norm of the limit is the limit of the norms. He made an extensive use of the space of L^2 functions with weak derivatives in L^2 two years before the celebrated work of Sobolev (1936). This space would later be called the Sobolev space H^1. Furthermore, Leray was the first to introduce the compactness method of solving partial differential equations (see the proof of Theorem 6.37); in fact this, together with his work with Juliusz Schauder opened a new branch in mathematics, the use of topological methods in PDEs.

The terms *Leray weak solution* (or *Leray–Hopf weak solution*) of the Navier–Stokes equations (see Definition 6.35), *Leray regularisation* (see (6.77)) and *Leray projection* (the projection of L^2 onto the space of weakly divergence-free vector fields) have become part of the mathematical lexicon in recognition of his seminal paper on the Navier–Stokes equations. We refer the interested reader to Lax (1998) for a broader description of Leray's work in the field of PDEs.

This article arose from series of lectures presented by the authors for a fluid mechanics reading group organised at the University of Warwick by James Robinson and José Rodrigo, and its purpose is to offer a modern exposition of Leray's work. We update the notation and we simplify some technical details by applying some modern methods; in particular we use the Fourier transform (see Theorem 6.8) and the distributional forms of the partial differential equations appearing in Leray's work. It is perhaps remarkable that these updates do not detract from the originality of Leray's arguments; rather they make them even more elegant.

We have also endeavoured to give a rigorous account of all non-trivial results that are left implicit in the original work, some of which we were not able to find in the literature. These include Leray's derivations of the blow-up rate of the norm $\|u(t)\|_p$ (with $p > 3$) of a strong solution u as t approaches the putative blow-up time (see Corollary 6.25), and a result on local existence of strong solutions for initial data $u_0 \in L^2 \cap L^p$ (with $p \in (3, \infty]$) that is weakly divergence free (see Corollary 6.31). In order to make the exposition self-contained we have also added appendices on relevant facts from the theory of the heat equation, integral inequalities, the Volterra equation and other topics.

For simplicity of notation we focus only on the case $\nu = 1$. The corresponding results for any $\nu > 0$ can be recovered using the following rescaling argument: if u, p is a solution of the Navier–Stokes equations with $\nu = 1$, then $\widetilde{u}(x, t) := u(x\nu, t\nu)$, $\widetilde{p} := p(x\nu, t\nu)$ is a solution for given $\nu > 0$.

The structure of the article is as follows. In the remainder of this section we describe some notation, recall some preliminary results, and introduce the Oseen kernel $\mathcal{T}$, which will be the main tool for solving the Stokes equations.

In Section 6.2 we study the Stokes equations,

$$\partial_t u - \nu \Delta u + \nabla p = F, \qquad \nabla \cdot u = 0.$$

We first show that if the forcing F is sufficiently smooth (see (6.28)) the equations can be solved classically using the representation formulae

$$u(t) := \Phi(t) * u_0 + \int_0^t \mathcal{T}(t-s) * F(s)\,\mathrm{d}s,$$

$$p(t) := -(-\Delta)^{-1}(\operatorname{div} F(t)),$$

see (6.22), (6.23), where $\Phi(t)$ denotes the heat kernel (Theorem 6.8). We also show some further properties of the representation formula for u in the case of less regular forcing F (Lemma 6.7). We then focus on a special form of the forcing

$$F = -(Y \cdot \nabla)Y,$$

see (6.31), which is reminiscent of the nonlinear term in the Navier–Stokes equations. This special form of F gives rise to the modified representation formulae,

$$u(t) := \Phi(t) * u_0 + \int_0^t \nabla \mathcal{T}(t-s) * [Y(s)Y(s)]\,\mathrm{d}s,$$

$$p(t) := \partial_k \partial_i (-\Delta)^{-1}(Y_i(t)Y_k(t)),$$

see (6.33) and (6.34). After deducing some properties of this modified representation formula (Lemma 6.9) we show that it gives a unique solution (in some wide sense) to the Stokes equations with the forcing F of the form above, in the sense of distributions (Theorem 6.11 and Theorem 6.12).

In Section 6.3 we study strong solutions of the Navier–Stokes equations. After defining strong solutions on an open time interval $(0, T)$ (Definition 6.14) we use the theory of the Stokes equations developed in Section 6.2 to deduce the smoothness of such solutions (Corollary 6.16), as well as other interesting properties, such as the energy equality

$$\|u(t_2)\|^2 + 2 \int_{t_1}^{t_2} \|\nabla u(s)\|^2\,\mathrm{d}s = \|u(t_1)\|^2,$$

(Theorem 6.17) and the comparison of strong solutions

$$\|(u-v)(t_2)\|^2 \le \|(u-v)(t_1)\|^2 \mathrm{e}^{\frac{1}{2}\int_{t_1}^{t_2} \|u(s)\|_\infty^2\,\mathrm{d}s}, \quad t_1 < t_2$$

(Lemma 6.18), where u, v is a pair of strong solutions on a time interval $(0, T) \ni t_1, t_2$. We then define strong solutions on a half-closed time interval $[0, T)$ (Definition 6.20) and prove the local-in-time existence and uniqueness of such solutions with weakly divergence-free initial data $u_0 \in L^2 \cap L^\infty$ (Theorem 6.22). Next, we discuss the issue of the maximal time of existence T_0 of strong solutions (Lemma 6.23 and Corollary 6.24), from which we deduce the rates of blow-up of $u(t)$ in various norms as t approaches T_0 (if T_0 is finite):

$$\|u(t)\|_\infty \geq \frac{C}{\sqrt{T_0 - t}}, \quad \|\nabla u(t)\| \geq \frac{C}{(T_0 - t)^{1/4}},$$

and

$$\|u(t)\|_p \geq \frac{C^{(1-3/p)/2}(1 - 3/p)}{(T_0 - t)^{(1-3/p)/2}}$$

(Corollary 6.25). The study of strong solutions is concluded with an observation that less regular initial data u_0 also gives rise to a unique strong solution on the time interval $(0, T)$ for some $T > 0$. This motivates the definition of *semi-strong solutions* (Definition 6.28); we show that if $u_0 \in L^2$ is weakly divergence free, and either

$$\nabla u_0 \in L^2 \quad \text{or} \quad u_0 \in L^p \quad \text{(with } p > 3\text{)},$$

then there exists a unique local-in-time semi-strong solution starting from initial data u_0 (Theorem 6.30 and Corollary 6.31).

In Section 6.4 we study weak solutions of the Navier–Stokes equations. To this end, for each $\varepsilon > 0$ we consider regularised equations, where a mollification operator is applied in the nonlinear term,

$$\partial_t u - \Delta u + ((J_\varepsilon u) \cdot \nabla)u + \nabla p = 0$$

(see Definition 6.32). We show that for each $\varepsilon > 0$ the regularised equations can be analysed in a similar way to strong solutions of the Navier–Stokes equations in Section 6.3, the difference being that the maximal time of existence of the solution u_ε is infinite (Theorem 6.33). In order to take the limit $\varepsilon \to 0^+$ we first show that the kinetic energy of $u_\varepsilon(t)$ outside a ball can be estimated independently of ε,

$$\int_{|x|>R_2} |u_\varepsilon(t)|^2 \, dx \leq \int_{|x|>R_1} |u_0|^2 \, dx + \frac{C(u_0, t)}{R_2 - R_1}, \qquad R_2 > R_1 > 0,$$

(Lemma 6.34). Thanks to this separation of energy result we can let $\varepsilon \to 0^+$ (along a carefully chosen subsequence) to obtain a global-in-time weak solution of the Navier–Stokes equations (Theorem 6.37). We then show the so-called weak-strong uniqueness result (Lemma 6.39) and we deduce that the weak solution admits a particular structure, namely that it is (locally) a strong solution at times

$$t \in \bigcup_i (a_i, b_i),$$

where the intervals $(a_i, b_i) \subset (0, \infty)$ are pairwise disjoint (Theorem 6.41). Finally we show that the complement of their union

$$\Sigma := (0, \infty) \setminus \bigcup_i (a_i, b_i),$$

(the set of putative singular times) is bounded, its box-counting dimension is bounded above by $1/2$ and that the weak solution satisfies certain decay for sufficiently large times (Theorem 6.42 and Corollary 6.43).

Unless specified otherwise, each proof follows Leray (1934b), possibly with minor modifications. We also comment on Leray's methodology throughout the text, in footnotes and in the "Notes" at the end of each section. Equation numbers marked in italics correspond to expressions in Leray's paper.

6.1.1 Preliminaries

The letter C denotes a numerical constant, whose value may change at each occurrence. Occasionally we write C' (or C'') to denote a constant that has the same value wherever it appears within a given section. Also, C_m (or c_m) denotes a numerical constant for each m. Throughout the article (unless specified otherwise) we consider function spaces on $\mathbb{R}^3$, for example $L^p := L^p(\mathbb{R}^3)$, $H^m := H^m(\mathbb{R}^3)$, $C_0^\infty := C_0^\infty(\mathbb{R}^3)$. Given an interval I and a Banach space B we denote by $C(I; B)$ the space of continuous functions from I with values in B; we similarly define the spaces $L_{\text{loc}}^1(I; B)$ and $L^p(I; B)$, $p \in [1, \infty]$, see Section 1.5 in Roubíček (2013) for details.

We also define $\int := \int_{\mathbb{R}^3}$, $\| \cdot \|_p := \| \cdot \|_{L^p}$ and we reserve the notation $\| \cdot \|$ for the L^2 norm. We say that a velocity field v is *weakly divergence free* if $\operatorname{div} v = 0$ in the sense of distributions, that is, if

$$\int v \cdot \nabla g = 0 \tag{6.1}$$

for all $g \in C_0^\infty$; we set

$$\begin{aligned} H &:= \{f \in L^2 : f \text{ is weakly divergence free}\}. \\ V &:= \{f \in H^1 : \operatorname{div} f = 0\}. \end{aligned} \tag{6.2}$$

We understand $\mathbb{Q}^+$ as the nonnegative rational numbers and we define $\partial_j := \partial/\partial x_j$ and $\nabla^m := D^m$, where all derivatives are understood in the weak sense. We use the convention of summing over repeated indices.

For example, we write $v_j \partial_j u_i$ to denote the i-th component of vector $(v \cdot \nabla)u$ (here $i = 1, 2, 3$). For an interval I we define

$$\mathcal{H}^{1/2}(I) := \{ f : \mathbb{R}^3 \times I \to \mathbb{R} : \exists C : I \to \mathbb{R}_{\geq 0} \text{ continuous, such that}$$

$$|f(x,t) - f(y,t)| \leq C(t)|x - y|^{1/2} \text{ for all } t \in I,\ x, y \in \mathbb{R}^3 \},$$
$$(6.3)$$

that is, $\mathcal{H}^{1/2}(I)$ is the space of functions such that $\|f(t)\|_{C^{0,1/2}} \leq C(t)$ for some continuous $C(t)$ (where $\|\cdot\|_{C^{0,1/2}}$ denotes $1/2$-Hölder seminorm). Note that $\mathcal{H}^{1/2}(I)$ is defined in the same way for vector-valued functions.

We recall the *integral Minkowski inequality*

$$\|f(t)\|_p \leq \int_0^t \|g(s)\|_p \, \mathrm{d}s, \quad \text{whenever } f(x,t) = \int_0^t g(x,s)\mathrm{d}s \quad (6.4)$$

and $p \in [1, \infty]$. More generally, for integrable and nonnegative ξ, η, if

$$F(x,t) = \int_{\mathbb{R}} \xi(s) \int \eta(y)f(x, y, t, s) \, \mathrm{d}y \, \mathrm{d}s,$$

then

$$\|F(t)\|_p \leq \int_{\mathbb{R}} \xi(s) \int \eta(y)\|f(\cdot, y, t, s)\|_p \, \mathrm{d}y \, \mathrm{d}s. \quad (6.5)$$

Now let $p, q, r \geq 1$ satisfy $1/q = 1/p + 1/r - 1$ and let $f \in L^p$, $g \in L^r$. Then

$$\|f * g\|_q \leq \|f\|_p \|g\|_r. \quad (6.6)$$

This is *Young's inequality for convolutions* (see e.g. Stein & Weiss (1971), p. 178, for the proof). Here "$*$" denotes the *convolution*, that is,

$$(f * g)(x) := \int f(x - y)g(y) \, \mathrm{d}y.$$

If f and g are also functions of time t we omit x and simply write $u(t) = f(t)*g(t)$. We apply this notation in the statement of the following extension of Young's inequality to the case of space-time convolutions, whose proof we give in Appendix 6.5.2 (see Lemma 6.46).

Lemma 6.1 *If $p, q, r \geq 1$ satisfy*

$$\frac{1}{q} = \frac{1}{p} + \frac{1}{r} - 1,$$

$A \in L^1_{\mathrm{loc}}([0, T); L^p)$ *and* $B \in C((0, T); L^r)$ *with* $\|B(t)\|_r$ *bounded as* $t \to 0^+$, *then* u *defined by*

$$u(t) := \int_0^t A(t - s) * B(s) \, \mathrm{d}s$$

belongs to $C([0,T]; L^q)$ and

$$\|u(t)\|_q \le \int_0^t \|A(t-s)\|_p \|B(s)\|_r \,\mathrm{d}s.$$

Let J_ε denote the *mollification* operator, that is

$$J_\varepsilon v := \eta_\varepsilon * v,$$

where η_ε is a standard mollifier, i.e. $\eta_\varepsilon(x) := \varepsilon^{-3}\eta(x/\varepsilon)$, where $\eta \in C_0^\infty(\mathbb{R}^3)$ is non-negative, radial, and $\int \eta = 1$.

Lemma 6.2 (properties of mollification) *The mollification operator J_ε (on $\mathbb{R}^3$) enjoys the following properties:*

  (i) $\|J_\varepsilon v\|_p \le \|v\|_p$ *for all $p \in [1,\infty]$, $\varepsilon > 0$,*
  (ii) $\partial_k J_\varepsilon v = J_\varepsilon(\partial_k v)$ *for every $k = 1,2,3$, whenever $\nabla v \in L_{\mathrm{loc}}^1$,*
 (iii) $\|J_\varepsilon v\|_\infty \le C\varepsilon^{-3/2}\|v\|$,
  (iv) *if $v \in L^2$ then $J_\varepsilon v \in H^m$ for all m, with $\|J_\varepsilon v\|_{H^m} \le C_m\varepsilon^{-m}\|v\|$,*
   (v) *if $v \in L_{\mathrm{loc}}^1$ then $J_\varepsilon v \to v$ almost everywhere as $\varepsilon \to 0$, and*
  (vi) *if $v \in L^p$, where $p \in [1,\infty)$, then $J_\varepsilon v \to v$ in L^p as $\varepsilon \to 0$.*

The proofs of the above properties are elementary (and can be found in Appendix C in Evans (2010), Section 3.5.2 in Majda & Bertozzi (2002) or Appendix A.3 in Robinson, Rodrigo, & Sadowski, 2016).

We will use the notation $\widehat{f}$ for the Fourier transform of f, defined for $f \in L^1(\mathbb{R}^3)$ by

$$\widehat{f}(\xi) = \int_{\mathbb{R}^3} f(x)\mathrm{e}^{-2\pi i x \cdot \xi} \,\mathrm{d}x \tag{6.7}$$

and extended by a density argument (using the Plancherel Theorem) to $f \in L^2$.

For $f \in C^\infty(\mathbb{R}^3)$ such that $\nabla^m f \in L^1 \cap L^\infty$ for every $m \ge 0$ we define

$$(-\Delta)^{-1} f(x) := \int \frac{f(y)}{4\pi|x-y|} \,\mathrm{d}y. \tag{6.8}$$

The symbol $(-\Delta)^{-1}$ relates to the fact that $g := (-\Delta)^{-1}f$ (is infinitely differentiable and) satisfies the Poisson equation $-\Delta g = f$ in $\mathbb{R}^3$ in the classical sense (see Theorem 1 in Section 2.2.1 in Evans (2010)).

Since we will often estimate terms similar to the right-hand side of (6.8), we formulate the following lemma (which is Leray's *(1.14)*).

Lemma 6.3 *If $f \in H^1$ then for any $y \in \mathbb{R}^3$*

$$\int \frac{|f(x)|^2}{|x-y|^2} \,\mathrm{d}x \le 4\|\nabla f\|^2. \tag{6.9}$$

Proof   It is enough to prove the claim when $y = 0$ and when f is a scalar function. If $f \in C_0^\infty$ then integration by parts, the Cauchy–Schwarz inequality and Young's inequality give

$$\int \frac{(f(x))^2}{|x|^2} \, \mathrm{d}x = -\int \frac{x}{|x|^2} \cdot \nabla(f(x))^2 \, \mathrm{d}x = -2 \int \frac{x}{|x|^2} \cdot \nabla f(x) f(x) \, \mathrm{d}x$$

$$\leq 2\|\nabla f\| \sqrt{\int \frac{(f(x))^2}{|x|^2} \, \mathrm{d}x} \leq 2\|\nabla f\|^2 + \frac{1}{2} \int \frac{(f(x))^2}{|x|^2} \, \mathrm{d}x.$$

For $f \in H^1$ the claim follows from the density of C_0^∞ functions in H^1 and Fatou's lemma.   □

We now show that the operator $\partial_i \partial_k (-\Delta)^{-1}$, $i, k = 1, 2, 3$, which is defined for f's as in (6.8), extends uniquely to a bounded operator from H^m to H^m for every $m \geq 0$.

Lemma 6.4   *The operator*

$$f \mapsto \partial_i \partial_k (-\Delta)^{-1} f$$

extends (uniquely) to a bounded operator from L^2 to L^2 for every i, k. More generally, it is a bounded operator from H^m to H^m for every $m \geq 0$.

Proof   Suppose that $f \in C_0^\infty$. Since the Fourier transform of $(4\pi|x|)^{-1}$ is $(2\pi|\xi|)^{-2}$ (for a proof of this fact see example 5 in Section 4.2 in Strichartz (1994)) we see that the Fourier transform of $\partial^\alpha(-\Delta)^{-1} f$ is $(2\pi|\xi|)^{-2}(2\pi i)^{|\alpha|}\xi^\alpha \widehat{f}(\xi)$ for any multiindex α. In particular, using the Plancherel Theorem

$$\|\partial_i \partial_k (-\Delta)^{-1} f\|^2 = \int \left| \frac{\xi_i \xi_k}{|\xi|^2} \widehat{f}(\xi) \right|^2 \mathrm{d}\xi \leq \int \left| \widehat{f}(\xi) \right|^2 \mathrm{d}\xi = \|f\|^2,$$

and similarly $\|\partial^\alpha \partial_i \partial_k (-\Delta)^{-1} f\| \leq \|\partial^\alpha f\|$ for any multiindex α.

Thus, since the space C_0^∞ is dense in L^2 and in H^m (for any $m \geq 1$) the operator $\partial_i \partial_k (-\Delta)^{-1}$ extends uniquely to a bounded operator from L^2 to L^2 as well as from H^m to H^m.   □

Consider the *heat equation* in $\mathbb{R}^3 \times (0, T)$:

$$v_t - \Delta v = 0,$$

with initial condition $v(0) = v_0$ (understood in the sense of L^2 limit as

$t \to 0^+$) for some $v_0 \in L^2$. Then the classical solution v of the heat equation is given by the convolution

$$v(t) = \Phi(t) * v_0,$$

where

$$\Phi(x, t) := \frac{1}{(4\pi t)^{3/2}} e^{-|x|^2/4t} \tag{6.10}$$

is the *heat kernel*. In what follows, we will rely on some well-known properties of the heat equation and the heat kernel, which we discuss in Appendix 6.5.1.

Finally, we will often use an integral version of an elementary fact from the theory of ordinary differential equations: let $f, \varphi \colon [0, T) \to \mathbb{R}^+$ be C^1 functions such that

$$\begin{cases} f' \le g\,f^k + a, \\ \varphi' \ge g\,\varphi^k + b \end{cases} \quad \text{on } [0, T),$$

with $f(0) < \varphi(0)$ and $k > 0$. Assume that g, a, b are continuous functions such that $g > 0$ and $a \le b$. Then $f < \varphi$ on $[0, T)$. In particular, we will use the following result, which is a stronger (integral) version of the case $k = 2$.

Lemma 6.5 (Integral inequalities) *Suppose $g > 0$ is a continuous function on $(0, T)$ that is locally integrable $[0, T)$, and that the functions $f, \varphi : (0, T) \to \mathbb{R}^+$ satisfy*

$$f(t) \le \int_0^t g(t - s)f(s)^2 \,\mathrm{d}s + a(t),$$

$$\varphi(t) \ge \int_0^t g(t - s)\varphi(s)^2 \,\mathrm{d}s + b(t)$$

for all $t \in (0, T)$, where a, b are continuous functions satisfying $a \le b$, φ is continuous, and f^2 and φ^2 are integrable near 0. Then $f \le \varphi$ on $(0, T)$.

Note that no assumption on the continuity of f is needed. We prove this lemma, along with a few related results, in Appendix 6.5.5 (see Lemma 6.48).

6.1.2 The Oseen kernel $\mathcal{T}$

The Oseen kernel is the main tool in solving the Stokes equations in $\mathbb{R}^3$ (which we discuss in the next section; see also the comment following

Theorem 6.8). It is a 3×3 matrix-valued function $\mathcal{T} = [\mathcal{T}_{ij}]$ defined by

$$\mathcal{T}_{ij}(x,t) := \delta_{ij}\Phi(x,t) + \partial_i\partial_j P(x,t), \qquad x \in \mathbb{R}^3, t > 0, \qquad (6.11)$$

where δ_{ij} denotes the Kronecker delta, Φ is given by (6.10), and

$$P(x,t) := \frac{1}{4\pi^{3/2}t^{1/2}|x|} \int_0^{|x|} e^{-\xi^2/4t} \, d\xi. \qquad (6.12)$$

It was first introduced by Oseen (1911) (see pages 3, 19 and 41 therein).

Note that $P(\cdot, t)$ is a smooth function for each $t > 0$. Indeed, for fixed t the function $\widetilde{P} : \mathbb{R} \to \mathbb{R}$ defined by $\widetilde{P}(0) := 1$,

$$\widetilde{P}(s) := \frac{1}{4\pi^{3/2}s} \int_0^s \frac{e^{-r^2/4t}}{t^{1/2}} \, dr$$

is even and smooth. Notice that a change of coordinates yields

$$\frac{1}{4\pi^{3/2}s} \int_0^s \frac{e^{-r^2/4t}}{t^{1/2}} \, dr = \frac{1}{4\pi^{3/2}} \int_0^1 \frac{e^{-s^2r^2/4t}}{t^{1/2}} \, dr$$

from which we immediately obtain the smoothness. Therefore, since $P(x,t) = \widetilde{P}(|x|)$, we obtain that $P(\cdot, t)$ is smooth as well (for each $t > 0$). A direct computation shows that

$$-\Delta P = \Phi. \qquad (6.13)$$

This yields an equivalent definition of $\mathcal{T}_{ij}$:

$$
\begin{aligned}
&\mathcal{T}_{1,1} = -(\partial_2^2 + \partial_3^2)P &\quad &\mathcal{T}_{1,2} = \partial_1\partial_2 P &\quad &\mathcal{T}_{1,3} = \partial_1\partial_3 P \\
&\mathcal{T}_{2,1} = \partial_1\partial_2 P &\quad &\mathcal{T}_{2,2} = -(\partial_1^2 + \partial_3^2)P &\quad &\mathcal{T}_{2,3} = \partial_2\partial_3 P \\
&\mathcal{T}_{3,1} = \partial_1\partial_3 P &\quad &\mathcal{T}_{3,2} = \partial_2\partial_3 P &\quad &\mathcal{T}_{3,3} = -(\partial_1^2 + \partial_2^2)P.
\end{aligned}
$$
$$(6.14)$$

Since the derivatives $\nabla^m P$ satisfy the pointwise estimate

$$|\nabla^m P(x,t)| \leq \frac{C_m}{(|x|^2 + t)^{(m+1)/2}}, \qquad m \geq 0$$

(see Theorem 6.47), we obtain the following pointwise estimates on the Oseen kernel

$$|\nabla^m \mathcal{T}(x,t)| \leq \frac{C_m}{(|x|^2 + t)^{(m+3)/2}} \qquad \text{for } x \in \mathbb{R}^3, t > 0, m \geq 0. \qquad (6.15)$$

Using these bounds we can easily deduce the integral estimates

$$\|\mathcal{T}(t)\| \leq C\, t^{-3/4} \quad \text{and} \quad \|\nabla\mathcal{T}(t)\|_1 \leq C\, t^{-1/2} \qquad (6.16)$$

for $t > 0$, where we used the facts that $\int(|x|^2 + t)^{-3} \, dx = C/t^{3/2}$ and $\int(|x|^2 + t)^{-2} \, dx = C/t^{1/2}$. Moreover, since $\mathcal{T}$ is pointwise continuous on $(\mathbb{R}^3\backslash\{0\}) \times (0,\infty)$, by (6.15) and an application of the Dominated Convergence Theorem we obtain

$$\mathcal{T} \in C((0,\infty); L^2) \quad \text{and} \quad \nabla\mathcal{T} \in C((0,\infty); L^1). \tag{6.17}$$

Finally $\mathcal{T}$ enjoys a certain integral continuity property, which we will use later to show Hölder continuity of solutions to the Stokes equations (see Lemma 6.9 (ii)).

Lemma 6.6 (1/2-Hölder continuity of $\nabla\mathcal{T}$ in an L^1 sense)   *For every* $x, y \in \mathbb{R}^3$, $t > 0$ *we have*

$$\int |\nabla\mathcal{T}(x - z, t) - \nabla\mathcal{T}(y - z, t)|\, \mathrm{d}z \leq C|x - y|^{1/2} t^{-3/4}.$$

Leray mentions this inequality on page 213, and he frequently uses it in his arguments (in *(2.14)*, *(2.18)*, the inequality following *(3.3)*, and the first inequality on page 219) to show Hölder continuity of functions given by a representation formulae involving $\nabla\mathcal{T}$. We provide a proof for the sake of completeness.

Proof Let $R := |x - y|$ and

$$\Omega := B(x, 2R) \cup B(y, 2R).$$

Given that $\Omega \subset B(x, 3R)$ we can use (6.15) to write

$$\int_\Omega |\nabla\mathcal{T}(x - z, t)|\, \mathrm{d}z$$

$$\leq C \int_{B(x,3R)} \frac{\mathrm{d}z}{(|x - z|^2 + t)^2} = C \int_0^{3R} \frac{r^2}{(r^2 + t)^2}\, \mathrm{d}r \tag{6.18}$$

$$\leq C \int_0^{3R} \frac{r^2 + t}{(r^2 + t)^2}\, \mathrm{d}r = \frac{C}{t^{1/2}} \tan^{-1}\left(\frac{3R}{t^{1/2}}\right) \leq C \frac{R^{1/2}}{t^{3/4}},$$

since $\tan^{-1}\alpha \leq \alpha^{1/2}$ for $\alpha > 0$. Analogously

$$\int_\Omega |\nabla\mathcal{T}(y - z, t)|\, \mathrm{d}z \leq C R^{1/2} t^{-3/4}. \tag{6.19}$$

As for $z \in \mathbb{R}^3 \setminus \Omega$ note that $|z - x| \geq 2R$. Hence for any point ξ on the line segment $[x, y]$

$$|z - x| \leq |z - \xi| + |\xi - x| \leq |z - \xi| + R \leq |z - \xi| + |z - x|/2,$$

and so $|z - \xi| \geq |z - x|/2$. Thus, using the Mean Value Theorem and the bound on $\nabla^2 \mathcal{T}$ in (6.15) we obtain

$$
\begin{aligned}
\int_{\mathbb{R}^3 \setminus \Omega} |\nabla \mathcal{T}(x-z,t) - \nabla \mathcal{T}(y-z,t)| \, \mathrm{d}z &= R \int_{\mathbb{R}^3 \setminus \Omega} |\nabla^2 \mathcal{T}(\xi(z) - z, t)| \, \mathrm{d}z \\
&\leq CR \int_{\mathbb{R}^3 \setminus \Omega} \frac{\mathrm{d}z}{(|\xi(z) - z|^2 + t)^{5/2}} \\
&\leq CR \int_{\mathbb{R}^3 \setminus B(x,R)} \frac{\mathrm{d}z}{(|x - z|^2 + 2t)^{5/2}} \\
&= CR \int_R^\infty \frac{r^2}{(r^2 + 2t)^{5/2}} \, \mathrm{d}r \leq CR^{1/2} \int_R^\infty \frac{r^{5/2}}{(r^2 + 2t)^{5/2}} \, \mathrm{d}r \\
&\leq CR^{1/2} t^{-3/4} \int_{R/\sqrt{t}}^\infty \frac{\rho^{5/2}}{(\rho^2 + 1)^{5/2}} \, \mathrm{d}\rho \leq CR^{1/2} t^{-3/4},
\end{aligned}
$$

where $\xi(z) \in [x, y]$ for each z. This together with (6.18) and (6.19) proves the claim. $\qquad\square$

6.2 The Stokes equations

In this section we consider the Stokes equations,

$$\partial_t u - \Delta u + \nabla p = F, \tag{6.20}$$
$$\operatorname{div} u = 0, \tag{6.21}$$

in $\mathbb{R}^3 \times (0, T)$, where $T > 0$ and $F(x, t)$ is a vector-valued forcing. Leray calls these equations the *infinitely slow motion*. The Stokes equations model a drift-diffusion flow of an incompressible velocity field u. Here p denotes the pressure function. One can think of the appearance of the pressure function as providing the extra freedom necessary to impose the incompressibility constraint (6.21) for an arbitrary F, see the comment after Theorem 6.8. As usual, we denote the initial condition for (6.20), (6.21) by $u_0 \in H$, which we will see is attained in the sense of the L^2 limit, that is $\|u(t) - u_0\| \to 0$ as $t \to 0^+$.

In his paper, Leray includes an essentially complete analysis of the Stokes initial value problem in $\mathbb{R}^3$. The results for this problem are fundamental in the analysis of the Navier–Stokes equations that follows; the arguments are at times somewhat technical, we therefore present them here, but with some details relegated to Appendix 6.5.4.

The Stokes equations with the general form of the forcing F can be solved using the representation formulae[1],

$$u(t) = u_1(t) + u_2(t) := \Phi(t) * u_0 + \int_0^t \mathcal{T}(t - s) * F(s)\, ds, \quad (6.22)$$

$$p(t) := -(-\Delta)^{-1}(\operatorname{div} F(t)), \quad (6.23)$$

see Theorem 6.8 below (in which we focus only on the case of regular F). See (6.8) for the definition of $(-\Delta)^{-1}$, and recall that $\Phi(t)$ denotes the heat kernel (6.10). Here the convolution of the matrix function $\mathcal{T}(t - s)$ and a vector function $F(s)$ is understood as a matrix-vector operation, that is

$$u_{2,i}(x, t) = \int_0^t \int \mathcal{T}_{ij}(x - y, t - s) F_j(y, s)\, dy\, ds.$$

In this section we study the representation formulae (6.22), (6.23) and certain modified representation formulae (that is (6.33), (6.34)) in the case when F is of the special form $F = -(Y \cdot \nabla)Y$ for some vector field Y. Of the two cases:

$$F \text{ of general form} \quad \text{and} \quad F = -(Y \cdot \nabla)Y \text{ for some } Y,$$

Leray considers[2] mainly the former case; studying the formula (6.22) given appropriate regularity of F, and only mentioning briefly the latter case[3]. Here we treat the two cases separately. We treat the former case briefly, and we focus more on the latter case. An advantage of this approach is that it makes our results for each of the two cases directly applicable in the analysis of the Navier–Stokes equations. Moreover, in this slight refinement of Leray's approach, we construct the solution using the representation formulae, rather than deducing the representation as a property of the solution. As a result, we obtain a simple existence and uniqueness theorem for the Stokes equations (Theorem 6.12).

6.2.1 A general forcing F

Consider a forcing $F \in C([0, T); L^2)$ and let u be given by the representation formula (6.22) above.

Lemma 6.7 *If $u_0 \in H$ and $F \in C([0, T); L^2)$ then the function u given by (6.22) satisfies*

---

[1] These formulae are stated by Leray in *(2.2)* and *(2.9)*.
[2] This corresponds to Sections 11–13.
[3] See Lemma 8 in his work.

(i) $u \in C((0,T); L^\infty)$ with[4],

$$\|u(t)\|_\infty \leq C \int_0^t \frac{\|F(s)\|}{(t-s)^{3/4}} \, ds + C\|u_0\| t^{-3/4}. \tag{6.24}$$

(ii) $\nabla u \in C((0,T); L^2)$ with[5]

$$\|\nabla u(t)\| \leq C \int_0^t \frac{\|F(s)\|}{(t-s)^{1/2}} \, ds + C\|u_0\| t^{-1/2}. \tag{6.25}$$

More generally, if $F, \ldots, \nabla^m F \in C([0,T); L^2)$ *then*

$$\nabla^{m+1} u \in C((0,T); L^2),$$

with

$$\|\nabla^{m+1} u(t)\| \leq C \int_0^t \frac{\|\nabla^m F(s)\|}{(t-s)^{1/2}} \, ds + C_{m+1}\|u_0\| t^{-(m+1)/2}.$$

(iii) $u \in C([0,T); L^2)$ with[6]

$$\|u(t)\| \leq \int_0^t \|F(s)\| \, ds + \|u_0\| \qquad \text{for all } t \in (0,T). \tag{6.26}$$

Moreover u *satisfies the energy dissipation equality*

$$\|u(t)\|^2 - \|u_0\|^2 + 2 \int_0^t \|\nabla u(s)\|^2 \, ds = 2 \int_0^t \int u \cdot F \, dx \, ds \tag{6.27}$$

for all $t \in (0,T)$.

The properties (i) and (ii) of the lemma follow from Lemma 6.1, the integral bounds on the Oseen kernel in (6.16) and the corresponding property of the heat kernel (see (ii), (iii) in Appendix 6.5.1). As for (iii), assuming first that F is smooth, the functions u, p constitute a classical solution to the Stokes equations (6.20), (6.21), which is proved in the following theorem. The required estimates are straightforward for classical solutions. If F is not smooth, one obtains (iii) by a density argument. See Appendix 6.5.4 for the detailed proof of (iii).

Theorem 6.8 (Classical solution for smooth forcing F)
Suppose that $u_0 \in H$ *and for some* $R > 0$

$$F \in C^\infty(\mathbb{R}^3 \times [0,T); \mathbb{R}^3) \quad \text{and} \quad \text{supp } F(t) \subset B(0,R) \text{ for } t \in [0,T). \tag{6.28}$$

---

[4] Leray does not state this bound (we state it as a tool for proving (iii)).
[5] This is Leray's *(2.8)*, *(2.12)* and *(2.19)*.
[6] This is Section 13 in Leray (1934b).

Then u, p given by (6.22)–(6.23) *constitute a classical solution of the Stokes equations* (6.20),(6.21) *with $u(0) = u_0$. Also $u \in C([0,T]; L^2)$.*

In fact the functions u, p constitute a unique solution in a much wider class, namely the class of distributional solutions u, p for which we have $u \in C([0,T]; L^2)$ and $p \in L^1_{loc}(\mathbb{R}^n \times [0,T))$, see Theorem 6.11 in Section 6.2.2.

Theorem 6.8 follows by showing that u_1, u_2 (recall (6.22)) satisfy the equations[7]

$$\begin{cases} \partial_t u_1 - \Delta u_1 = 0, \\ \operatorname{div} u_1 = 0, \\ u_1(0) = u_0, \end{cases} \qquad \begin{cases} \partial_t u_2 - \Delta u_2 + \nabla p = F, \\ \operatorname{div} u_2 = 0, \\ u_2(0) = 0. \end{cases}$$
(6.29)

The part of the claim for u_1 follows directly from the analysis of the heat equation, see Appendix 6.5.1. As for u_2, using the Fourier transform one can see that it is enough to prove the claim in Fourier space. It therefore suffices to use the Fourier transform of the Oseen kernel,

$$\mathcal{F}[\mathcal{T}(t)] = \left(I - \frac{\xi \otimes \xi}{|\xi|^2} \right) e^{-4\pi^2 t|\xi|^2} \quad t > 0, \tag{6.30}$$

obtained from (6.11) and (6.13) (see (6.128)), where I denotes the identity matrix and $\xi \otimes \xi$ denotes the 3×3 matrix with components $\xi_i \xi_j$. A detailed proof can be found in Appendix 6.5.4.

At this point it is interesting to note that the Stokes equations (6.20) are in fact a nonhomogeneous heat equation for u under the incompressibility constraint $\operatorname{div} u = 0$. Since $\Delta p = \operatorname{div} F$ we see that p appearing in the Stokes equations acts as a modification of the forcing F to make it divergence free (that is $\operatorname{div}(F - \nabla p) = 0$). In other words, $F - \nabla p$ is the projection of F onto the space of weakly divergence-free vector fields (which is often called the *Leray projection*). Since any solution of a nonhomogeneous heat equation with divergence-free forcing and initial data remains divergence free for positive times, one can think that the role of p is to guarantee that $u(t)$ remains divergence free for $t > 0$.

Moreover, from (6.23) we see that the Fourier transform of the modified forcing $F - \nabla p$ is

$$\left(I - \frac{\xi \otimes \xi}{|\xi|^2} \right) \widehat{F}(\xi, t).$$

Thus we see from (6.30) that the Oseen kernel is precisely the modifi-

---

[7] The study of u_1 and u_2 corresponds to Leray's Sections 11 and 12, respectively.

cation of the heat kernel that accounts for this new form of the forcing. This is particularly clear from a calculation in the Fourier space in Appendix 6.5.4.

6.2.2 A forcing of the form $F = -(Y \cdot \nabla)Y$

Here we assume that F is of a particular form, namely

$$F = -(Y \cdot \nabla)Y \qquad (6.31)$$

(given in components by $F_k = -Y_i \partial_i Y_k$) for some weakly divergence-free $Y \in C((0,T), L^\infty)$ such that $\|Y(t)\|_\infty$ remains bounded as $t \to 0^+$. Note that, since the derivatives $\partial_i Y_k$ are not well-defined for such Y, we understand (6.31) in a formal sense and we will consider the Stokes equations (6.20), (6.21) in the sense of distributions. More precisely, we say that u, p is a *distributional solution* of (6.20), (6.21) with F of the form (6.31) if $u(t)$ is weakly divergence free for $t \in (0,T)$ and

$$\int u_0 \cdot \varphi(0) \, \mathrm{d}x + \int_0^T \int (u \cdot (\varphi_t + \Delta\varphi) + p \operatorname{div}\varphi) = -\int_0^T \int Y \cdot (Y \cdot \nabla)\varphi \qquad (6.32)$$

for $\varphi \in C_0^\infty(\mathbb{R}^3 \times [0,T); \mathbb{R}^3)$. We will also consider a modified form of the representation formulae (6.22), (6.23) that accounts for this special form of the forcing,

$$u(t) = u_1(t) + u_2(t) := \Phi(t) * u_0 + \int_0^t \nabla\mathcal{T}(t-s) * [Y(s)Y(s)] \, \mathrm{d}s, \qquad (6.33)$$

$$p(t) := \partial_k \partial_i (-\Delta)^{-1}(Y_i(t)Y_k(t)), \qquad (6.34)$$

where we write

$$(\nabla\mathcal{T}(t-s) * [Y(s)Y(s)])_j(x) := \int \partial_i T_{jk}(x-y, t-s) Y_i(y,s) Y_k(y,s) \, \mathrm{d}y. \qquad (6.35)$$

Clearly, such u, p are well defined since no derivatives fall on Y in these modified representation formulae. If Y is regular (as in (6.28)) then the above definition of u, p is equivalent to (6.22), (6.23), and so Theorem 6.8 implies that such u, p constitute a classical solution of the Stokes equations (and hence also a distributional solution). In this section we show that u, p constructed above constitute the unique distributional solution in a wide class if we have $Y \in C([0,T); L^2)$ in addition to the assumptions on Y mentioned just after (6.31), see Theorem 6.12 below[8].

For this purpose we derive several properties of such u, p. In view of Lemma 6.7, we now prove refined bounds on $\|u(t)\|_\infty$ and $\|u(t)\|$, and show that $\nabla u \in C((0,T); L^\infty)$ and $u \in \mathcal{H}^{1/2}((0,T))$ (as defined in (6.3)).

Lemma 6.9 (Properties of u, p)   *Let $u_0 \in L^2$ and u,p be given by (6.33) and (6.34) for some weakly divergence-free $Y \in C((0,T); L^\infty)$ such that $\|Y(t)\|_\infty$ remains bounded as $t \to 0^+$.*

(i) *If u_0 is bounded then $u \in C((0,T); L^\infty)$ with[9]*

$$\|u(t)\|_\infty \leq C \int_0^t \frac{\|Y(s)\|_\infty^2}{\sqrt{t-s}}\, ds + \|u_0\|_\infty.$$

Moreover $u \in C([0,T); L^\infty)$ if $u_0 \in L^\infty$ is uniformly continuous.

(ii) *$u \in \mathcal{H}^{1/2}((0,T))$ and the corresponding Hölder constant $C_0(t)$ satisfies[10]*

$$C_0(t) \leq c_0 \int_0^t \frac{\|Y(s)\|_\infty^2}{(t-s)^{3/4}}\, ds + c_0 \frac{\|u_0\|}{t}$$

for some $c_0 > 0$.

More generally, if $Y, \nabla Y, \ldots, \nabla^m Y \in C((0,T); L^\infty)$ with the respective L^∞ norms bounded as $t \to 0^+$ then $\nabla^m u \in \mathcal{H}^{1/2}((0,T))$ and the corresponding constant $C_m(t)$ satisfies

$$C_m(t) \leq c_m \sum_{\alpha+\beta=m} \int_0^t \frac{\|D^\alpha Y(s)\|_\infty \|D^\beta Y(s)\|_\infty}{(t-s)^{3/4}}\, ds + c_m \frac{\|u_0\|}{t^{(m+2)/2}}.$$

(iii) *If additionally we have $Y \in C([0,T); L^2)$ then $p \in C([0,T); L^2)$ and $u \in C([0,T); L^2)$ with[11]*

$$\|u(t)\| \leq C \int_0^t \frac{\|Y(s)\|_\infty \|Y(s)\|}{\sqrt{t-s}}\, ds + \|u_0\|. \qquad (6.36)$$

---

[8] This corresponds to Leray's Lemma 8, in which he states that the representation formula (6.33) is a property of the solution.

[9] Leray does not state this bound explicitly, but he uses it in later sections during the study of the Navier–Stokes equations (for instance in *(3.5)*, at the bottom of p. 222, and at the top of p. 232).

[10] Leray shows a similar property of ∇u in the case of F of general form (which he obtains in *(2.18)* and as a consequence of *(2.7)* and *(2.8)*). We translate this result to the case of F of the form $F = -(Y \cdot \nabla)Y$.

[11] Leray does not state this property, but he uses it in showing existence of strong solutions to the Navier–Stokes equations (in the inequality he states at the bottom of page 223). We will apply it in a similar way (see Theorem 6.22) and also in the existence and uniqueness theorem for the Stokes equations (Theorem 6.12).

Moreover if $T' < T$ and $\{Y^{(n)}\}$ is a sequence such that $Y^{(n)} \to Y$ in $C([0, T']; L^2)$ and $\max_{t \in [0,T']} \|Y^{(n)}(t)\|_\infty \leq \max_{t \in [0,T']} \|Y(t)\|_\infty$ then

$$u^{(n)} \to u \quad \text{and} \quad p^{(n)} \to p \quad \text{in} \quad C([0, T']; L^2).$$

(iv) *If additionally $Y \in \mathcal{H}^{1/2}((0, T))$ with the corresponding constant $C_0(t)$ bounded as $t \to 0^+$ then $\nabla u \in C((0, T); L^\infty)$ with[12]*

$$\|\nabla u(t)\|_\infty \leq C \int_0^t \frac{\|Y(s)\|_\infty C_0(s)}{(t-s)^{3/4}} \, ds + C \frac{\|u_0\|}{t^{5/4}}. \tag{6.37}$$

More generally if for every multi-index α with $|\alpha| \leq m - 1$ we have $D^\alpha Y \in \mathcal{H}^{1/2}((0, T)) \cap C((0, T); L^\infty)$ with the corresponding Hölder constant $C_\alpha(t)$ and $\|D^\alpha Y(t)\|_\infty$ bounded as $t \to 0^+$ then $\nabla^m u \in C((0, T); L^\infty)$, with

$$\|\nabla^m u(t)\|_\infty \leq C_m \int_0^t \frac{\sum_{\alpha+\beta=m-1} \|D^\alpha Y(s)\|_\infty C_\beta(s)}{(t-s)^{3/4}} \, ds$$
$$+ C_m t^{-\frac{m}{2} - \frac{3}{4}} \|u_0\|.$$

Proof Since $Y \in C((0, T); L^\infty)$ with $\|Y(t)\|_\infty$ bounded as $t \to 0^+$ the same is true of $Y_i Y_k$ for each pair i, k and so claim (i) follows from Lemma 6.1 and from the properties of the heat kernel (see Appendix 6.5.1; note also that Lemma 6.44 verifies the comment in (i)). In a similar way one obtains (iii), where the claim for p follows directly from Lemma 6.4, and the limiting property follows from (6.36) and the same Lemma. Property (ii) is a consequence of the Hölder continuity of the heat kernel (see (v) in Appendix 6.5.1) and the Hölder continuity of $\nabla \mathcal{T}$ in L^1 (see Lemma 6.6). Indeed we have

$$|u_2(x,t) - u_2(y,t)|$$
$$\leq \int_0^t \int |\nabla \mathcal{T}(x-z, t-s) - \nabla \mathcal{T}(y-z, t-s)| \, dz \|Y(s)\|_\infty^2 \, ds$$
$$\leq C|x-y|^{1/2} \int_0^t \frac{\|Y(s)\|_\infty^2}{(t-s)^{3/4}} \, ds.$$

As for property (iv), note that the bound on ∇u_1 in (6.37) (that is $\|\nabla u_1(t)\|_\infty \leq C \|u_0\| t^{-5/4}$) and the continuity $\nabla u_1 \in C((0, T); L^\infty)$ follow from properties of the heat kernel (see (iv) in Appendix 6.5.1). As

---

[12] This corresponds to Leray's property of ∇u (which he obtains in *(2.7)* and *(2.20)*).

for u_2, the bound on ∇u_2 in (6.37) can be shown using the following argument. Recalling that Y is weakly divergence free, we obtain

$$\int \partial_{li} \mathcal{T}_{kj}(x - y, t) Y_i(y, s) \, dy = 0$$

for all j, k, l and $s, t \in (0, T)$, $x \in \mathbb{R}^3$, where the integral exists due to (6.15). Hence, for each j, l and $t \in (0, T)$

$$\partial_l u_{2,j}(x, t) = -\int_0^t \int \partial_{li} \mathcal{T}_{jk}(x - y, t - s) Y_i(y, s) Y_k(y, s) \, dy \, ds$$

$$= -\int_0^t \int \partial_{li} \mathcal{T}_{jk}(x - y, t - s) Y_i(y, s) \left[Y_k(y, s) - Y_k(x, s) \right] \, dy \, ds,$$

$$(6.38)$$

and so using the bound $|\nabla^2 \mathcal{T}(x, t)| \leq C(|x|^2 + t)^{-5/2}$ (see (6.15)) and the assumption $Y \in \mathcal{H}^{1/2}((0, T))$ we obtain

$$|\nabla u_2(x, t)| \leq C \int_0^t \int \frac{C_0(s) \|Y(s)\|_\infty |x - y|^{1/2}}{(|x - y|^2 + (t - s))^{5/2}} \, dy \, ds$$

$$= C \int_0^t \frac{\|Y(s)\|_\infty C_0(s)}{(t - s)^{3/4}} \, ds,$$

where we used the fact $\int |y|^{1/2}/(|y|^2 + t)^{5/2} \, dy = Ct^{-3/4}$. Thus (6.37) follows. One can also employ a similar argument to show the continuity $\nabla u_2 \in C((0, T); L^\infty)$, see Appendix 6.5.4 for the details.

Finally, claims (ii) and (iv) for higher derivatives $\nabla^m u$ follow in a similar way. Indeed, the claims corresponding to u_1 follow from the properties of the heat kernel (see (iv) in Appendix 6.5.1). For u_2, we write any $D^\gamma u_{2,j}$ with $|\gamma| = m$ as a sum of terms of the form

$$\int_0^t \int \partial_l \partial_i \mathcal{T}_{jk}(x - y, t - s) Y_{\alpha,i}(y, s) Y_{\beta,k}(y, s) \, dy \, ds$$

where Y_α, Y_β denote appropriate derivatives of Y of orders α, β, respectively, where $\alpha + \beta = m - 1$, $l, j = 1, 2, 3$, and we repeat the reasoning above. $\square$

Corollary 6.10   *The results of the above lemma extend to the case* $F_i = -\partial_k(Y_i Z_k)$ *for some weakly divergence-free* $Y, Z \in C((0, T); L^\infty)$ *with the* L^∞ *norms bounded as* $t \to 0^+$. *In particular, if we also have* $Y, Z \in C([0, T); L^2)$ *and* v *is given by*

$$v(t) := \Phi(t) * u_0 + \int_0^t \nabla \mathcal{T}(t - s) * [Y(s) Z(s)] \, ds \qquad (6.39)$$

then $v \in C((0,T); L^{\infty}) \cap C([0,T); L^2)$, with

$$\|v(t)\|_{\infty} \leq C \int_0^t \frac{\|Y(s)\|_{\infty}\|Z(s)\|_{\infty}}{\sqrt{t-s}} \, ds + \|u_0\|_{\infty},$$

$$\|v(t)\| \leq C \int_0^t \frac{\|Y(s)\|_{\infty}\|Z(s)\|}{\sqrt{t-s}} \, ds + \|u_0\|.$$

The existence and uniqueness theorem for distributional solutions to the Stokes equations is based on the following uniqueness result.

Theorem 6.11 (Uniqueness of distributional solutions to the Stokes equations[13]) *If u, p are such that $u \in C([0,T); L^2)$ is weakly divergence-free, $p \in L^1_{\text{loc}}(\mathbb{R}^3 \times [0,T))$, and*

$$\int_0^T \int ((\varphi_t + \Delta\varphi) \cdot u + p \operatorname{div} \varphi) \, dx \, dt = 0 \qquad (6.40)$$

for all $\varphi \in C_0^{\infty}(\mathbb{R}^3 \times [0,T); \mathbb{R}^3)$, then $u \equiv 0$.

Note that in this statement it is essential that the supports of the test functions φ are allowed to include $t = 0$. Indeed, if the test functions φ were supported away from $t = 0$ then the uniqueness result would be false (as one could take a classical solution of the homogeneous Stokes equations starting from a non-zero initial data).

Observe that it follows from $u \equiv 0$ that $\int_0^T \int p \operatorname{div} \varphi \, dx \, dt = 0$, and so integration by parts and the Fundamental Lemma of the Calculus of Variations give $\nabla p \equiv 0$; that is p is a function of t only. Since both the Stokes equations and the Navier–Stokes equations are invariant under addition to the pressure function any function of time, we will identify two solutions u_1, p_1 and u_2, p_2 of the Stokes equations (or the Navier–Stokes equations) if $u_1 \equiv u_2$ and p_1 differs from p_2 by a function of time.

Proof (sketch; see Appendix 6.5.4 for details). The proof of the theorem is based on considering the regularisations of u, p,

$$v(x,t) := \int_0^t (J_\varepsilon u)(x,s) \, ds, \qquad q(x,t) := \int_0^t (J_\varepsilon p)(x,s) \, ds, \qquad \varepsilon > 0.$$

Such a regularisation is still a solution of (6.40) and one can show that $\Delta q \equiv 0$. Thus Δv satisfies the homogeneous heat equation in a distributional sense and the uniqueness of the solution to the heat equation gives $\Delta v \equiv 0$. An application of Liouville's theorem and the assumption

---

[13] This is a version of the argument from Section 14 of Leray (1934b).

$\|u(t)\| < \infty$ for all t then gives $v \equiv 0$ for all $\varepsilon > 0$, and consequently $u \equiv 0$. $\qquad\qquad\qquad\qquad\qquad\qquad\qquad\qquad\qquad\qquad\qquad\square$

We are now ready to prove the existence and uniqueness of distributional solutions, a central result of the study of the Stokes equations.

Theorem 6.12 (Distributional solution for F of the form (6.31)) *Let $Y \in C([0,T); L^2) \cap C((0,T); L^\infty)$ be weakly divergence free with $\|Y(t)\|_\infty$ bounded as $t \to 0^+$. Then u, p given by (6.33), (6.34) are a distributional solution of (6.20), (6.21) with initial data u_0 and $F = -(Y \cdot \nabla)Y$. Also the solution is unique in the class $u \in C([0,T); L^2)$, $p \in L^1_{\text{loc}}(\mathbb{R}^3 \times [0,T))$.*

Proof The uniqueness follows from the theorem above. The fact that $u \in C([0,T); L^2)$ and the L^2 continuity at $t = 0$, $\|u(t) - u_0\| \to 0$ as $t \to 0$, is clear from Lemma 6.9 (iii).

To show that (u,p) indeed solves (6.32), fix $\varphi \in C_0^\infty(\mathbb{R}^3 \times [0,T); \mathbb{R}^3)$ and let $T' \in (0,T)$ be such that $\varphi = 0$ for $t \geq T'$. Let $\{Y^{(n)}\}$ be a sequence of functions $Y^{(n)} \in C^\infty(\mathbb{R}^3 \times [0,T))$ such that $\operatorname{supp} Y^{(n)}(t) \subset B(0, R_n)$ for some $R_n > 0$,

$$\|Y - Y^{(n)}\|_{C([0,T'];L^2)} \to 0 \qquad \text{as } n \to \infty$$

and $\max_{t \in [0,T']} \|Y^{(n)}(t)\|_\infty \leq \max_{t \in [0,T']} \|Y(t)\|_\infty$. Note that the above convergence means that also

$$\|Y_i Y_k - Y_i^{(n)} Y_k^{(n)}\|_{C([0,T'];L^2)} \to 0 \qquad \text{as } n \to \infty$$

for all i, k. The existence of such $Y^{(n)}$ is guaranteed by Lemma 6.55. Let (u_n, p_n) be given by (6.22), (6.23) with F replaced by $F^{(n)}$, where $F_k^{(n)} := -\partial_i(Y_i^{(n)} Y_k^{(n)})$. By Theorem 6.8 (u_n, p_n) satisfies the equations (6.20), (6.21) with F replaced by $F^{(n)}$ in the classical sense, and so also in the sense of distributions, that is, u_n is weakly divergence free and

$$\int u_0 \cdot \varphi \, \mathrm{d}x + \int_0^T \int (u_n \cdot (\varphi_t + \Delta\varphi) + p_n \operatorname{div}\varphi) = \int_0^T \int Y^{(n)} \cdot (Y^{(n)} \cdot \nabla)\varphi$$

$$(6.41)$$

for any $\varphi \in C_0^\infty(\mathbb{R}^3 \times [0,T); \mathbb{R}^3)$. By Lemma 6.9 (iii), we have

$$\|u_n - u\|_{C([0,T'];L^2)} \to 0, \quad \|p_n - p\|_{C([0,T'];L^2)} \to 0 \qquad \text{as } n \to \infty$$

and so we can take the limit as n goes to infinity to obtain that u is weakly divergence free and, from (6.41), satisfies

$$\int u_0 \cdot \varphi \, \mathrm{d}x + \int_0^T \int (u \cdot (\varphi_t + \Delta\varphi) + p \operatorname{div}\varphi) = \int_0^T \int Y \cdot (Y \cdot \nabla)\varphi,$$

that is u, p is indeed the distributional solution. $\qquad\qquad\qquad\qquad\square$

Corollary 6.13 *The conclusion of Theorem 6.12 also holds if F is of the form $F_i = -\partial_i(Y_iZ_k)$, where $Y, Z \in C([0,T), L^2) \cap C((0,T), L^\infty)$ are weakly divergence free with $\|Y(t)\|_\infty$, $\|Z(t)\|_\infty$ bounded as $t \to 0^+$, and the representation formula (6.33) is replaced by (6.39).*

Notes

As remarked in the beginning of the section, we focused on the forcing of the form $F = -(Y \cdot \nabla)Y$ more directly than Leray. In particular Lemma 6.9 is not stated explicitly by Leray. Thanks to the use of the distributional form of the Stokes equations (6.32) and the limiting property of the representation formulae (6.33), (6.34) (that is Lemma 6.9 (iii)) the main results of the section can be encapsulated in Theorem 6.12.

In the next section we will follow Leray in applying the results for the Stokes equations to study the Navier–Stokes equations. In particular we will employ the other properties of the modified representation formulae (6.33), (6.34), that is Lemma 6.9 (i), (ii), and (iv). These properties were presented by Leray either implicitly during the study of the Navier–Stokes equations or by showing a related result for a general form of the forcing F, as we pointed out in the footnotes in Lemma 6.9.

6.3 Strong solutions of the Navier–Stokes equations

We now consider the Navier–Stokes equations

$$\partial_t u - \Delta u + \nabla p = -(u \cdot \nabla)u, \qquad (6.42)$$

$$\operatorname{div} u = 0 \qquad (6.43)$$

in $\mathbb{R}^3 \times (0, T)$. We will consider a weak form of these equations,

$$\int_0^T \int (u \cdot (\varphi_t + \Delta\varphi) + p \operatorname{div}\varphi) = -\int_0^T \int u \cdot (u \cdot \nabla)\varphi \qquad (6.44)$$

for $\varphi \in C_0^\infty(\mathbb{R}^3 \times (0,T))$, where $u(t)$ is weakly divergence free. We first define solutions on the open time interval $(0,T)$ (see below) and we study their properties in Section 6.3.1. In Section 6.3.2 we equip the problem (6.42)–(6.43) with initial data, and for this reason we extend the definition of strong solutions to the half-closed time interval $[0,T)$ (Definition 6.20). We then show existence and uniqueness of local-in-time strong solutions (Theorem 6.22). In Section 6.3.3 we study the maximal time of existence for strong solutions and the rate of blow-up of strong solutions if the maximal time is finite. In Section 6.3.4 we study local existence and uniqueness of strong solutions with less regular initial data.

Definition 6.14   A function $u \in C((0,T); L^2) \cap C((0,T); L^\infty)$ is a *strong solution of the Navier–Stokes equations on the time interval* $(0,T)$ if it satisfies the weak form of the equations (6.44) for some function $p \in L^1_{\text{loc}}(\mathbb{R}^3 \times (0,T))$.

This is how Leray defines a strong solution, except that he requires the continuity (in both x and t) of all terms appearing in the Navier–Stokes equations (6.42) (see p. 217). Here we make use of the weak form of equations and thus we avoid specifying any conditions on derivatives of u. However, smoothness of strong solutions (Corollary 6.16) implies that the two definitions are equivalent to each other. Note also that if u is a strong solution of the Navier–Stokes equations then (6.44) is equivalent to requiring that

$$\int u(t_1) \cdot \varphi(t_1) - \int u(t_2) \cdot \varphi(t_2) + \int_{t_1}^{t_2} \int (u \cdot (\varphi_t + \Delta\varphi) + p \operatorname{div} \varphi)$$

$$= -\int_{t_1}^{t_2} \int u \cdot (u \cdot \nabla)\varphi \qquad (6.45)$$

for all $\varphi \in C_0^\infty(\mathbb{R}^3 \times (0,T))$ and $t_1, t_2 \in (0,T)$ with $t_1 < t_2$. While the "$\Leftarrow$" part this equivalence is trivial, the "$\Rightarrow$" part is not immediate but can be obtained by a simple cut-off procedure, which we now explain.

For $h > 0$ let $F_h(x,s) := \varphi(x,s)\theta_h(s)$, where $\theta_h \in C^\infty(\mathbb{R})$ is a non-increasing function such that $\theta_h(s) = 1$ for $s \le t_2$, $\theta_h(s) = 0$ for $s \ge t_2 + h$. Using F_h as a test function in (6.44) we obtain

$$\int_0^{t_2+h} \int (u \cdot (\varphi_t + \Delta\varphi) + p \operatorname{div} \varphi)\theta_h + \int_{t_2}^{t_2+h} \int u \cdot \varphi\theta_h'$$

$$= -\int_0^{t_2+h} \int u \cdot (u \cdot \nabla)\varphi\theta_h. \qquad (6.46)$$

Since $u, \varphi \in C((0,T); L^2)$ the function $s \mapsto \int u(s) \cdot \varphi(s)$ is continuous. Thus, since θ_h is non-increasing and $\int_{\mathbb{R}} \theta_h' = -1$ we obtain

$$\int_{t_2}^{t_2+h} \int u \cdot \varphi\, \theta_h' \to -\int u(t_2) \cdot \varphi(t_2) \quad \text{as } h \to 0^+.$$

Thus taking the limit as $h \to 0^+$ in (6.46) (via the Dominated Convergence Theorem) gives

$$-\int u(t_2)\cdot\varphi(t_2) + \int_0^{t_2}\int (u\cdot(\varphi_t + \Delta\varphi) + p\operatorname{div}\varphi) = -\int_0^{t_2}\int u\cdot(u\cdot\nabla)\varphi.$$

Applying a similar cut-off procedure at time t_1 gives (6.45).

From (6.45) and the theorem about the existence and uniqueness of distributional solutions to Stokes equations (Theorem 6.12) we see that a strong solution of the Navier–Stokes equations admits the representation formulae[14],

$$u(t_2) = \Phi(t_2 - t_1) * u(t_1) + \int_{t_1}^{t_2}\int \nabla\mathcal{T}(t_2 - s) * [u(s)u(s)]\,\mathrm{d}s,$$

$$p(t_2) = \partial_i\partial_k(-\Delta)^{-1}(u_i(t_2)u_k(t_2)),$$

$$(6.47)$$

for all $t_1, t_2 \in (0, T)$ with $t_1 < t_2$. Recall that we employ the notation

$$(\nabla\mathcal{T}(t - s) * [u(s)u(s)])_j(x) := \int \partial_i\mathcal{T}_{jk}(x - y, t - s)u_i(y, s)u_k(y, s)\,\mathrm{d}y.$$

Note this representation formula also determines uniquely (up to a function of time) the pressure function p (which is not specified (uniquely) in the definition of strong solutions, Definition 6.14).

6.3.1 Properties of strong solutions

In this section we study the properties of strong solutions of the Navier–Stokes equations on the open time interval $(0, T)$. We will show that if u is a strong solution and p is the corresponding pressure function then u and p are smooth and u satisfies an energy equality, along with some other useful results. The theorem below and the following corollary show that u, p are smooth.

Theorem 6.15 *If u is a strong solution of the Navier–Stokes equations on $(0, T)$ then*

$$\nabla^m u \in C((0, T); L^2) \cap C((0, T); L^\infty) \quad \textit{for all } m \ge 0.$$

In particular, using the Sobolev embedding[15] $H^2 \hookrightarrow C(\mathbb{R}^3)$ (see Theorem 4.12 in Adams & Fournier (2003)), a strong solution u admits a continuous representative with $\nabla^m u \in C((0, T); C(\mathbb{R}^3))$ for $m \ge 0$.

---

[14] These are *(3.2)* and *(3.3)* in Leray (1934b).

[15] The need to use Sobolev embeddings here is an artefact of our chosen definition of strong solutions i.e. so that the continuity of solutions Leray assumes in his definition can be recovered.

The proof of the theorem (as well as the corollary that follows) is a simplification[16] of Leray's arguments (which he presents on pages 218-219).

Proof The proof proceeds by a double use of induction. First, we show that

$$\nabla^m u \in C((0,T); L^\infty) \cap \mathcal{H}^{1/2}((0,T)) \quad \text{for } m \geq 0.$$

Here the base case follows from the definition of a strong solution and from Lemma 6.9 (ii), and the induction step follows from the same lemma, properties (iv) and (ii).

Next, we show that

$$\nabla^m u \in C((0,T); L^2), \ \nabla^m\left[(u \cdot \nabla)u\right] \in C((0,T); L^2) \quad \text{for } m \geq 0.$$

Here the base case follows from the definition of a strong solution and by deducing that $(u \cdot \nabla)u \in C((0,T); L^2)$ by using Hölder's inequality,

$$\|(u(t) \cdot \nabla)u(t) - (u(s) \cdot \nabla)u(s)\|$$
$$\leq \|u(t)\| \, \|\nabla u(t) - \nabla u(s)\|_\infty + \|\nabla u(s)\|_\infty \|u(t) - u(s)\|.$$

The induction step follows from Lemma 6.7 (ii) and from a similar use of Hölder's inequality. $\square$

Note that for each $s \in (0,T)$ we have bounded the norms $\|\nabla^m u(s)\|$, $\|\nabla^m u(s)\|_\infty$, $m \geq 0$ using only $\|u(t)\|$, $\|u(t)\|_\infty$, $t \in (0,T)$.

Corollary 6.16 (Smoothness of strong solutions) *If u is a strong solution of the Navier–Stokes equations on $(0,T)$ and p is the corresponding pressure then*

$$\partial_t^k \nabla^m u, \partial_t^k \nabla^m p \in C((0,T); L^2) \cap C((0,T); L^\infty) \quad \text{for all } m,k \geq 0.$$

In particular $u,p \in C^\infty(\mathbb{R}^3 \times (0,T))$ and u,p constitute a classical solution of the Navier–Stokes equations on $\mathbb{R}^3 \times (0,T)$.

Proof From the representation of p, (6.47), and Lemma 6.4, we obtain that $\nabla^m p \in C((0,T); L^2)$ for all m. By the Sobolev embedding $H^2 \hookrightarrow C(\mathbb{R}^3)$, it follows that[17]

$$\nabla^m p \in C((0,T); C(\mathbb{R}^3)) \qquad \text{for all } m \geq 0. \tag{6.48}$$

---

[16] This simplification is due to our choice to organise the properties of the representation formulae (6.22), (6.23) and (6.33), (6.34) into Lemmas 6.7 and 6.9, see the Notes at the end of this section.

[17] Leray is not explicit in showing this property (see page 219). It can be verified in an elementary way, which we suspect to be Leray's approach (rather than using the Sobolev embedding), and we discuss it in the Notes at the end of the section.

From the distributional form of the Navier–Stokes equations (6.44) we see that u_t (understood as the weak derivative of $u \in C(\mathbb{R}^3 \times (0, T); \mathbb{R}^3)$) is given by $\Delta u - (u \cdot \nabla)u - \nabla p \in C(\mathbb{R}^3 \times (0, T); \mathbb{R}^3)$. Therefore u_t is in fact the classical derivative and the Navier–Stokes equations are satisfied in the classical sense. Furthermore $\nabla^m u_t \in C((0, T); L^2) \cap C((0, T); L^\infty)$ for all m.

The regularity of higher derivatives in time follows by induction: regularity of $\partial_t^k u$ follows from the regularity of $u, \partial_t u, \ldots, \partial_t^{k-1} u$ and $\partial_t^{k-1} p$, and by taking $(k-1)$-th time derivative of the Navier–Stokes equations (6.42), and the regularity of $\partial_t^k p$ follows by taking k time derivatives of the representation formula (6.47) of p. □

Theorem 6.17 (Energy equality for strong solutions[18])   *A strong solution u of the Navier–Stokes equations on $(0, T)$ satisfies*

$$\|u(t_2)\|^2 + 2 \int_{t_1}^{t_2} \|\nabla u(s)\|^2 \, \mathrm{d}s = \|u(t_1)\|^2 \qquad (6.49)$$

for all $t_1, t_2 \in (0, T)$.

Proof Since Theorem 6.15 gives in particular $(u \cdot \nabla)u \in C((0, T); L^2)$, Lemma 6.7 (iii) gives

$$\|u(t_2)\|^2 - \|u(t_1)\|^2 + 2 \int_{t_1}^{t_2} \|\nabla u(s)\|^2 \, \mathrm{d}s = 2 \int_{t_1}^{t_2} \int u \cdot (u \cdot \nabla)u \, \mathrm{d}x \, \mathrm{d}s.$$

The theorem follows by noting that the right-hand side vanishes: integration by parts and the incompressibility constraint, $\partial_k u_k = 0$, give

$$\int u \cdot (u \cdot \nabla)u = \int u_i \, u_k \, \partial_k u_i = - \int \partial_k u_i \, u_k \, u_i, \qquad (6.50)$$

that is $\int u \cdot (u \cdot \nabla)u = - \int u \cdot (u \cdot \nabla)u = 0$. □

We now show that we can control the separation of two strong solutions.

Lemma 6.18 (Comparison of two strong solutions[19])   *Suppose that u, v are strong solutions to the Navier–Stokes equations on $(0, T)$ and let*

$$w := u - v.$$

Then

$$\|w(t_2)\|^2 \leq \|w(t_1)\|^2 e^{\frac{1}{2} \int_{t_1}^{t_2} \|u(s)\|_\infty^2 \, \mathrm{d}s} \qquad (6.51)$$

---

[18] This is *(3.4)* in Leray (1934b).
[19] This is Section 18 of Leray (1934b).

for $t_1, t_2 \in (0, T)$ with $t_1 < t_2$.

In particular, if u, v coincide at time t_1 then they continue to coincide for later times. We will extend this uniqueness property to account for the initial data in Section 6.3.2 (Lemma 6.21).

Proof Since both u and v satisfy the Navier–Stokes equations pointwise, subtracting the corresponding equations gives

$$\partial_t w - \Delta w + \nabla q = -(u \cdot \nabla)u + (v \cdot \nabla)v = -(v \cdot \nabla)w - (w \cdot \nabla)u.$$

As in (6.50) we have $\int w \cdot (v \cdot \nabla)w = 0$ and hence multiplying the above equality by w, integrating by parts in spatial variables and using the incompressibility constraint, $\partial_k w_k = 0$, we obtain for $t \in (0, T)$

$$\frac{1}{2}\frac{d}{dt}\|w(t)\|^2 + \|\nabla w(t)\|^2 = -\int w \cdot ((v \cdot \nabla)w + (w \cdot \nabla)u)$$

$$= \int w_i \, w_k \, \partial_k u_i = -\int \partial_k w_i \, w_k \, u_i$$

$$\leq \|\nabla w(t)\| \|w(t)\| \|u(t)\|_\infty$$

$$\leq \|\nabla w(t)\|^2 + \frac{1}{4}\|w(t)\|^2 \|u(t)\|_\infty^2,$$

where we also used the Cauchy-Schwarz and Young inequalities (and we omitted the argument "t" under the integrals). Hence

$$\frac{d}{dt}\|w(t)\|^2 \leq \frac{1}{2}\|w(t)\|^2 \|u(t)\|_\infty^2,$$

and the claim follows by applying Gronwall's inequality. $\qquad\square$

Finally, Theorem 6.15 implies the following convergence property of a family of strong solutions of the Navier–Stokes equations.

Lemma 6.19 (Convergence lemma[20]) *Suppose $\{u_\varepsilon\}_{\varepsilon>0}$ is a family of strong solutions of the Navier–Stokes equations such that*

$$\|u_\varepsilon(t)\|_\infty, \|u_\varepsilon(t)\| \leq f(t) \qquad \text{for } t \in (0, T),$$

where f is a continuous function on $(0, T)$. Then there exists a sequence $\varepsilon_k \to 0^+$ and a function u, such that $u_{\varepsilon_k} \to u$, $\nabla u_{\varepsilon_k} \to \nabla u$ uniformly on compact sets in $\mathbb{R}^3 \times (0, T)$ as $\varepsilon_k \to 0^+$.

Moreover u is a strong solution of the Navier–Stokes equations in $(0, T)$ and satisfies $\|u(t)\|_\infty, \|u(t)\| \leq f(t)$ for all $t \in (0, T)$.

---

[20] This is Lemma 9 in Leray (1934b).

Proof Let $p_\varepsilon(t)$ denote the pressure function corresponding to u_ε (which is determined by the representation formula (6.47) with u replaced by u_ε). Fix a bounded domain $\Omega \subset \mathbb{R}^3$ and $\delta > 0$, we define $\Omega_\delta := \Omega + B(0, \delta)$. By Theorem 6.15 and associated remarks, we see that for any m and a multi-index α such that $m, |\alpha| \leq 3$

$$\partial_t^m D^\alpha u_\varepsilon, \partial_t^m D^\alpha p_\varepsilon$$

are bounded in $C((\delta/2, T - \delta/2); L^2(\Omega_\delta))$ and $C((\delta/2, T - \delta/2); L^\infty(\Omega_\delta))$ in terms of only $\|u_\varepsilon(t)\|_\infty$, $\|u_\varepsilon(t)\|$, $t \in (\delta/4, T - \delta/4)$, and thus (by assumption) are bounded by a constant $C(\Omega, \delta, \max_{t \in (\delta/4, T - \delta/4)} |f(t)|)$. Having obtained a bound independent of ε we can apply the Arzelà–Ascoli theorem to obtain a sequence $\{\varepsilon_k\}$ such that the derivatives $\partial_t^m D^\alpha u_{\varepsilon_k}$, $\partial_t^m D^\alpha p_{\varepsilon_k}$ with $m, |\alpha| \leq 2$ converge to the respective derivatives of u and p uniformly on $\Omega \times (\delta, T - \delta)$, for some functions u, p. In particular the Navier–Stokes equations

$$\partial_t u_{\varepsilon_k} - \Delta u_{\varepsilon_k} + \nabla p_{\varepsilon_k} = -(u_{\varepsilon_k} \cdot \nabla) u_{\varepsilon_k},$$
$$\operatorname{div} u_{\varepsilon_k} = 0$$

converge uniformly on $\Omega \times (\delta, T - \delta)$ to the Navier–Stokes equations for u (in the sense that all terms converge). Now consider a sequence of bounded sets $\Omega_n \nearrow \mathbb{R}^3$, a sequence $\delta_n \to 0^+$ and apply a diagonal argument to obtain a subsequence $\{\varepsilon_k\}$ (which we relabel) such that for $m, |\alpha| \leq 2$

$$\partial_t^m D^\alpha u_{\varepsilon_k} \to \partial_t^m D^\alpha u, \quad \partial_t^m D^\alpha p_{\varepsilon_k} \to \partial_t^m D^\alpha p \qquad (6.52)$$

uniformly on compact sets in $\mathbb{R}^3 \times (0, T)$. In particular u, p satisfy the Navier–Stokes equations pointwise, and thus also in the sense of distributions (6.44). The fact that $\|u(t)\|_\infty \leq f(t)$ holds for all t is clear, and the inequality $\|u(t)\| \leq f(t)$ follows by a simple application of Fatou's lemma.

Therefore, according to Definition 6.14 it remains to verify that $u \in C((0, T); L^2) \cap C((0, T); L^\infty)$. For this let I be a closed interval in $(0, T)$ and note that there exists $M > 0$ such that $\|\partial_t u_{\varepsilon_k}(t)\|_\infty, \|\partial_t u_{\varepsilon_k}(t)\| \leq M$ for $t \in I$, $k \geq 0$ (see Corollary 6.16). Thus the Mean Value Theorem gives

$$\|u_{\varepsilon_k}(t) - u_{\varepsilon_k}(s)\|_\infty, \|u_{\varepsilon_k}(t) - u_{\varepsilon_k}(s)\| \leq M|t - s| \quad \text{for } s, t \in I, \ k \geq 0.$$

Thus taking the limit in k (and applying Fatou's lemma) gives the required continuity. $\qquad \square$

6.3.2 Local existence and uniqueness of strong solutions

In this section we study the Navier–Stokes initial value problem, that is we consider the equations (6.42), (6.43) with given initial data. For this reason we extend the definition of strong solutions (Definition 6.14) to the half-closed time interval $[0, T)$.

Definition 6.20 A function u is a *strong solution of the Navier–Stokes equations on* $[0, T)$ if for some $p \in L^1_{\text{loc}}(\mathbb{R}^3 \times [0, T))$

$$\int u(0) \cdot \varphi(0) + \int_0^T \int (u \cdot (\varphi_t + \Delta\varphi) + p \operatorname{div} \varphi) = \int_0^T \int u \cdot (u \cdot \nabla)\varphi \quad (6.53)$$

for every $\varphi \in C_0^\infty(\mathbb{R}^3 \times [0, T); \mathbb{R}^3)$, $u(t)$ is weakly divergence free for $t \in (0, T)$, and

$$u \in C([0, T); L^2) \cap C((0, T); L^\infty), \quad (6.54)$$

with $\|u(t)\|_\infty$ bounded as $t \to 0^+$.

The regularity (6.54) is part of Leray's definition of solutions on the time interval $[0, T)$, but he also requires $\nabla u \in C([0, T); L^2)$ and that u and its spatial derivatives are continuous at $t = 0$ (see pp. 220-221 of his paper). It is remarkable that by use of the weak formulation (6.53), these additional assumptions are not necessary for showing local well-posedness (see Theorem 6.22 below). Moreover, the above definition is much less restrictive than the commonly used definition of strong solutions, which usually requires

$$u \in L^\infty_{\text{loc}}([0, T); H^1) \cap L^2_{\text{loc}}([0, T); H^2),$$

and so consequently $u \in C([0, T); H^1)$ see, for example, Definition 6.1 and the discussion that follows in Robinson et al. (2016). In particular, Definition 6.20 makes no assumption on the regularity of $\nabla u(t)$ for times t near 0.

Since Definition 6.20 is an extension of the definition of the strong solution on the open time interval $(0, T)$ (Definition 6.14), we see that u, p admit the representation formulae (6.47). Moreover, now the representation formula also holds for $t_1 = 0$,

$$u(t) = \Phi(t) * u(0) + \int_0^t \nabla \mathcal{T}(t - s) * [u(s) u(s)] \, ds \quad (6.55)$$

for $t \in (0, T)$, a consequence of the definition above and Theorem 6.12.

We also see that (6.53) is equivalent to

$$\int u(0) \cdot \varphi(0) - \int u(t) \cdot \varphi(t)$$

$$+ \int_0^t \int (u \cdot (\varphi_t + \Delta\varphi) + p \operatorname{div} \varphi) = - \int_0^t \int u \cdot (u \cdot \nabla)\varphi$$

$$(6.56)$$

being satisfied for all $t \in (0, T)$, $\varphi \in C_0^\infty(\mathbb{R}^3 \times [0, T); \mathbb{R}^3)$ (see the discussion following Definition 6.14). A consequence of this fact is that a strong solution on the time interval $[0, T)$ is also a strong solution on the time interval $[\tau, T)$ for any $\tau \in (0, T)$.

Given the definition of strong solutions on the half-closed time interval $[0, T)$, we immediately obtain the energy equality (6.49) with $t_1 = 0$ and the uniqueness of strong solutions.

Lemma 6.21 (Uniqueness and the energy equality for local strong solution) *A strong solution u to the Navier–Stokes equations on $[0, T)$ satisfies the energy equality*

$$\|u(t)\|^2 + 2 \int_0^t \|\nabla u(s)\|^2 \, ds = \|u_0\|^2 \qquad (6.57)$$

for $t \in [0, T)$.

Moreover, if v is another strong solution to the Navier–Stokes equations on $[0, T)$ with $v(0) = u(0)$ then $u \equiv v$.

Proof The claim follows by taking the limit $t_1 \to 0^+$ in (6.49) and (6.51) and applying the Monotone Convergence Theorem. □

From the lemma we obtain the semigroup property: if $\tau \in (0, T_1)$ and u, $\widetilde{u}$ are strong solutions on the intervals $[0, T_1)$, $[\tau, T_2)$ respectively with $\widetilde{u}(\tau) = u(\tau)$, then $\widetilde{u} = u$ on the time interval $[\tau, \min\{T_1, T_2\})$. We now state the central theorem regarding strong solutions to the Navier–Stokes equations.

Theorem 6.22 (Local existence of strong solutions) *If $u_0 \in H \cap L^\infty$ (that is $u_0 \in L^2 \cap L^\infty$ is weakly divergence free, see (6.2)) then there exists a unique strong solution u of the Navier–Stokes equations on $[0, T)$ with $u(0) = u_0$, where $T > C/\|u_0\|_\infty^2$.*

This theorem is proved by Leray in Section 19 in the case in which $u_0 \in C^1 \cap H^1 \cap L^\infty$. Here, we use the distributional form of the equations to relax this regularity requirement to only $u_0 \in H \cap L^\infty$, and to demonstrate how Leray's original proof can be simplified while exposing his main ideas.

Proof Uniqueness is guaranteed by Lemma 6.21. As for existence, we consider the following iterative definition of $u^{(n)}$:

$$u^{(0)}(t) := \Phi(t) * u_0$$

and

$$u^{(n+1)}(t) := \int_0^t \nabla \mathcal{T}(t-s) * \left[u^{(n)}(s)\, u^{(n)}(s) \right] \mathrm{d}s + u^{(0)}(t),$$

using the notation from (6.35). From properties of the heat kernel (see (iii) in Appendix 6.5.1) we have

$$u^{(0)} \in C([0,\infty); L^2) \cap C((0,\infty); L^\infty),$$

$\|u^{(0)}(t)\|_\infty \le \|u_0\|_\infty$ and $\|u^{(0)}(t)\| \le \|u_0\|$ for $t \ge 0$. Moreover, using induction we deduce from Lemma 6.9 (i) and (iii) (applied with $Y := u^{(n)}$) that for all $n \ge 0$, $t \ge 0$,

$$u^{(n+1)} \in C([0,\infty); L^2) \cap C((0,\infty); L^\infty),$$

$$\|u^{(n+1)}(t)\|_\infty \le C' \int_0^t \frac{\|u^{(n)}(s)\|_\infty^2}{\sqrt{t-s}}\, \mathrm{d}s + \|u_0\|_\infty, \qquad (6.58)$$

and

$$\|u^{(n+1)}(t)\| \le C' \int_0^t \frac{\|u^{(n)}(s)\|_\infty \|u^{(n)}(s)\|}{\sqrt{t-s}}\, \mathrm{d}s + \|u_0\|. \qquad (6.59)$$

Theorem 6.12 guarantees that if

$$p^{(n+1)}(t) := \partial_k \partial_i (-\Delta)^{-1} \left(u_i^{(n)}(t)\, u_k^{(n)}(t) \right)$$

(recall (6.8) for the notation) then for each $n \ge 0$ the pair $u^{(n+1)}, p^{(n+1)}$ is a distributional solution of the problem

$$\partial_t u^{(n+1)} - \Delta u^{(n+1)} + \nabla p^{(n+1)} = -(u^{(n)} \cdot \nabla) u^{(n)}, \qquad \mathrm{div}\, u^{(n+1)} = 0$$

with initial condition $u^{(n+1)}(0) = u_0$, that is $u^{(n+1)}$ is weakly divergence free and

$$\int u_0 \cdot \varphi(0)\, \mathrm{d}x + \int_0^\infty \int \left(u^{(n+1)} \cdot (\varphi_t + \Delta\varphi) + p^{(n+1)}\, \mathrm{div}\, \varphi \right)$$

$$= -\int_0^\infty \int u^{(n)} \cdot (u^{(n)} \cdot \nabla)\varphi \qquad (6.60)$$

for all $\varphi \in C_0^\infty(\mathbb{R}^3 \times [0, \infty))$, $n \geq 0$. In order to take the limit in n we will find a uniform bound on $\|u^{(n+1)}\|_\infty$ on the finite time interval $[0, T]$, where

$$T := \frac{1}{32(1 + C')^4 \|u_0\|_\infty^2}.$$

For such choice of $T > 0$ the constant function $\psi(t) := (1 + C')\|u_0\|_\infty$ satisfies the integral inequality

$$\psi(t) \geq C' \int_0^t \frac{\psi^2(s)}{\sqrt{t - s}} \, ds + \|u_0\|_\infty \tag{6.61}$$

for all $t \in [0, 8T)$. A use of induction and (6.58) thus gives

$$\|u^{(n)}(t)\|_\infty \leq \psi(t) \tag{6.62}$$

for all such t and all $n \geq 0$. Now noting that for all $j, k = 1, 2, 3$, $n \geq 1$

$$u_k^{(n)} u_j^{(n)} - u_k^{(n-1)} u_j^{(n-1)} = u_k^{(n)} \left(u_j^{(n)} - u_j^{(n-1)} \right) + u_j^{(n-1)} \left(u_k^{(n)} - u_k^{(n-1)} \right)$$

we use Corollary 6.10 twice (first with $Y := u^{(n)}$, $Z := u^{(n)} - u^{(n-1)}$ and then with $Y := u^{(n-1)}$, $Z := u^{(n)} - u^{(n-1)}$) to obtain

$$\|u^{(n+1)}(t) - u^{(n)}(t)\|_\infty$$
$$\leq C' \int_0^t \frac{\left(\|u^{(n)}(s)\|_\infty + \|u^{(n-1)}(s)\|_\infty \right) \|u^{(n)}(s) - u^{(n-1)}(s)\|_\infty}{\sqrt{t - s}} \, ds \tag{6.63a}$$

and

$$\|u^{(n+1)}(t) - u^{(n)}(t)\|$$
$$\leq C' \int_0^t \frac{\left(\|u^{(n)}(s)\|_\infty + \|u^{(n-1)}(s)\|_\infty \right) \|u^{(n)}(s) - u^{(n-1)}(s)\|}{\sqrt{t - s}} \, ds. \tag{6.63b}$$

Applying (6.62) to the second of the above inequalities gives for $t \in [0, T]$

$$\|u^{(n+1)}(t) - u^{(n)}(t)\|$$
$$\leq 2C'(1 + C')\|u_0\|_\infty \|u^{(n)} - u^{(n-1)}\|_{C([0,T];L^2)} \int_0^t \frac{1}{\sqrt{t - s}} \, ds$$
$$= 4C'(1 + C')\|u_0\|_\infty \sqrt{t} \|u^{(n)} - u^{(n-1)}\|_{C([0,T];L^2)}$$
$$\leq \lambda \|u^{(n)} - u^{(n-1)}\|_{C([0,T];L^2)},$$

where $\lambda := 4C'(1 + C')\|u_0\|_\infty \sqrt{T}$. Hence

$$\|u^{(n+1)} - u^{(n)}\|_{C([0,T];L^2)} \leq \lambda \|u^{(n)} - u^{(n-1)}\|_{C([0,T];L^2)}.$$

Because the definition of T implies $\lambda \in (0,1)$ we see that $\{u^{(n)}\}$ is a Cauchy sequence in $C([0,T];L^2)$ and so

$$u^{(n)} \to u \quad \text{in } C([0,T];L^2)$$

for some $u \in C([0,T];L^2)$ such that $u(0) = u_0$ and $u(t)$ is weakly divergence free for each $t \in [0,T]$ (since $u^{(n)}(0) = u_0$ and $u^{(n)}(t)$ is weakly divergence free for each n). Similarly, applying (6.62) to the first inequality in (6.63) gives

$$\|u^{(n+1)} - u^{(n)}\|_{C([0,T];L^\infty)} \leq \lambda \|u^{(n)} - u^{(n-1)}\|_{C([0,T];L^\infty)}.$$

Although this does not imply that $\{u^{(n)}\}$ is Cauchy in $C([0,T];L^\infty)$ (recall each $u^{(n)}$ need not belong to this space; it is continuous into L^∞ only on the open time interval $(0,\infty)$), it does follow that $\{u^{(n)}\}$ is a Cauchy sequence in $C([\delta,T];L^\infty)$ for any $\delta \in (0,T)$, and therefore

$$u^{(n)} \to u \quad \text{in } C([\delta,T];L^\infty) \qquad (6.64)$$

for each δ. Note that the limit function is u since L^2 convergence implies convergence almost everywhere on a subsequence. Therefore

$$u \in C([0,T];L^2) \cap C((0,T];L^\infty)$$

and (6.62) gives $\|u(t)\|_\infty \leq (1 + C')\|u_0\|_\infty$ for all $t \in [0,T]$. Letting

$$p(t) := \partial_i \partial_k (-\Delta)^{-1} (u_i(t) u_k(t))$$

we see that Lemma 6.4 implies

$$p^{(n)} \to p \quad \text{in } C([0,T];L^2).$$

Therefore, taking the limit as n tends to infinity in (6.60) we obtain

$$\int u_0 \cdot \varphi \, dx + \int_0^T \int (u \cdot (\varphi_t + \Delta \varphi) + p \operatorname{div} \varphi) = -\int_0^T \int u \cdot (u \cdot \nabla)\varphi$$

for all $\varphi \in C_0^\infty(\mathbb{R}^3 \times [0,T))$. Thus u is a strong solution of the Navier–Stokes equations on $[0,T)$. $\qquad \square$

Note that if u_0 is more regular then the continuity as $t \to 0^+$ of the corresponding strong solution u on $[0,T)$ may be stronger. This is, in essence, the issue of continuity as $t \to 0^+$ of the solution of the heat equation (cf. Lemma 6.44). If u_0 is uniformly continuous then the representation formula (6.55) gives that $u(t) \to u_0$ in L^∞ as $t \to 0^+$ (cf. the proof of Lemma 6.9 (i)). Furthermore, for such u_0 the proof above simplifies since each $u^{(n)}$ belongs to $C([0,\infty); L^\infty)$ (cf. the same lemma) and so they converge in $C([0,T]; L^\infty)$ rather than in $C([\delta, T]; L^\infty)$ for all δ as in (6.64).

6.3.3 Characterisation of singularities

Here we investigate the maximal time of existence of strong solutions and derive the rates of blow-up of u in various norms at a (putative) blow-up time.

Let $u_0 \in V \cap L^\infty$ (that is $u_0 \in H^1 \cap L^\infty$ is divergence free, see (6.2)), u be the strong solution of the Navier–Stokes equations starting from u_0 and T_0 be its maximal time of existence, that is u cannot be extended to a solution on $[0, T')$ for any $T' > T_0$. Note that Theorem 6.22 gives that $T_0 \geq C/\|u_0\|_\infty^2$ and, if T_0 is finite,

$$\|u(t)\|_\infty \quad \text{blows up as} \quad t \to T_0^-,$$

as otherwise we could extend u beyond T_0 and hence obtain a contradiction.

In this section we will apply the theory of integral inequalities (see Lemma 6.5) to bound the L^∞ norm of a strong solution u on some time interval starting from 0 and thus obtain lower bounds on T_0 as well as lower bounds on $\|u(t)\|_\infty$, $\|\nabla u(t)\|$ and $\|u(t)\|_p$ with $p > 3$ when $t \to T_0^-$ (if T_0 is finite). In this section (and the next) p will always denote an exponent, and should not be confused with the pressure.

Since (6.55) holds for $t \in (0, T_0)$, Lemma 6.9 (i) gives that for all $t \in [0, T_0)$, $p > 3$

$$\|u(t)\|_\infty \leq C' \int_0^t \frac{\|u(s)\|_\infty^2}{\sqrt{t-s}} \, ds + \min\left(\|u_0\|_\infty, C' \frac{\|\nabla u_0\|}{t^{1/4}}, C' \frac{\|u_0\|_p}{t^{3/2p}} \right).$$
$$(6.65)$$

Here, the minimum on the right-hand side is obtained by applying Young's inequality for convolutions (6.6) to $u_1(t) = \Phi(t) * u_0$ (with exponents $(1, \infty)$, $(6/5, 6)$ and $(p/(p-1), p)$ respectively), the fact that $\|\Phi(t)\|_p \leq C/t^{-3(p-1)/2p}$, and the Gagliardo-Nirenberg-Sobolev[21] in-

---

[21] Note that this inequality was not available in the 1930s. Instead, Leray used Lemma 6.3 to obtain

$$\left| \int \Phi(x-y,t) u_0(y) \, dy \right| \leq 2 \left| \int |\Phi(x-y,t)|^2 |x-y|^2 \right|^{1/2} \|\nabla u_0\| \leq C \|\nabla u_0\| \, t^{-1/4}.$$

equality $\|u_0\|_6 \leq C\|\nabla u_0\|$. Similarly, the convolution under the time integral in the representation formula (6.55) can be bounded in terms of $\|\nabla T(t-s)\|_\infty \|u(s)\|^2$ (rather than by $\|\nabla T(t-s)\|_1 \|u(s)\|^2_\infty$ as in Lemma 6.9 (i)), and this gives for $t \in [0, T_0)$,

$$\|u(t)\|_\infty \leq C'' \int_0^t \min\left(\frac{\|u(s)\|^2_\infty}{\sqrt{t-s}}, \frac{\|u_0\|^2}{(t-s)^2} \right) ds + \|u_0\|_\infty, \quad (6.66)$$

where we have also used the facts $\|\nabla T(t)\|_\infty \leq Ct^{-2}$ (see (6.15)) and $\|u(t)\| \leq \|u_0\|$ (see the energy equality (6.57)). The two inequalities above[22] allow us to obtain bounds on $\|u(t)\|_\infty$ in terms of various norms of the initial data.

Lemma 6.23 [23] *If u is the strong solution with initial data $u_0 \in V \cap L^\infty$ then*

(i) $\|u(t)\|_\infty \leq C\|u_0\|_\infty$ *for* $t \leq C/\|u_0\|^2_\infty$,
(ii) $\|u(t)\|_\infty \leq C\|\nabla u_0\|t^{-1/4}$ *for* $t \leq C/\|\nabla u_0\|^4$,
(iii) $\|u(t)\|_\infty \leq C\|u_0\|_p t^{-3/2p}$ *for* $t \leq (C(1 - 3/p)/\|u_0\|_p)^{2p/(p-3)}$, *and* $p > 3$.

Moreover, there exists $\varepsilon > 0$ such that $\|u(t)\|_\infty \leq C\|u_0\|_\infty$ for all $t \geq 0$ if $\|u_0\|^2 \|u_0\|_\infty < \varepsilon$.

Proof As in the proof of Theorem 6.22, the function $\varphi(t) = C\|u_0\|_\infty$ satisfies the integral inequality (6.61) for $t \in [0, C/\|u_0\|^2_\infty]$ and so (i) follows from this, together with (6.65) and from the theory of integral inequalities (see Lemma 6.5). Similarly, a direct calculation shows that the function $\psi(t) = C\|\nabla u_0\|t^{-1/4}$ satisfies the integral inequality

$$\psi(t) \geq C' \int_0^t \frac{\psi(s)^2}{\sqrt{t-s}} ds + C'\|\nabla u_0\|t^{-1/4}$$

for $t \in (0, C/\|\nabla u_0\|^4]$, and so (ii) follows. One can also check that the function $\psi(t) = C\|u_0\|_p t^{-3/2p}$ (where $\|u_0\|_p < \infty$ due to Lebesgue interpolation inequality) satisfies the integral inequality

$$\psi(t) \geq C' \int_0^t \frac{\psi(s)^2}{\sqrt{t-s}} ds + C'\|u_0\|_p t^{-3/2p}$$

for $t \in (0, (C(1 - 3/p)/\|u_0\|_p)^{2p/(p-3)}]$, and (iii) follows.

---

[22] These comprise *(3.5)* in Leray (1934b).
[23] This lemma and the two following corollaries correspond to Sections 21 and 22 in Leray (1934b).

The last claim follows from (6.66) and the fact that a constant function $\psi(t) = C\|u_0\|_\infty$ with $C > 1$ satisfies the integral inequality

$$\psi(t) > C'' \int_0^t \min\left(\frac{\psi(s)^2}{\sqrt{t-s}}, \frac{\|u_0\|^2}{(t-s)^2} \right) ds + \|u_0\|_\infty, \qquad (6.67)$$

for all $t > 0$ if

$$\|u_0\|_\infty > \tilde{C} \int_0^\infty \min\left(\frac{\|u_0\|_\infty^2}{\sqrt{s}}, \frac{\|u_0\|^2}{s^2} \right) ds,$$

for a constant $\tilde{C}$ depending on C, C''.

One can check that this last condition is equivalent to a smallness condition $\|u_0\|^2 \|u_0\|_\infty < \varepsilon$ for some $\varepsilon > 0$. Therefore, the integral inequalities (6.67) and (6.66) show that, given the smallness condition, $\|u(t)\|_\infty \leq \psi(t)$ for all $t \geq 0$ (where we apply another fact from the theory of integral inequalities, see Corollary 6.50). $\qquad \square$

From the lemma above we immediately obtain lower bounds on the maximal time of existence T_0, the rates of blow-up of the norms $\|u(t)\|_\infty$, $\|\nabla u(t)\|$ and $\|u(t)\|_p$, where $p > 3$, as $t \to T_0^-$, as well as global existence result for small data, which we formulate in the following three corollaries.

Corollary 6.24 (Lower bounds on the existence time T_0) *If T_0 is the maximal time of existence of the strong solution u with initial data $u_0 \in V \cap L^\infty$ then*

(i) $T_0 > C/\|u_0\|_\infty^2$,

(ii) $T_0 > C/\|\nabla u_0\|^4$,

(iii) $T_0 > \left(C \left(1 - \frac{3}{p} \right) / \|u_0\|_p \right)^{2p/(p-3)}$ *for all $p > 3$.*

Proof These are direct consequences of Lemma 6.23 (i)-(iii). $\qquad \square$

Corollary 6.25 (Blow-up rates) *If u is a strong solution of the Navier–Stokes equations on the time interval (T, T_0), where $T_0 < \infty$ is the maximal existence time, then for $t \in (T, T_0)$,*

$$\|u(t)\|_\infty \geq \frac{C}{\sqrt{T_0 - t}}, \quad \|\nabla u(t)\| \geq \frac{C}{(T_0 - t)^{1/4}},$$

and, for $p > 3$,

$$\|u(t)\|_p \geq \frac{C^{(1-3/p)/2}(1 - 3/p)}{(T_0 - t)^{(1-3/p)/2}}.$$

Proof Let $t \in (T, T_0)$. Since $u(t) \in H \cap L^\infty$, the local existence and uniqueness theorem (Theorem 6.22) gives that $(T_0 - t) \geq C/\|u(t)\|_\infty^2$, which gives the first bound. The other two follow in a similar way using Corollary 6.24 (ii) and (iii) respectively.   □

Corollary 6.26 (Global existence for small initial data) *There exists $\varepsilon > 0$ such that if either $\|u_0\|^2 \|u_0\|_\infty < \varepsilon$, $\|u_0\| \|\nabla u_0\| < \varepsilon$ or*

$$(C\|u_0\|)^{2(p-3)} \|u_0\|_p^p < C\left(1 - \frac{3}{p}\right) \varepsilon^{p-3} \qquad \text{for any } p > 3 \qquad (6.68)$$

then $T_0 = \infty$, that is the strong solution with initial data u_0 exists for all times.

Observe that each of the above criteria on small initial data u_0 is scaling invariant (that is invariant with respect to the scaling $\lambda u_0(\lambda \cdot)$ for $\lambda > 0$).

Proof The first claim follows directly from Lemma 6.23. As for the smallness condition on $\|u_0\| \|\nabla u_0\|$, let $t_0 := C/\|\nabla u_0\|^4$, the endpoint time in Lemma 6.23 (ii). Then

$$\|u(t_0)\|_\infty \leq C\|\nabla u_0\| t_0^{-1/4} = C\|\nabla u_0\|^2,$$

and so, using the energy equality (6.57), we obtain

$$\|u(t_0)\|^2 \|u(t_0)\|_\infty \leq C\|u_0\|^2 \|\nabla u_0\|^2.$$

In other words, the condition $\|u(t_0)\|^2 \|u(t_0)\|_\infty < \varepsilon$ holds if $\|u_0\| \|\nabla u_0\|$ is sufficiently small, as required.

As for the smallness condition on $(C\|u_0\|)^{2(p-3)} \|u_0\|_p^p$, take $t_0 := (C(1 - 3/p)/\|u_0\|_p)^{2p/(p-3)}$, the endpoint time in Lemma 6.23 (iii). Then

$$\|u(t_0)\|_\infty \leq C\|u_0\|_p t_0^{-3/2p} = C\|u_0\|_p^{p/(p-3)} \left(C\left(1 - \frac{3}{p}\right)\right)^{-3/(p-3)}.$$

Thus the energy equality (6.57) together with (6.68) gives

$$\|u(t_0)\|^2 \|u(t_0)\|_\infty \leq \|u_0\|^2 \|u(t_0)\|_\infty < \varepsilon,$$

as required.   □

Finally, we deduce the following result[24], which we will use later in analysing the structure of a weak solution (Theorem 6.42).

---

[24] This corollary is a consequence of Leray's *(3.19)*.

Corollary 6.27 *Let u be a strong solution of the Navier–Stokes equations on the time interval (T, T_0), let $t_1 \in (T, T_0)$ and $t_2 > t_1$. If $t_2 - t_1 \le C\|\nabla u(t_1)\|^{-4}$ then*

$$\|u(t_2)\|_\infty \le C \frac{\|\nabla u(t_1)\|}{(t_2 - t_1)^{1/4}}, \quad and \quad \|\nabla u(t_2)\| \le C\|\nabla u(t_1)\|.$$

Proof The first inequality follows directly from Lemma 6.23 (ii). For the second one, note that from the smoothness of strong solutions (see Corollary 6.16) we have $(u \cdot \nabla)u \in C([t_1, T_0); L^2)$, so by the representation formula (6.47) and Lemma 6.7 (ii), we obtain[25]

$$\|\nabla u(t)\| \le C \int_{t_1}^t \frac{\|\nabla u(s)\|\|u(s)\|_\infty}{\sqrt{t-s}} \, ds + \|\nabla u(t_1)\|$$

$$\le C'''\|\nabla u(t_1)\| \int_{t_1}^t \frac{\|\nabla u(s)\|}{\sqrt{t-s}(s-t_1)^{1/4}} \, ds + \|\nabla u(t_1)\|$$

for all $t \in [t_1, T_0)$. Now, a direct calculation shows that the constant function $\psi(t) := C\|\nabla u(t_1)\|$ satisfies the integral inequality

$$\psi(t) \ge C'''\|\nabla u(t_1)\| \int_{t_1}^t \frac{\psi(s)}{\sqrt{t-s}(s-t_1)^{1/4}} \, ds + \|\nabla u(t_1)\|$$

for $t \in [t_1, t_1 + C/\|\nabla u(t_1)\|^4]$. Therefore,

$$\|\nabla u(t)\| \le \psi(t) \quad \text{for } t \in [t_1, t_1 + C/\|\nabla u(t_1)\|^4],$$

where we also used the theory of integral inequalities, see Corollary 6.49. Thus we obtain the second of the required inequalities. $\square$

6.3.4 Semi-strong solutions

In this section we focus on the regularity required from u_0 in order to generate a unique strong solution. In Section 6.3.2 we have shown that $u_0 \in H \cap L^\infty$ generates a solution that is strong on $[0, T)$ (see Definition 6.20) for some $T > 0$ (see Theorem 6.22). We also observed that the high regularity of u_0 guarantees some further properties of the solution; in particular the representation formula (6.55).

It turns out that relaxing the regularity of u_0 still gives a unique strong solution for (sufficiently small) positive times. This motivates the following definition[26].

---

[25] This is *(3.6)* in Leray (1934b).
[26] Definition 6.28 and the uniqueness result in Lemma 6.29 are stated in Section 23 in Leray's paper.

Definition 6.28   A function u is a *semi-strong solution of the Navier–Stokes equations* (6.42), (6.43) on the time interval $[0, T)$ if it is a strong solution on the open time interval $(0, T)$ (see Definition 6.14) such that $u \in C([0, T); L^2)$ and

$$\int_0^t \|u(s)\|_\infty^2 \, \mathrm{d}s < \infty \quad \text{for all } t \in (0, T). \tag{6.69}$$

Note this definition is less restrictive than the definition of strong solutions on the time interval $[0, T)$ (Definition 6.20). Namely, we replace the weak form of the equations (6.53) by the weak form (6.44), which does not include the initial data u_0, and we replace the boundedness of $\|u(t)\|_\infty$ as $t \to 0^+$ by the integral condition (6.69). Note that the initial condition $u(0) = u_0$ is now incorporated in the assumption that $u \in C([0, T); L^2)$.

Lemma 6.29   *Semi-strong solutions to the Navier–Stokes equations on $[0, T)$ (that satisfy a given initial condition) are unique and satisfy the energy equality*

$$\|u(t)\|^2 + 2 \int_0^t \|\nabla u(s)\|^2 \, \mathrm{d}s = \|u(0)\|^2$$

for all $t \in [0, T)$.

Proof   As in Lemma 6.21, the claim follows by taking the limit as $t_1 \to 0^+$ in (6.49) and (6.51) and applying the Monotone Convergence Theorem (note that the integral condition (6.69) guarantees that the exponent in (6.51) remains finite as $t_1 \to 0^+$). □

We now use the notion of semi-strong solutions to obtain a local-in-time well-posedness for initial data $u_0 \in V$ (rather than $u_0 \in H \cap L^\infty$ as in Theorem 6.22)[27].

Theorem 6.30   *If $u_0 \in V$ (that is $u_0 \in H^1$ is divergence free) then there exists a unique semi-strong solution u on the time interval $[0, T)$, where $T \geq C/\|\nabla u_0\|^4$, such that $u(0) = u_0$.*

Proof   Note that $J_\varepsilon u_0 \in H \cap L^\infty$ and $\mathrm{div}(J_\varepsilon u_0) = 0$ (see Section 6.1.1). Hence Theorem 6.22 implies the existence of a unique strong solution $u_\varepsilon(t)$ to the Navier–Stokes equations on some time interval $[0, T_\varepsilon)$ such that $u_\varepsilon(0) = J_\varepsilon u_0$. Denote the associated pressure by $p_\varepsilon \in L^1_{\mathrm{loc}}(\mathbb{R}^3 \times$

---

[27]   This is Section 24 in Leray (1934b).

$[0, T)$). The energy equality (6.57) and the properties of mollification (see Lemma 6.2) give

$$\|u_\varepsilon(t)\| \le \|u_\varepsilon(0)\| = \|J_\varepsilon u_0\| \le \|u_0\|, \tag{6.70}$$

$$\|\nabla u_\varepsilon(0)\| = \|\nabla(J_\varepsilon u_0)\| = \|J_\varepsilon(\nabla u_0)\| \le \|\nabla u_0\|. \tag{6.71}$$

The last bound and Corollary 6.24 let us bound the existence time T_ε from below independently of ε,

$$T_\varepsilon \ge C/\|\nabla u_0\|^4 =: T.$$

Moreover, Lemma 6.23 (ii) gives

$$\|u_\varepsilon(t)\|_\infty \le C\|\nabla u_\varepsilon(0)\|t^{-1/4} \le C\|\nabla u_0\|t^{-1/4} \tag{6.72}$$

for $t \in [0, T)$. By (6.70) and (6.72) we may apply the convergence lemma (Lemma 6.19) to extract a sequence $\{\varepsilon_k\}$ such that for all $t \in (0, T)$, $u_{\varepsilon_k}(t) \to u(t)$ uniformly on compact sets in $\mathbb{R}^3$. Here u is a strong solution of the Navier–Stokes equations on $\mathbb{R}^3 \times (0, T)$ with

$$\|u(t)\| \le \|u_0\| \tag{6.73}$$

and

$$\|u(t)\|_\infty \le C\|\nabla u_0\|t^{-1/4} \tag{6.74}$$

for $t \in (0, T)$.

By (6.52) we may also assume that for all $t \in (0, T)$, $p_{\varepsilon_k}(t) \to p(t)$ uniformly on compact sets in $\mathbb{R}^3$, where $p \in C((0, T); C(\mathbb{R}^3))$ is the pressure function corresponding to u. Furthermore, Lemma 6.4, (6.70), and (6.72) imply that

$$\|p_{\varepsilon_k}(t)\| \le C\|u_{\varepsilon_k}(t)\|_\infty \|u_{\varepsilon_k}(t)\| \le C(u_0)t^{-1/4} \quad \text{for } t \in (0, T). \tag{6.75}$$

It follows from (6.74) that $\int_0^t \|u(s)\|_\infty^2 \, ds$ is finite for all $t \in (0, T)$. Thus, in order to verify that u is a semi-strong solution, it remains to check that $u(t) \to u_0$ in L^2 as $t \to 0$. Since u_{ε_k} is a strong solution to the Navier–Stokes equations on $[0, T)$ (6.56) gives

$$0 = \int u_{\varepsilon_k}(t) \cdot \varphi - \int (J_{\varepsilon_k} u_0) \cdot \varphi - \int_0^t \int u_{\varepsilon_k} \cdot \Delta\varphi$$

$$- \int_0^t \int u_{\varepsilon_k} \cdot (u_{\varepsilon_k} \cdot \nabla)\varphi - \int_0^t \int p_{\varepsilon_k} \nabla \cdot \varphi \tag{6.76}$$

for $t \in [0, T)$ and $\varphi \in C_0^\infty(\mathbb{R}^3)$. By (6.52) and the fact that $J_\varepsilon u_0 \to u_0$ in

L^2 as $\varepsilon \to 0$, we see that each of the spatial integrals in (6.76) converge to the corresponding integral with u_ε, p_ε replaced by u, p, respectively.

By (6.70), (6.75), and the Dominated Convergence Theorem (applied to the time integrals) we can therefore pass to the limit in the above equation to obtain

$$0 = \int u(t) \cdot \varphi - \int u_0 \cdot \varphi - \int_0^t \int u \cdot \Delta\varphi - \int_0^t \int u \cdot (u \cdot \nabla)\varphi,$$

from which it follows that

$$\int u(t) \cdot \varphi \to \int u_0 \cdot \varphi \qquad \text{as } t \to 0^+$$

for all $\varphi \in C_0^\infty$, which, by the L^2 boundedness (6.73), gives that $u(t) \rightharpoonup u_0$ weakly in L^2 as $t \to 0^+$. In order to show that $u(t) \to u_0$ strongly in L^2 it is enough to show the convergence of the norms, $\|u(t)\| \to \|u_0\|$ as $t \to 0^+$. This convergence follows from properties of weak limits and (6.73) by writing

$$\|u_0\| \le \liminf_{t\to 0} \|u(t)\| \le \limsup_{t\to 0} \|u(t)\| \le \|u_0\|. \qquad \square$$

Similarly, we obtain that the notion of semi-strong solutions gives local-in-time well-posedness for $u_0 \in H \cap L^p$, where $p > 3$.[28]

Corollary 6.31   *Given $u_0 \in H \cap L^p$ with $p \in (3, \infty)$ there exists a semi-strong solution u of the Navier–Stokes equations on $[0, T)$ that satisfies $u(0) = u_0$, where $T \ge (C(1 - 3/p)/\|u_0\|_p)^{2p/(p-3)}$.*

Proof   The proof is identical to the that of the previous Theorem, with the following modification. Replace (6.71) by $\|u_\varepsilon(0)\|_p \le \|u_0\|_p$, (6.72) by $\|u_\varepsilon(t)\|_\infty \le C\|u_0\|_p t^{-3/2p}$, (6.74) by $\|u(t)\|_\infty \le C\|u_0\|_p t^{-3/2p}$, and T by $(C(1 - 3/p)/\|u_0\|_p)^{2p/(p-3)}$. $\qquad \square$

Notes

This section corresponds to Chapters III and IV of Leray (1934b).

In Section 20 Leray (1934b) considers the issue of the existence of solutions that blow up. He points out that such a solution exists if

$$\begin{cases} \Delta U(x) - \alpha U(x) - \alpha(x \cdot \nabla)U(x) - \nabla P(x) = (U(x) \cdot \nabla)U(x), \\ \operatorname{div} U(x) = 0, \end{cases}$$

---

[28] This is Section 25 of Leray (1934b). Note that the case $u_0 \in H \cap L^\infty$, for which Leray states well-posedness of semi-strong solutions, was covered in this article in the well-posedness result for strong solutions, see Theorem 6.22.

has a nontrivial solution in $\mathbb{R}^3$ for some $\alpha > 0$. In that case there would exist a strong solution of the Navier–Stokes equations on the time interval $(-\infty, T)$ of the following self-similar form

$$u(x,t) := \frac{1}{\sqrt{2\alpha(T-t)}} U\left(\frac{x}{\sqrt{2\alpha(T-t)}}\right),$$

which would blow up at time T. However, Nečas, Růžička, & Šverák (1996) have shown that this system of equations has no nontrivial L^3 solutions.

In fact, it is rather remarkable that the issue of the existence of solutions that blow up is one of the most important open problems in mathematics to this day; it was announced by Clay Mathematics Institute as one of seven Millennium Problems (see Fefferman, 2006).

Leray showed the smoothness of strong solutions via similar bounds as in the analysis of the Stokes equations (pp. 218-219). Since in our presentation the properties of the representation formulae (6.22), (6.23) and (6.33), (6.34) are organised in Lemmas 6.7 and 6.9, we were able to prove the smoothness of strong solutions by induction.

Furthermore, Section 6.3.3 shows that the analysis of the maximal time of existence and the blow-up rates can be done without the use of Leray's *(3.6)*,

$$\|\nabla u(t)\| \leq C \int_0^t \frac{\|\nabla u(s)\|\,\|u(s)\|_\infty}{\sqrt{t-s}}\,\mathrm{d}s + \|\nabla u_0\|,$$

which we use only in the proof of Corollary 6.27.

Finally, we prove (6.48) in an elementary way. Using (6.8), the Cauchy-Schwarz inequality and Lemma 6.3 we see that for any $x \in \mathbb{R}^3$

$$|p(x,t) - p(x,s)| \leq \int \frac{|\partial_i u_j(y,t)(\partial_j u_i(y,t) - \partial_j u_i(y,s))|}{4\pi|x-y|}\,\mathrm{d}y$$

$$+ \int \frac{|\partial_j u_i(y,s)(\partial_i u_j(y,t) - \partial_i u_j(y,s))|}{4\pi|x-y|}\,\mathrm{d}y$$

$$\leq C\|\nabla u(t) - \nabla u(s)\| \left(\sqrt{\int \frac{|\nabla u(y,t)|^2}{|x-y|^2}\,\mathrm{d}y} + \sqrt{\int \frac{|\nabla u(y,s)|^2}{|x-y|^2}\,\mathrm{d}y}\right)$$

$$\leq C\|\nabla u(t) - \nabla u(s)\| \left(\|\nabla^2 u(t)\| + \|\nabla^2 u(s)\|\right).$$

Hence $p \in C((0,T); C(\mathbb{R}^3))$ and performing a similar calculation for each of the spatial derivatives of p shows that $\nabla^m \in C((0,T); C(\mathbb{R}^3))$ for all $m \geq 0$.

6.4 Weak solutions of the Navier–Stokes equations

In this section we show the global existence of a weak solution (which Leray termed a *turbulent solution*) of the Navier–Stokes equations. His approach is characterised by considering, for $\varepsilon > 0$, the following modified system:

$$\partial_t u - \Delta u + ((J_\varepsilon u) \cdot \nabla) u + \nabla p = 0,$$
$$\operatorname{div} u(t) = 0, \qquad\qquad\qquad\qquad (6.77)$$

which is often called the *Leray regularisation*. We will see that this regularisation of the nonlinear term gives, for each $\varepsilon > 0$, a unique, global in time, strong solution. We then study the limit $\varepsilon \to 0$ of the solutions of the above equations.

6.4.1 Well-posedness for the regularised equations

Definition 6.32   A function u_ε is a *strong solution of the regularised equations* (6.77) *on the interval* $[0, T)$ if for some $p_\varepsilon \in L^1_{\mathrm{loc}}(\mathbb{R}^3 \times [0, T))$

$$\int u_\varepsilon(0) \cdot \varphi(0) + \int_0^T \int (u_\varepsilon \cdot (\varphi_t + \Delta\varphi) + p_\varepsilon \operatorname{div}\varphi) = -\int_0^T \int u_\varepsilon \cdot (J_\varepsilon u_\varepsilon \cdot \nabla)\varphi$$
$$(6.78)$$

for all $\varphi \in C_0^\infty(\mathbb{R}^3 \times [0, T); \mathbb{R}^3)$, $u_\varepsilon(t)$ is weakly divergence free for every $t \in (0, T)$, and

$$u_\varepsilon \in C([0, T); L^2) \cap C((0, T); L^\infty),$$

with $\|u_\varepsilon(t)\|_\infty$ bounded as $t \to 0^+$.

Note this definition follows the lines of the definition of a strong solution of the Navier–Stokes equations on the time interval $[0, T)$ (Definition 6.20), the difference appearing only in the form of the distributional equations (6.78). Moreover, we see that a solution u_ε and the corresponding pressure p_ε are given by

$$u_\varepsilon(t) = \Phi(t) * u_\varepsilon(t_1) + \int_{t_1}^t \int \nabla T(t - s) * [(J_\varepsilon u_\varepsilon)(s)\, u_\varepsilon(s)]\, ds,$$
$$(6.79)$$
$$p_\varepsilon(t) = \partial_i \partial_k (-\Delta)^{-1} ((J_\varepsilon u_{\varepsilon,i})(t)\, u_{\varepsilon,k}(t))$$

for all $0 \le t_1 < t < T$, cf. (6.47) and (6.55), see also Corollary 6.13.

Now let $u_0 \in H^1 \cap L^\infty$ be divergence free; we will show existence and uniqueness of global-in-time strong solution of the regularised equations with initial data u_0.

Theorem 6.33 (Global well-posedness of the regularised equations[29])
For each $\varepsilon > 0$ there exists a unique strong solution u_ε of the regularised equations (6.77) on the time interval $[0, \infty)$ such that $u_\varepsilon(0) = u_0$, u_ε is smooth on the time interval $(0, \infty)$ (in the sense of Corollary 6.16) and the energy equality

$$\|u_\varepsilon(t)\|^2 + 2 \int_0^t \|\nabla u_\varepsilon(s)\|^2 \, ds = \|u_0\|^2 \qquad (6.80)$$

holds for all $t \geq 0$.

Proof We note that the analysis from Section 6.3 can be applied to the regularised equations (6.77). In particular, by noting that $\|J_\varepsilon v\| \leq \|v\|$ and $\|J_\varepsilon v\|_\infty \leq \|v\|_\infty$ for any $v \in L^2 \cap L^\infty$ (see Lemma 6.2 (i)), we can prove a local existence and uniqueness theorem following Theorem 6.22, to obtain a unique strong solution u_ε of the system (6.77) on the time interval $[0, T)$ for some $T \geq C/\|u_0\|_\infty^2$. Now following the arguments in Section 6.3.1 we note that (using the representation formulae (6.79) instead of (6.47)) u_ε is smooth on the time interval $(0, T)$, and so following Theorem 6.17 and Lemma 6.21 we obtain the energy equality

$$\|u_\varepsilon(t)\|^2 + 2 \int_0^t \|\nabla u_\varepsilon(s)\|^2 \, ds = \|u_0\|^2$$

for $t \in [0, T)$. It remains to show that T, the maximal time of existence, is infinite.

As in Section 6.3.3 we see that $\|u_\varepsilon(t)\|_\infty$ must blow-up as $t \to T^-$ if $T < \infty$. We also obtain

$$\|u(t)\|_\infty \leq C \int_0^t \frac{\|(J_\varepsilon u_\varepsilon)(s)\|_\infty \|u(s)\|_\infty}{\sqrt{t-s}} \, ds + \|u_0\|_\infty$$

for $t \in [0, T)$, following the approach that led to (6.65). This inequality, however, is fundamentally different from (6.65) in the sense that we can now apply the bound

$$\|(J_\varepsilon u_\varepsilon)(s)\|_\infty \leq C\varepsilon^{-3/2} \|u_\varepsilon(s)\| \leq C\varepsilon^{-3/2} \|u_0\|,$$

(see Lemma 6.2 (iii) and the energy equality (6.80)). Moving this bound outside of the integral, we obtain a linear integral inequality,

$$\|u_\varepsilon(t)\|_\infty \leq C'\varepsilon^{-3/2} \|u_0\| \int_0^t \frac{\|u_\varepsilon(s)\|_\infty}{\sqrt{t-s}} \, ds + \|u_0\|_\infty.$$

---

[29] This is Section 26 of Leray (1934b).

Letting $\varphi_\varepsilon \in C([0, \infty))$ be the unique solution to the corresponding linear integral equation[30],

$$\varphi_\varepsilon(t) = C'\varepsilon^{-3/2}\|u_0\| \int_0^t \frac{\varphi_\varepsilon(s)}{\sqrt{t-s}}\, \mathrm{d}s + \|u_0\|_\infty, \quad t \geq 0, \qquad (6.81)$$

we see that $\|u_\varepsilon(t)\|_\infty \leq \varphi_\varepsilon(t)$ for all $t \geq 0$ (see Corollary 6.49). Therefore $\|u_\varepsilon(t)\|_\infty$ remains bounded on every finite interval and hence $T = \infty$, as required. $\qquad\square$

In order to study the limit as $\varepsilon \to 0$ of the solutions u_ε to the system (6.77), we first show that the kinetic energy of $u_\varepsilon(t)$ outside of a ball can be bounded independently of ε.

Lemma 6.34 (Separation of energy[31]) *Let $\varepsilon > 0$, $0 < R_1 < R_2$ and let u_ε be the solution of (6.77) with initial condition u_0. Then for $t \geq 0$*

$$\int_{|x|>R_2} |u_\varepsilon(t)|^2\, \mathrm{d}x \leq \int_{|x|>R_1} |u_0|^2\, \mathrm{d}x + \frac{\bar{C}(u_0, t)}{R_2 - R_1},$$

where $\bar{C}(u_0, t) := C\|u_0\|^2\sqrt{t} + C\|u_0\|^3 t^{1/4}$.

Proof For simplicity we will write u in place of u_ε. Let

$$\varphi(x) := \begin{cases} 0 & |x| < R_1, \\ \frac{|x|-R_1}{R_2-R_1} & R_1 \leq |x| \leq R_2, \\ 1 & |x| > R_2. \end{cases}$$

Taking the scalar product of the regularised equations (6.77) with the vector $-2\varphi(x)u(x,t)$, integrating in space and time and using $\operatorname{div} u = 0$ yields

$$2\int_0^t \int \varphi|\nabla u|^2 + \int \varphi|u(t)|^2$$

$$= \int \varphi|u_0|^2 - \int_0^t \int \left(2\partial_k\varphi\, u_i\, \partial_k u_i + 2\partial_i\varphi\, p\, u_i + \partial_k\varphi(J_\varepsilon u_k)|u|^2 \right).$$

$$(6.82)$$

Bounding below the second term on the left-hand side by $\int_{|x|>R_2} |u(t)|^2$

---

and using the nonnegativity of the first term yields

$$\int_{|x|>R_2} |u(t)|^2$$

$$\leq \int \varphi |u_0|^2 - \int_0^t \int \left(2\partial_k \varphi u_i \partial_k u_i + 2\partial_i \varphi \, p \, u_i + \partial_k \varphi (J_\varepsilon u_k) |u|^2 \right)$$

$$\leq \int_{|x|>R_1} |u_0|^2 + \frac{1}{R_2 - R_1} \int_0^t \left(2\|u\| \|\nabla u\| + 2\|u\| \|p\| + \|J_\varepsilon u\| \|u\|_4^2 \right)$$

$$\leq \int_{|x|>R_1} |u_0|^2 + \frac{\|u_0\|}{R_2 - R_1} \left(2\int_0^t \|\nabla u\| + 2\int_0^t \|p\| + \int_0^t \|u\|_4^2 \right).$$

Let us denote the last three integrals on the right hand side by I_1, I_2 and I_3 respectively. We have

$$I_1 \leq \sqrt{t} \left(\int_0^t \|\nabla u(s)\|^2 \, ds \right)^{1/2} \leq \frac{\sqrt{t}}{2} \|u_0\| \tag{6.83}$$

by the energy equality (6.80). To treat I_2 and I_3, observe that the Gagliardo-Nirenberg-Sobolev inequality[32] gives

$$\|u\|_4^4 \leq C \|\nabla u\|^3 \|u\|, \tag{6.84}$$

(see, for example, Theorem 6.1.1 in Giga, Giga, & Saal (2010)). Note that this estimate is independent of t (which we omitted in the notation). Moreover, the representation formula (6.79) for p together with Lemma 6.4 and the bound $\|J_\varepsilon u\|_4 \leq \|u\|_4$ (see Lemma 6.2 (i)) give

$$\|p\| \leq C \|\,|J_\varepsilon u|\,|u|\,\| \leq C \|u\|_4^2 \leq C \|\nabla u\|^{3/2} \|u\|^{1/2}. \tag{6.85}$$

Therefore $\|p\|$ and $\|u\|_4^2$ enjoy the same bound $C \|\nabla u\|^{3/2} \|u\|^{1/2}$. Thus, by the energy equality (6.80) and Hölder's inequality, we obtain

$$I_2, I_3 \leq C \int_0^t \|\nabla u(s)\|^{3/2} \|u(s)\|^{1/2} \, ds \leq C \|u_0\|^{1/2} \int_0^t \|\nabla u(s)\|^{3/2} \, ds$$

$$\leq C \|u_0\|^{1/2} t^{1/4} \left(\int_0^t \|\nabla u(s)\|^2 \, ds \right)^{3/4} \leq C \|u_0\|^2 t^{1/4}.$$

---

[32] Note that this inequality was not available in the 1930s. Instead, Leray proved (6.84) in an elementary way using (6.8), Lemma 6.3 and the Cauchy-Schwarz inequality, see the Notes at the end of this section for details.

Hence, finally

$$\int_{|x|>R_2} |u(t)|^2 \leq \int_{|x|>R_1} |u_0|^2 + \frac{\|u_0\|}{R_2 - R_1} (2I_1 + 2I_2 + I_3)$$

$$\leq \int_{|x|>R_1} |u_0|^2 + \frac{C}{R_2 - R_1} \left(t^{1/2}\|u_0\|^2 + t^{1/4}\|u_0\|^3 \right). \quad \square$$

Remark It is interesting to note that Leray (1934b) presented a way of deriving the bound (6.85) that does not use Lemma 6.4 (in fact Leray does not mention the use of the Fourier transform). See Appendix 6.5.7 for details.

6.4.2 Global existence of a weak solution

Here we study the limit as ε tends to zero of the solutions u_ε of the regularised equations (6.77) to obtain a global-in-time weak solution of the Navier–Stokes equations, as defined below.

In this section we use test functions that are divergence-free and not necessarily have compact support. That is we set

$$E := \{\varphi \in C^\infty(\mathbb{R}^3 \times [0,\infty); \mathbb{R}^3) : \operatorname{div} \varphi = 0$$

$$\text{and } \partial_t^m \nabla^k \varphi \in C([0,\infty); L^2) \cap C([0,\infty); L^\infty) \text{ for all } k, m \geq 0\}.$$

Definition 6.35 [33] A function u is a *weak solution of the Navier–Stokes equations* if there exists a set $S \subset (0,\infty)$ of measure zero such that $u(t)$ is weakly divergence free for all $t \in (0,\infty) \setminus S$,

$$\int u(t) \cdot \varphi(t) = \int u_0 \cdot \varphi(0) + \int_0^t \int u \cdot (\partial_t \varphi + \Delta \varphi) + \int_0^t \int u \cdot (u \cdot \nabla) \varphi \quad (6.86)$$

for all $t > 0$, $\varphi \in E$, and

$$\|u(t)\|^2 + 2 \int_s^t \|\nabla u(\tau)\|^2 \, d\tau \leq \|u(s)\|^2 \quad (6.87)$$

for every $s \in [0,\infty) \setminus S$ and every $t \geq s$.

Such a solution is often called a *Leray–Hopf weak solution* (Eberhard Hopf (1951) considered weak solutions in a similar sense on a bounded domain), while *weak solutions* are functions (belonging to the spaces $L^\infty((0,T); H) \cap L^2((0,T); V)$) satisfying the first part of Definition 6.35, but not necessarily the energy inequality, see, for example, Definitions 4.9 and 3.3 in Robinson et al. (2016) (see also Lemma 6.6 therein for the

---

[33] This is the definition in Section 31 of Leray (1934b).

equivalence of the spaces of test functions). The set S is often called the *set of singular times.*

Corollary 6.36   *A weak solution u satisfies*

$$u \in L^\infty((0,\infty); L^2) \cap L^2((0,\infty); H^1).$$

Moreover, it is L^2 weakly continuous in time and, for $s \in [0,\infty) \setminus S$, $u(t) \to u(s)$ in L^2 as $t \to s^+$.

Proof The first property is a consequence of the energy inequality (6.87), the L^2 weak continuity is a consequence of (6.86) and the boundedness of $\|u(t)\|$ in t, and the last property is a consequence of the weak continuity and the convergence of the norms $\|u(t)\| \to \|u(s)\|$ for $s \in [0,\infty) \setminus S$ (which follows from (6.87)). $\square$

Theorem 6.37 (Global existence of a weak solution[34])   *If $u_0 \in H$ then there exists a weak solution u of the Navier–Stokes equations such that $\|u(t) - u_0\| \to 0$ as $t \to 0^+$.*

Proof For each $\varepsilon > 0$ let u_ε be the unique strong solution of (6.77) with $u_\varepsilon(0) = J_\varepsilon u_0$. The existence of such u_ε is guaranteed by Theorem 6.33, which also gives

$$\|u_\varepsilon(t)\|^2 + 2 \int_0^t \|\nabla u_\varepsilon(s)\|^2 \, ds = \|J_\varepsilon u_0\|^2 \leq \|u_0\|^2 \tag{6.88}$$

for $t \geq 0$, $\varepsilon > 0$ (see (6.80); the last inequality is a property of the mollification operator, see Lemma 6.2 (i)). The weak solution is constructed in the following four steps.

Step 1. Construct the sequence $\{\varepsilon_n\}$.

Taking the scalar product of (6.77) with a test function $\varphi \in E$, integrating in space and time, and integrating by parts we obtain, for $t \geq 0$,

$$\int u_\varepsilon(t) \cdot \varphi(t)$$

$$= \int J_\varepsilon u_0 \cdot \varphi(0) + \int_0^t \int u_\varepsilon \cdot (\partial_t \varphi + \Delta \varphi) + \int_0^t \int u_\varepsilon \cdot (J_\varepsilon u_\varepsilon \cdot \nabla)\varphi. \tag{6.89}$$

Inequality (6.88) implies that, for each t, $\|u_\varepsilon(t)\|$ is bounded independently of ε, and so using a diagonal argument we extract a sequence

---

[34] This theorem corresponds to Sections 28-31 of Leray (1934b).

$\{\varepsilon_n\}$ such that for all $t \in \mathbb{Q}^+$, $\|u_{\varepsilon_n}(t)\| \to W(t)$ as $n \to \infty$, for some function $W \colon \mathbb{Q}^+ \to [0, \infty)$. Extend W to the whole of $\mathbb{R}_+$ by letting $W(t) := \liminf_{s \to t^-} W(s)$ for $t \in \mathbb{R}_+ \setminus \mathbb{Q}^+$. Using the following fact, we see that $\|u_{\varepsilon_n}(t)\| \to W(t)$ as $n \to \infty$ for times t at which W is continuous.

Theorem 6.38 (Helly's theorem[35])   *Let $g_n \in C([0,1])$ be a family of non-increasing functions such that $g_n(t) \to g(t)$ for all $t \in \mathbb{Q} \cap [0,1]$. Then $g_n(t) \to g(t)$ at each continuity point t of g.*

Since $\|u_{\varepsilon_n}(t)\|$ is a non-increasing function of t for each n (see the energy equality (6.80)), the same is true of the limit function $W(t)$. Therefore, since any non-increasing non-negative function has at most countably many points of discontinuity, we can apply a diagonal argument to account for such points and obtain

$$\|u_{\varepsilon_n}(t)\| \to W(t) \qquad \text{for } t \geq 0, \tag{6.90}$$

where we have also relabelled the sequence $\{\varepsilon_n\}$ and redefined W at its points of discontinuity. Note that since

$$u_{\varepsilon_n}(0) = J_{\varepsilon_n} u_0 \to u_0 \quad \text{in } L^2 \tag{6.91}$$

(see Lemma 6.2 (vi)) we obtain

$$W(0) = \lim_{n \to \infty} \|u_{\varepsilon_n}(0)\| = \lim_{n \to \infty} \|J_{\varepsilon_n} u_0\| = \|u_0\|.$$

We now want to take the limit of the functions themselves (rather than their norms). Since

$$\|J_{\varepsilon_n} u_{\varepsilon_n}(t)\| \leq \|u_{\varepsilon_n}(t)\| \leq \|u_0\|, \qquad \varepsilon_n > 0, t \geq 0 \tag{6.92}$$

(see Lemma 6.2 (i) and (6.88)) we can apply the diagonal argument once more (and relabel the sequence $\{\varepsilon_n\}$) to deduce that the quantities

$$\int_{t_1}^{t_2} \int_\Omega (u_{\varepsilon_n})_k \quad \text{and} \quad \int_{t_1}^{t_2} \int_\Omega (u_{\varepsilon_n})_k (J_{\varepsilon_n} u_{\varepsilon_n})_l \quad \text{converge as } n \to \infty \tag{6.93}$$

for all $t_1, t_2 \in \mathbb{Q}^+$, $k, l = 1, 2, 3$ and all cubes $\Omega \subset \mathbb{R}^3$ with rational vertices, that is, vertices whose coordinates are rational numbers. Moreover, from the timewise uniform continuity of the above integrals (that is from the bound $C(T)|t_2 - t_1|$ of the integrals whenever $t_1, t_2 \in (0, T)$ for some T) we obtain that in fact they converge as $n \to \infty$ for every pair $t_1, t_2 \geq 0$, every $k, l = 1, 2, 3$ and every cube Ω with rational vertices.

---

[35] This result is due to Helly (1912), see also Lemma 13.15 in Carothers (2000).

From the convergence in (6.93) we see that

$$\int_{t_1}^{t_2} \int u_{\varepsilon_n} \cdot (\partial_t \varphi + \Delta \varphi) \quad \text{and} \quad \int_{t_1}^{t_2} \int u_{\varepsilon_n} \cdot (J_{\varepsilon_n} u_{\varepsilon_n} \cdot \nabla) \varphi \quad (6.94)$$

converge as $n \to \infty$ for all $t_1, t_2 \geq 0$ and all test functions $\varphi \in E$. Indeed, fix $t_1, t_2 \geq 0$, $\varphi \in E$, and $\varepsilon > 0$. Without loss of generality we can assume $t_1 < t_2$. There exist $N \in \mathbb{N}$, $\{g_k\}_{k=1}^N \subset \mathbb{R}^3$, $\{\Omega_k\}_{k=1}^N$ (cubes with rational coordinates), and a family of intervals

$$\{(p_k, q_k): 0 \leq p_k < q_k \leq \infty, \ k = 1, \ldots, N\}$$

such that the (vector-valued) function

$$G_N(x, t) := \sum_{k=1}^N g_k \chi_{\Omega_k}(x) \chi_{(p_k, q_k)}(t),$$

satisfies

$$\|(\partial_t \varphi + \Delta \varphi) - G_N\|_{L^\infty((t_1, t_2); L^2)} \leq \frac{\varepsilon}{4\|u_0\|(t_2 - t_1)}. \quad (6.95)$$

Moreover, the first part of (6.93) gives that for sufficiently large n, m

$$\left| \int_{t_1}^{t_2} \int (u_{\varepsilon_n} - u_{\varepsilon_m}) \cdot G_N \right| \leq \frac{\varepsilon}{2}.$$

Thus

$$\left| \int_{t_1}^{t_2} \int (u_{\varepsilon_n} - u_{\varepsilon_m}) \cdot (\partial_t \varphi + \Delta \varphi) \right|$$

$$\leq \left| \int_{t_1}^{t_2} \int (u_{\varepsilon_n} - u_{\varepsilon_m}) \cdot G_N \right| + \int_{t_1}^{t_2} \int (|u_{\varepsilon_n}| + |u_{\varepsilon_m}|) |\partial_t \varphi + \Delta \varphi - G_N|$$

$$\leq \frac{\varepsilon}{2} + \frac{\varepsilon}{4\|u_0\|(t_2 - t_1)} \int_{t_1}^{t_2} (\|u_{\varepsilon_n}(s)\| + \|u_{\varepsilon_m}(s)\|) \, \mathrm{d}s$$

$$\leq \frac{\varepsilon}{2} + \frac{\varepsilon}{2\|u_0\|(t_2 - t_1)} \int_{t_1}^{t_2} \|u_0\| \, \mathrm{d}s = \varepsilon,$$

where we have also used the energy inequality (6.88). Thus we obtain the first part of (6.94). The second part follows in a similar way, by choosing a simple (matrix) function G_N such that

$$\|\nabla \varphi - G_N\|_{L^\infty((t_1, t_2); L^\infty)} \leq \frac{\varepsilon}{2\|u_0\|^2(t_2 - t_1)},$$

rather than (6.95). Therefore we obtain (6.94).

On the other hand, the convergence $J_{\varepsilon_n} u_0 \to u_0$ in L^2 (see (6.91)) implies, in particular, the convergence

$$\int J_{\varepsilon_n} u_0 \cdot \varphi(0) \to \int u_0 \cdot \varphi(0) \qquad \text{for all } \varphi \in E.$$

Combining this with (6.94) we can use the weak form of the regularised equations (6.89) to obtain that

$$\int u_{\varepsilon_n}(t) \cdot \varphi(t) \qquad \text{converges for all } \varphi \in E, t \geq 0.$$

Thus letting $\varphi(x,t) := \psi(x)$ for some divergence-free $\psi \in L^2$ such that $\psi \in H^m$ for all $m \geq 1$ and $\operatorname{div} \psi = 0$, we obtain that the numbers

$$\int u_{\varepsilon_n}(t) \cdot \psi \qquad \text{converge for all } t \geq 0, \text{ and } \psi. \tag{6.96}$$

This together with the fact that $\|u_{\varepsilon_n}(t)\| \leq \|u_0\|$ (see (6.88)) implies that [36]

$$u_{\varepsilon_n}(t) \rightharpoonup u(t) \qquad \text{as } n \to \infty \text{ in } L^2, \quad t \geq 0, \tag{6.97}$$

for some $u(t) \in H$. Note $u(0) = u_0$ by (6.91). We have thus constructed the sequence $\{\varepsilon_n\}$ and we obtained u as the weak limit of u_{ε_n}. Before showing that u is the required weak solution (Step 4), we first prove that in fact $u_{\varepsilon_n}(t) \to u(t)$ strongly in L^2 for almost all t (Step 3).

Step 2. Define the set of singular times S.

Fatou's lemma and (6.88) give

$$\int_0^\infty \liminf_{\varepsilon_n \to 0} \|\nabla u_{\varepsilon_n}(t)\|^2 \, dt \leq \liminf_{\varepsilon_n \to 0} \int_0^\infty \|\nabla u_{\varepsilon_n}(t)\|^2 \, dt \leq \frac{1}{2} \|u_0\|^2.$$

Hence $\liminf_{\varepsilon_n \to 0} \|\nabla u_{\varepsilon_n}(t)\|^2 < \infty$ for almost every $t > 0$. Defining the *set of singular times* as

$$S := \{t > 0 : \|\nabla u_{\varepsilon_n}(t)\| \to \infty \text{ as } n \to \infty\} \tag{6.98}$$

we see that $|S| = 0$.

Step 3. Show that $u_{\varepsilon_n}(t) \to u(t)$ strongly in L^2 for $t \in (0,\infty) \setminus S$.

---

[36] This functional analytical fact is one of Leray's remarkable contributions, which he discusses on page 209. To see it, note that if (6.97) does not hold then (using the boundedness in L^2) one can extract subsequences $\{n_k\}$, $\{m_k\}$ such that $u_{\varepsilon_{n_k}}(t) \rightharpoonup v(t)$, $u_{\varepsilon_{m_k}}(t) \rightharpoonup w(t)$ for some weakly divergence-free $v(t), w(t) \in L^2$, $v(t) \neq w(t)$. Thus (6.96) gives $\int (u_{\varepsilon_{n_k}}(t) - u_{\varepsilon_{m_k}}(t))\psi \to 0$ for all ψ, and so $\int (v(t) - w(t))\psi = 0$. Taking $\psi := J_\delta(v(t) - w(t))$ gives $\int (v(t) - w(t))J_\delta(v(t) - w(t)) = 0$, which in the limit $\delta \to 0^+$ implies $\|v(t) - w(t)\| = 0$, a contradiction.

Recalling that $W(t) = \lim_{\varepsilon_n \to 0} \|u_{\varepsilon_n}(t)\|$ (see (6.90)) we see that it is enough to show that

$$\|u(t)\| = W(t), \quad t \in (0, \infty) \setminus S, \tag{6.99}$$

since weak convergence (6.97) together with the convergence of the norms is equivalent to strong convergence. The estimate $\|u(t)\| \le W(t)$ holds for all $t \ge 0$ by the property of weak limits,

$$\|u(t)\| \le \liminf_{\varepsilon_n \to 0} \|u_{\varepsilon_n}(t)\| = W(t). \tag{6.100}$$

In order to prove the converse inequality, fix $t \in (0, \infty) \setminus S$. For such t let $\{\varepsilon_{n_k}\}$ be a subsequence along which $\liminf_{\varepsilon_n \to 0} \|\nabla u_{\varepsilon_n}(t)\|^2$ is attained, that is

$$\lim_{k \to \infty} \|\nabla u_{\varepsilon_{n_k}}(t)\| = \liminf_{\varepsilon_n \to 0} \|\nabla u_{\varepsilon_n}(t)\|. \tag{6.101}$$

It follows that the sequence $\{\nabla u_{\varepsilon_{n_k}}(t)\}$ is bounded in L^2 (as $t \notin S$) and therefore

$$\nabla u_{\varepsilon_{n_k}}(t) \rightharpoonup \nabla u(t) \text{ in } L^2 \tag{6.102}$$

on a subsequence (which we relabel back to ε_{n_k}; note on such a subsequence (6.101) still holds). The limit function is $\nabla u(t)$ by the definition of weak derivatives, and

$$\|\nabla u(t)\| \le \liminf_{\varepsilon_n \to 0} \|\nabla u_{\varepsilon_n}(t)\| \qquad \text{for } t \in (0, \infty) \setminus S. \tag{6.103}$$

At this point we need to apply the separation of energy result (Lemma 6.34). We fix $\eta > 0$, and we let $R_1(\eta) > 0$ be large enough that

$$\int_{|x| > R_1(\eta)} |u_0|^2 \le \frac{\eta}{2},$$

and set

$$R_2(\eta, t) := R_1(\eta) + \frac{4}{\eta} \bar{C}(u_0, t),$$

where $\bar{C}(u_0, t)$ is the constant from Lemma 6.34, which implies that

$$\int_{|x| > R_2(\eta, t)} |u_\varepsilon(t)|^2 \le \eta, \qquad \varepsilon > 0. \tag{6.104}$$

Since $u_{\varepsilon_{n_k}}(t) \rightharpoonup u(t)$ in H^1 (see (6.97) and (6.102)) we have

$$u_{\varepsilon_{n_k}}(t) \rightharpoonup u(t) \quad \text{in } H^1(B(R_2(\eta, t))),$$

and so the compact embedding[37] $H^1(B(R_2(\eta,t))) \hookrightarrow L^2(B(R_2(\eta,t)))$ implies

$$u_{\varepsilon_{n_k}}(t) \to u(t) \quad \text{in } L^2(B(R_2(\eta,t))).$$

Therefore, from (6.104), we obtain

$$\limsup_{\varepsilon_{n_k} \to 0} \|u_{\varepsilon_{n_k}}(t)\|^2$$

$$\leq \limsup_{\varepsilon_{n_k} \to 0} \int_{|x| \leq R_2(\eta,t)} |u_{\varepsilon_{n_k}}(t)|^2 + \limsup_{\varepsilon_{n_k} \to 0} \int_{|x| > R_2(\eta,t)} |u_{\varepsilon_{n_k}}(t)|^2$$

$$\leq \int_{|x| \leq R_2(\eta,t)} |u(t)|^2 + \eta \leq \|u(t)\|^2 + \eta, \qquad \eta > 0.$$

And hence, taking $\eta \to 0$,

$$W(t) = \limsup_{\varepsilon_{n_k} \to 0} \|u_{\varepsilon_{n_k}}(t)\| \leq \|u(t)\|.$$

Thus we obtained the $\geq$ inequality in (6.99), as required.

Step 4. Verify that u is a weak solution.

We already established (after (6.97)) that $u(t)$ is weakly divergence free. Step 3 gives

$$\int u_{\varepsilon_n}(t) \cdot (\partial_t \varphi(t) + \Delta \varphi(t)) \to \int u(t) \cdot (\partial_t \varphi(t) + \Delta \varphi(t)),$$

$$\int u_{\varepsilon_n}(t) \cdot (J_{\varepsilon_n} u_{\varepsilon_n}(t) \cdot \nabla) \varphi(t) \to \int u(t) \cdot (u(t) \cdot \nabla) \varphi(t)$$

for almost every $t > 0$. Thus, using (6.92), we obtain (6.86) by taking the limit $\varepsilon_n \to 0^+$ in (6.89) and applying the Dominated Convergence Theorem to the time integrals. As for the energy inequality (6.87), let $s \in (0, \infty) \setminus S$ and $t > s$. Since $\|\nabla u(\tau)\| \leq \liminf_{\varepsilon_n \to 0} \|\nabla u_{\varepsilon_n}(\tau)\|$ for almost every $\tau \geq 0$ (see (6.103)), we can use (6.100), Fatou's lemma, the identity

$$\|u_\varepsilon(t)\|^2 + 2 \int_s^t \|\nabla u_\varepsilon(\tau)\|^2 \, \mathrm{d}\tau = \|u_\varepsilon(s)\|^2, \qquad \varepsilon > 0$$

(see (6.88)) and the fact that $\lim_{\varepsilon_n \to 0} \|u_{\varepsilon_n}(s)\| = \|u(s)\|$ (see (6.99)) to

---

[37] Leray's Lemma 2 provides a similar compactness result.

obtain

$$\|u(t)\|^2 + 2 \int_s^t \|\nabla u(\tau)\|^2 \, \mathrm{d}\tau$$

$$\leq \liminf_{\varepsilon_n \to 0} \|u_{\varepsilon_n}(t)\|^2 + 2 \int_s^t \liminf_{\varepsilon_n \to 0} \|\nabla u_{\varepsilon_n}(\tau)\|^2 \, \mathrm{d}\tau$$

$$\leq \liminf_{\varepsilon_n \to 0} \left(\|u_{\varepsilon_n}(t)\|^2 + 2 \int_0^t \|\nabla u_{\varepsilon_n}(\tau)\|^2 \, \mathrm{d}\tau \right)$$

$$= \liminf_{\varepsilon_n \to 0} \|u_{\varepsilon_n}(s)\|^2 = \|u(s)\|^2.$$

The case $s = 0$ follows similarly by using (6.91) (rather than (6.99)) in the last equality (recall also $u(0) = u_0$). Finally $\|u(t) - u_0\| \to 0$ as $t \to 0$ since $u(t) \rightharpoonup u_0$ in L^2 (a consequence of (6.86)) and $\|u(t)\| \to \|u_0\|$ (a consequence of Fatou's lemma and the energy inequality (6.87) with $s = 0$). $\qquad\square$

6.4.3 Structure of the weak solution

Let $u_0 \in H$ and consider the weak solution u given by Theorem 6.37. We first show that such solution enjoys the following *weak-strong uniqueness* property[38].

Lemma 6.39 (Weak-strong uniqueness) *If $\|\nabla u(t_0)\| < \infty$ for some $t_0 \geq 0$ then $u = v$ on the time interval $[t_0, T)$ for some $T > t_0$, where v is the semi-strong solution corresponding to the initial data $u(t_0)$ (recall Definition 6.28).*

Proof The assumption gives that $u(t_0) \in V$ (that is $u(t_0) \in H^1$ and $u(t_0)$ is divergence free, recall (6.2)) and so Theorem 6.30 gives the existence of a unique semi-strong solution $v(t)$ on $[t_0, T)$, for some $T > t_0$. We will show that $u = v$ on time interval $[t_0, T)$.

From the energy equality for v (see Lemma 6.29) and the energy inequality for u (see (6.87)), we obtain for a.e. $t_1 \in (t_0, T)$ and every $t \in (t_1, T)$

$$\|v(t)\|^2 + 2 \int_{t_1}^t \|\nabla v(s)\|^2 \, \mathrm{d}s = \|v(t_1)\|^2,$$

$$\|u(t)\|^2 + 2 \int_{t_1}^t \|\nabla u(s)\|^2 \, \mathrm{d}s \leq \|u(t_1)\|^2.$$

---

[38] Leray calls this *Comparison of a regular solution and a turbulent solution*, see pp. 242–244.

Adding these together and letting $w := u - v$ gives

$$\|u(t_1)\|^2 + \|v(t_1)\|^2 \geq \|w(t)\|^2 + 2 \int_{t_1}^{t} \|\nabla w(s)\|^2 \, ds + 2 \int u(t) \cdot v(t)$$

$$+ 4 \int_{t_1}^{t} \int \nabla u : \nabla v,$$

where $A : B := A_{ij} B_{ij}$ denotes the inner product of matrices. Since v is a strong solution on (t_0, T) (see Definition 6.28), it satisfies the equation $\partial_t v - \Delta v + (v \cdot \nabla) v + \nabla p = 0$ in $(t_0, T) \times \mathbb{R}^3$ and it can be used as a test function for u on the time interval (t_1, t) (recall (6.86) and the definition of E) to write

$$\int u(t_1) \cdot v(t_1) = \int u(t) \cdot v(t) - \int_{t_1}^{t} \int u \cdot (\partial_t v + \Delta v + (u \cdot \nabla) v)$$

$$= \int u(t) \cdot v(t) - \int_{t_1}^{t} \int u \cdot (2 \Delta v + (w \cdot \nabla) v - \nabla p)$$

$$= \int u(t) \cdot v(t) + \int_{t_1}^{t} \int v \cdot (w \cdot \nabla) w + 2 \int_{t_1}^{t} \int \nabla u : \nabla v,$$

where in the last step we integrated by parts all three terms under the last integral and used the facts that $u(t)$ and $v(t)$ are weakly divergence free and that $\int v \cdot (w \cdot \nabla) v = 0$ (cf. (6.50)). Hence, using the inequality above and the Cauchy-Schwarz inequality we obtain

$$\|w(t_1)\|^2 = \|u(t_1)\|^2 + \|v(t_1)\|^2 - 2 \int u(t_1) \cdot v(t_1)$$

$$\geq \|w(t)\|^2 + 2 \int_{t_0}^{t} \|\nabla w(s)\|^2 \, ds - 2 \int_{t_1}^{t} \int v \cdot (w \cdot \nabla) w$$

$$\geq \|w(t)\|^2 + 2 \int_{t_1}^{t} \|\nabla w(s)\|^2 \, ds - 2 \int_{t_1}^{t} \|v(s)\|_{\infty} \|w(s)\| \, \|\nabla w(s)\| \, ds$$

$$\geq \|w(t)\|^2 - \frac{1}{2} \int_{t_1}^{t} \|v(s)\|_{\infty}^2 \|w(s)\|^2,$$

where in the last step we used Young's inequality $ab \leq a^2 + b^2/4$. Hence Gronwall's inequality implies

$$\|w(t)\|^2 \leq \|w(t_1)\|^2 \exp\left(\frac{1}{2} \int_{t_1}^{t} \|v(s)\|_{\infty}^2 \, ds \right).$$

Since $\int_{t_0}^{t} \|v(s)\|_{\infty} < \infty$ (see Definition 6.28) and both u and v are continuous in L^2 as $t_1 \to t_0^+$ (since $t_0 \notin S$, see Corollary 6.36) and therefore we can take the limit $t_1 \to t_0^+$ to obtain

$$\|w(t)\|^2 \le \|w(t_0)\|^2 \exp\left(\frac{1}{2}\int_{t_0}^t \|v(s)\|_\infty^2 \, ds\right) = 0 \qquad t \in (t_0, T).$$

Thus $u(t) = v(t)$ for $t \in [t_0, T)$, as required. $\qquad\square$

We can use the weak-strong uniqueness to obtain that u, the weak solution given by Theorem 6.37, is regular in certain time intervals.

Definition 6.40   We call an open interval $(a, b) \subset (0, \infty)$ a *maximal interval of regularity* if u is a strong solution on (a, b) (see Definition 6.14) and u is not a strong solution on any open interval I strictly containing (a, b).

Theorem 6.41 (The structure of the weak solution[39])   *If u is a weak solution given by Theorem 6.37 then there exists a family of pairwise disjoint maximal intervals of regularity $(a_i, b_i) \subset (0, \infty)$ of u such that the set*

$$\Sigma := (0, \infty) \setminus \bigcup_i (a_i, b_i)$$

has measure zero.

Clearly $S \subset (0, \infty) \setminus \bigcup_i (a_i, b_i)$ (where S is the set of singular times of u, see Definition 6.35), since a strong solution is divergence free for all times and satisfies the energy equality (6.49). Therefore this theorem asserts that any $t_0 \in (0, \infty) \setminus S$ is an initial point of an interval in the interior of which u coincides with a strong solution, and the coincidence continues as long as the strong solution exists. Moreover, the energy inequality (6.87) shows that $\|u(t)\|$ is strictly decreasing on every interval of regularity (a_i, b_i). Note also that the family $\{(a_i, b_i)\}_i$ is at most countable (as a family of pairwise disjoint open intervals on $\mathbb{R}$).

Proof   Since for the weak solution u constructed in the last theorem the set S is given by (6.98), we see that if $t_0 \notin S$ then $\|\nabla u(t_0)\| < \infty$ and so Lemma 6.39 gives that t_0 either belongs to a maximal interval of regularity (see Definition 6.40) or is a left endpoint of one such interval. That is $t_0 \in \bigcup_i [a_i, b_i)$, and so

$$\Sigma \subset S \cup \bigcup_i \{a_i\}.$$

[39] This theorem corresponds to Sections 32 and 33 in Leray (1934b).

Since there are at most countably many maximal intervals of regularity we obtain $|\Sigma| = 0$, as required. $\qquad\square$

Furthermore, it turns out that one of the intervals of regularity (a_i, b_i) of the weak solution u contains $(C\|u_0\|^4, \infty)$ and on this interval u enjoys a certain decay, which we make precise in the following theorem.

Theorem 6.42 (Supplementary information[40]) *Under the assumptions of the last theorem the set of singular times Σ is bounded above by $C\|u_0\|^4$, and for $t > C\|u_0\|^4$*

$$\|\nabla u(t)\| < C\|u_0\|t^{-1/2}, \tag{6.105}$$

$$\|u(t)\|_\infty < C\|u_0\|t^{-3/4}. \tag{6.106}$$

Moreover

$$\sum_{i:\, b_i < \infty} \sqrt{b_i - a_i} \le C\|u_0\|^2. \tag{6.107}$$

Proof If the union $\bigcup_i (a_i, b_i)$ has only one element $(0, \infty)$ then the boundedness of Σ is trivial. If not then let i be such that $b_i < \infty$ and let $t \in (a_j, b_j)$, where j is such that $b_j \le b_i$ (that is (a_i, b_i) does not precede (a_j, b_j) on the line). Since u becomes singular at b_j Corollary 6.25 gives

$$\|\nabla u(t)\| \ge C(b_j - t)^{-1/4} \ge C(b_i - t)^{-1/4}. \tag{6.108}$$

Applying this lower bound in the energy inequality (6.87) gives

$$\|u_0\|^2 \ge 2\int_0^{b_i} \|\nabla u(t)\|^2 \, dt \ge C\int_0^{b_i} (b_i - t)^{-1/2} \, ds = C(b_i)^{1/2}.$$

Thus all finite b_i's are bounded by $C\|u_0\|^4$, hence Σ is bounded. As for (6.107), apply the first inequality of (6.108) to the energy inequality (6.87) to obtain

$$\|u_0\|^2 \ge 2\sum_{j:\, b_j < \infty} \int_{a_j}^{b_j} \|\nabla u(t)\|^2 \, dt$$

$$\ge C\sum_{j:\, b_j < \infty} \int_{a_j}^{b_j} (b_j - t)^{-1/2} \, dt = C\sum_{j:\, b_j < \infty} \sqrt{b_j - a_j},$$

as required. It remains to show the decay estimates (6.105), (6.106). Let

[40] This is Section 34 in Leray (1934b).

$s \in \bigcup_i (a_i, b_i)$ and $t > s$. Since u is strong on an interval containing s, we can use Corollary 6.27 to write that if $t - s \leq C' \|\nabla u(s)\|^{-4}$ then

$$\|\nabla u(t)\| \leq C \|\nabla u(s)\|, \tag{6.109}$$

$$\|u(t)\|_\infty \leq C \|\nabla u(s)\|(t - s)^{-1/4}. \tag{6.110}$$

From the first of these two inequalities we see that either $t - s \geq C \|\nabla u(s)\|^{-4}$ or $\|\nabla u(t)\| \leq C \|\nabla u(s)\|$, and so

$$\|\nabla u(s)\| \geq C \min\left((t - s)^{-1/4}, \|\nabla u(t)\|\right).$$

Using this lower bound in the energy inequality (6.87) we obtain

$$\|u_0\|^2 \geq 2C \int_0^t \min\left((t - s)^{-1/2}, \|\nabla u(t)\|^2\right) ds$$

$$\geq C'' \int_0^t \min\left(t^{-1/2}, \|\nabla u(t)\|^2\right) ds$$

$$= C'' \min\left(t^{1/2}, t\|\nabla u(t)\|^2\right).$$

Therefore, since the first argument of the minimum tends to ∞ as $t \to \infty$, we see that for $t^{1/2} > \|u_0\|^2/C''$ we necessarily must have $\min\left(t^{1/2}, t\|\nabla u(t)\|^2\right) = t\|\nabla u(t)\|^2$ and hence

$$\|u_0\|^2 \geq C'' t \|\nabla u(t)\|^2, \tag{6.111}$$

from which (6.105) follows.

To obtain (6.106), fix $\alpha > 1/C''$ such that $2\alpha^2 C'(C'')^2 > 1$, let $s \geq \alpha^2 \|u_0\|^4$ and $t := 2s$. Then $t^{1/2} > s^{1/2} \geq \|u_0\|^2/C''$ and so

$$t - s = t/2 = \frac{1}{2}(t^{-1/4} t^{1/2})^4 \leq \frac{1}{2\alpha^2}(\|u_0\|^{-1} t^{1/2})^4$$

$$\leq \frac{1}{2\alpha^2 (C'')^2} \|\nabla u(t)\|^{-4} \leq C' \|\nabla u(t)\|^{-4},$$

where we also used (6.111) and our choice of α. Therefore we may apply (6.110) to obtain $\|u(t)\|_\infty \leq C \|\nabla u(s)\| t^{-1/4}$. Hence, since $s^{1/2} > \|u_0\|^2/C''$ we can apply (6.111) again to obtain

$$\|u(t)\|_\infty \leq C \|\nabla u(s)\| t^{-1/4} \leq C \|u_0\| t^{-3/4},$$

which gives (6.106). $\qquad \square$

Finally, we remark on a consequence of (6.107).

Corollary 6.43 *The set of singular times Σ satisfies*

$$d_H(\Sigma) \leq d_B(\Sigma) \leq 1/2.$$

Here d_H denotes the Hausdorff dimension and d_B denotes the upper box-counting dimension (see Sections 2.1 and 3.2 in Falconer (2014) for the definitions). This corollary does not appear in Leray's paper. Here we show it following the proof of a more general result by Lapidus & van Frankenhuijsen (2006) (see Theorem 1.10 therein); see also Besicovitch & Taylor (1954). An alternative approach can be found in the proof of Theorem 8.13 of Robinson et al. (2016).

Proof Note that due to the general inequality $d_H(K) \leq d_B(K)$ for any compact $K \subset \mathbb{R}$ (see Lemma 3.7 in Falconer (2014)) it is enough to show the bound $d_B(\Sigma) \leq 1/2$. In this context the upper box-counting dimension is equivalent to the Minkowski–Bouligand dimension, which (for subsets of $\mathbb{R}$) is given by

$$d_B(\Sigma) = 1 - \liminf_{\delta \to 0^+} \frac{\log |\Sigma_\delta|}{\log \delta}, \tag{6.112}$$

where Σ_δ denotes the δ-neighbourhood of Σ (see, for instance, Proposition 2.4 in Falconer (2014)). Since Σ is bounded there exists a unique index i_0 such that $b_{i_0} = \infty$, and so the set $\bigcup_{i \neq i_0}(a_i, b_i)$ is bounded by a_{i_0}. Thus

$$\Sigma = (0, a_{i_0}] \setminus \bigcup_i (a_i, b_i),$$

where now each b_i is finite. Thus we can renumber the intervals (a_i, b_i) such that the length of (a_i, b_i) does not increase with i, that is $b_i - a_i \geq b_{i+1} - a_{i+1}$ for all i.

For $\delta > 0$ let N_δ be such that

$$\begin{cases} b_i - a_i < 2\delta & i > N_\delta, \\ b_i - a_i \geq 2\delta & i \leq N_\delta. \end{cases}$$

Observe that

$$\Sigma_\delta = \Sigma \cup \left(\bigcup_{i=1}^{N_\delta} (a_i, a_i + \delta) \cup (b_i - \delta, b_i) \right) \cup \bigcup_{i > N_\delta} (a_i, b_i).$$

Thus, since $|\Sigma| = 0$,

$$|\Sigma_\delta| \leq 2\delta N_\delta + \sum_{i > N_\delta} (b_i - a_i) \leq 2\delta N_\delta + \sqrt{2\delta} \sum_{i > N_\delta} \sqrt{b_i - a_i} \leq 2\delta N_\delta + C\sqrt{\delta},$$

$$\tag{6.113}$$

where we used the assumption $\sum_i \sqrt{b_i - a_i} \leq C$ (see the last inequality in Theorem 6.42). Now since for each k

$$k\sqrt{b_k - a_k} = \sum_{j=1}^{k} \sqrt{b_k - a_k} \leq \sum_{j=1}^{k} \sqrt{b_j - a_j} \leq C,$$

we see that $b_k - a_k < 2\delta$ for $k > C/\sqrt{2\delta}$. In other words $N_\delta \leq C/\sqrt{\delta}$. Hence (6.113) gives $|\Sigma_\delta| \leq C\sqrt{\delta}$ and so (since $\log \delta < 0$ for small $\delta > 0$)

$$\frac{\log |\Sigma_\delta|}{\log \delta} \geq \frac{\log C + \frac{1}{2}\log \delta}{\log \delta} \to \frac{1}{2}$$

as $\delta \to 0^+$. Therefore $d_B(\Sigma) \leq 1/2$, as required.   □

It is interesting to note that apart from the bound $d_H(\Sigma) \leq 1/2$ a stronger property $\mathcal{H}_{1/2}(\Sigma) = 0$ holds, where $\mathcal{H}_{1/2}$ denotes the 1/2-dimensional Hausdorff measure. This can be shown by an (earlier) argument due to Scheffer (1976), which is similar to the above and is also presented by Robinson (2006), see Proposition 8 therein.

Notes

This section corresponds to Chapters V and VI of Leray (1934b), except that Leray did not discuss the dimension of the set of singular times. In fact the notion of the box-counting dimension was not properly unified in the 1930's, although some variants were already being studied at the time (see Bouligand (1928) and Pontrjagin & Schnirelmann (1932) for instance). On the other hand, the notion of the Hausdorff dimension was already quite well developed (see Hausdorff (1918), Besicovitch (1935)), but it is not apparent whether or not Leray was aware of these developments. In any case, it was Scheffer (1976) who was the first to study the dimension of the set of singular times for the Navier–Stokes equations.

Finally, we present Leray's proof of (6.84) (which he discusses on p. 234). Since $|u|^2$ solves (trivially) the Poisson equation $\Delta |u|^2 = \Delta |u|^2$, $|u|^2 = (-\Delta)^{-1}(-\Delta |u|^2)$. Indeed $|u|^2$ is the unique solution of this equation in the space $L^1 \cap L^\infty$ by Liouville's theorem. Hence, in the definition (6.8) of the symbol $(-\Delta)^{-1}$ (which is valid since $\nabla^m |u|^2 \in L^1 \cap L^\infty$ for all $m \geq 0$), we can integrate by parts to obtain

$$|u(x)|^2 = \frac{-1}{4\pi} \int \frac{1}{|x-y|} \Delta |u(y)|^2 \, dy = \frac{1}{4\pi} \int \frac{x-y}{|x-y|^3} \cdot \nabla |u(y)|^2 \, dy.$$

Thus

$$\|u\|_4^4 = \frac{1}{4\pi} \int \int |u(x)|^2 \frac{x-y}{|x-y|^3} \cdot \nabla |u(y)|^2 \, dx \, dy$$

$$\leq \frac{1}{2\pi} \int \int \frac{|u(x)|^2}{|x-y|^2} |u(y)| \, |\nabla u(y)| \, dx \, dy$$

$$\leq C\|\nabla u\|^2 \int |u(y)| \, |\nabla u(y)| \, dy \leq C\|\nabla u\|^3 \|u\|,$$

where we used Lemma 6.3 and the Cauchy-Schwarz inequality.

Acknowledgements

We would like to thank Robert Terrell for his English translation of Leray's original work, and Simon Baker, Tobias Barker, Antoine Choffrut and Kenneth Falconer for their helpful comments. We are particularly grateful to James Robinson and José Rodrigo for their encouragement and interest in this work.

WSO is supported by EPSRC as part of the MASDOC DTC at the University of Warwick, Grant No. EP/HO23364/1.

BCP was partially supported by an EPSRC Doctoral Training Award and partially by postdoctoral funding from ERC 616797.

6.5 Appendix

6.5.1 The heat equation and the heat kernel

Let $v_0 \in L^2$ and $v(t) := \Phi(t) * v_0$, where $\Phi = (4\pi t)^{-3/2} e^{-|x|^2/4t}$ is the heat kernel.

If v_0 is a vector-valued function which is weakly divergence free then $\operatorname{div} v(t) = 0$ for each $t > 0$ (which can be shown directly, see, for example, Proposition 4.1.4 (II) in Giga et al. (2010)). Furthermore for $m \geq 1$, v satisfies

(i) $v \in C([0,T); L^2)$ with $v(0) = v_0$, and $\|v(t)\| \leq \|v_0\|$;
(ii) $\nabla^m v \in C((0,T); L^2)$ and

$$\|\nabla^m v(t)\| \leq C_m \|v_0\| t^{-m/2}.$$

Moreover if $\nabla^m v_0 \in L^2$ then we have $\nabla^m v \in C([0,T); L^2)$ with $\|\nabla^m v(t)\| \leq C_m \min_{k \leq m} \left\{ \|\nabla^k v_0\| t^{-(m-k)/2} \right\};$

(iii) $v \in C((0,T); L^\infty)$ and $\|v(t)\|_\infty \leq C \min\{\|v_0\| t^{-3/4}, \|v_0\|_\infty\}$;

(iv) $\nabla^m v \in C((0,T); L^\infty)$ and

$$\|\nabla^m v\|_\infty \leq C_m \min\{\|v_0\| t^{-m/2-3/4}, \|v_0\|_\infty t^{-m/2}\};$$

(v) $v, \nabla^m v \in \mathcal{H}^{1/2}((0,T))$ with the corresponding constants satisfying $C_0(t) \leq c_0 \|v_0\|/t$, $C_m(t) \leq c_m \|v_0\| t^{-1-m/2}$.

The inequalities in (i)-(iv) follow directly from Young's inequality (6.6) and the bounds $\|\nabla^m \Phi(t)\|_1 \leq C_m t^{-m/2}$, $\|\nabla^m \Phi(t)\| \leq C_m t^{-m/2-3/4}$, $m \geq 0$ (which can be verified by a direct calculation). The claims regarding continuity in $(0,T)$ follow from Young's inequality (6.6) and the fact that Φ and all its spatial derivatives $\nabla^m \Phi$, $m \geq 1$, belong to $C((0,T); L^1) \cap C((0,T); L^2)$ (a consequence of pointwise continuity of Φ in time and the Dominated Convergence Theorem). The claims regarding continuity at $t = 0$ in (i) and (ii) are known as the *approximation of identity*, see e.g. Theorem 1.18 in Chapter 1 of Stein & Weiss (1971) for a proof. Finally, property (v) follows from Morrey's inequality (see, for example, Section 4.5.3 in Evans & Gariepy, 2015) and Young's inequality for convolutions (6.6) by writing

$$\|\Phi(t) * v_0\|_{C^{0,1/2}} \leq C\|\nabla \Phi(t) * v_0\|_6 \leq C\|\nabla \Phi(t)\|_{3/2}\|v_0\| = C\|v_0\|/t.$$
$$(6.114)$$

Recall that we denote by $\|\cdot\|_{C^{0,1/2}}$ the $1/2$-Hölder seminorm. The claim for the derivatives follows similarly by noting that

$$\|\nabla^{m+1} \Phi(t)\|_{3/2} = C_m t^{-1-m/2}.$$

The issue of pointwise convergence as $t \to 0^+$ is addressed in the following lemma.

Lemma 6.44 (pointwise convergence as $t \to 0^+$ of the heat flow) *If $x_0 \in \mathbb{R}^3$ is a continuity point of v_0 then*

$$v(x_0, t) \to v_0(x_0) \qquad \text{as } t \to 0^+.$$

If v_0 is bounded and uniformly continuous then

$$\|v(t) - v_0\|_\infty \to 0 \qquad \text{as } t \to 0^+.$$

See Section 4.2 in Giga et al. (2010) for a proof.

One can verify that $v(t)$ is a solution of the heat equation $v_t = \Delta v$ in $\mathbb{R}^3 \times (0, \infty)$ by a direct calculation. It is the unique such solution in the class of functions $C([0, \infty); L^2)$ satisfying $v(0) = v_0$ for some $v_0 \in L^2$, which we make precise in the following lemma.

Lemma 6.45   *Let $w \in C([0,T]; L^2)$ be a weak solution of the heat equation with $w(0) = 0$, that is*

$$\int_0^T \int (\varphi_t + \Delta\varphi) w \, dx \, dt = 0 \qquad (6.115)$$

for all $\varphi \in C_0^\infty(\mathbb{R}^3 \times [0,T))$. Then $w \equiv 0$.

Proof   We modify the argument from Section 4.4.2 of Giga et al. (2010). We focus on the case $T < \infty$ (the case $T = \infty$ follows trivially by applying the result for all $T > 0$).

We first show that assumption $w \in C([0,T]; L^2)$ implies that (6.115) holds also for all $\varphi \in C^\infty(\mathbb{R}^3 \times [0,T))$ such that

$$\begin{cases} \sup_{t\in[0,T)} \left(\|\partial_t\varphi(t)\| + \|D^\alpha\varphi(t)\| \right) \leq C & \text{for some } C > 0, \\ \operatorname{supp}\varphi \subset \mathbb{R}^3 \times [0,T') & \text{for some } T' < T, \end{cases} \qquad (6.116)$$

where $|\alpha| \leq 2$. Indeed, let $\theta \in C^\infty(\mathbb{R}; [0,1])$ be such that $\theta(\tau) = 1$ for $\tau \leq 1$ and $\theta(\tau) = 0$ for $\tau \geq 2$ (take for instance $\theta(\tau) := q(2 - \tau)/(q(2 - \tau) + q(1 - \tau))$, where $q(s) := e^{-1/s}$ for $s > 0$ and $q(s) := 0$ otherwise). For $x \in \mathbb{R}^3$ let $\theta_j(x) := \theta(|x|/j)$. Then $\theta_j \in C_0^\infty$ and $\|\nabla\theta_j\|_\infty \leq C/j$ and $\|D^2\theta_j\|_\infty \leq C/j^2$, as well as

$$\theta_j(x) = \begin{cases} 1 & |x| < j, \\ 0 & |x| > 2j. \end{cases}$$

Now for $\varphi \in C^\infty(\mathbb{R}^3 \times [0,T))$ satisfying (6.116), let $T' < T$ be such that $\varphi(t) \equiv 0$ for $t \geq T'$ and let

$$\varphi_j(x,t) := \theta_j(x)\varphi(x,t).$$

Then $\varphi_j \in C_0^\infty(\mathbb{R}^3 \times [0,T))$ and so (6.115) gives

$$\int_0^T \int (\partial_t\varphi_j + \Delta\varphi_j) w \, dx \, dt = 0,$$

or equivalently

$$\int_0^T \int \theta_j(\partial_t\varphi + \Delta\varphi) w \, dx \, dt + \int_0^T \int w \left(2\nabla\theta_j \cdot \nabla\varphi + \varphi\Delta\theta_j \right) dx \, dt = 0. \qquad (6.117)$$

Since θ_j converges pointwise to 1 and

$$\int_0^T \int |(\partial_t\varphi + \Delta\varphi) w| \, dx \, dt \leq \int_0^{T'} \|\partial_t\varphi(t) + \Delta\varphi(t)\| \|w(t)\| \, dt$$

$$\leq T'C \sup_{t\in[0,T']} \|w(t)\| < \infty$$

the Dominated Convergence Theorem gives that the first integral on the left-hand side of (6.117) converges to $\int_0^T \int (\partial_t \varphi + \Delta \varphi) w \, \mathrm{d}x \, \mathrm{d}t$. The second integral is bounded by

$$\int_0^T \int |w| \, |2\nabla\theta_j \cdot \nabla\varphi + \varphi\Delta\theta_j| \, \mathrm{d}x \, \mathrm{d}t$$

$$\leq T' \sup_{t\in[0,T']} \|w(t)\| (2C/j + C/j^2) \xrightarrow[j\to\infty]{} 0.$$

Therefore taking the limit $j \to \infty$ in (6.117) gives

$$\int_0^T \int (\partial_t \varphi + \Delta \varphi) w \, \mathrm{d}x \, \mathrm{d}t = 0$$

as required.

We will show that

$$\int_0^T \int \Psi w \, \mathrm{d}x \, \mathrm{d}t = 0$$

for all $\Psi \in C_0^\infty(\mathbb{R}^3 \times (0,T))$. This finishes the proof as C_0^∞ is dense in L^2 and so $\int_0^T \int vw \, \mathrm{d}x \, \mathrm{d}t = 0$ for all $v \in L^2(\mathbb{R}^3 \times (0,T))$. Hence $w \equiv 0$.

Given $\Psi \in C_0^\infty(\mathbb{R}^3 \times (0,T))$, let $T' < T$ be such that $\Psi(t) \equiv 0$ for $t \geq T'$. Extend Ψ by zero for $t \leq 0$ and $t \geq T$ and let φ be a solution of the backward heat equation with forcing Ψ, that is let φ be a classical solution to the problem

$$\begin{cases} \varphi_t + \Delta\varphi = \Psi & \text{in } \mathbb{R}^3 \times (-\infty, T), \\ \varphi(x, T) = 0 & \text{for } x \in \mathbb{R}^3. \end{cases} \tag{6.118}$$

As for the existence of such a φ, writing $\widetilde{\varphi}(x,t) := \varphi(x, T - t)$ and $\widetilde{\Psi}(x,t) := \Psi(x, T - t)$ the above problem becomes

$$\begin{cases} \widetilde{\varphi}_t - \Delta\widetilde{\varphi} = -\widetilde{\Psi} & \text{in } \mathbb{R}^3 \times (0, \infty), \\ \widetilde{\varphi}(x, 0) = 0 & \text{for } x \in \mathbb{R}^3, \end{cases}$$

which admits a smooth classical solution $\widetilde{\varphi}$, given by an application of the Duhamel principle,

$$\widetilde{\varphi}(t) := -\int_0^t \Phi(t - s) * \widetilde{\Psi}(s) \, \mathrm{d}s,$$

see, for example, Section 4.3.2 of Giga et al., 2010. Observe that, since $\Psi(s) \equiv 0$ for $s \geq T'$, $\widetilde{\varphi}(t) \equiv 0$ for $t \in [0, T - T']$. Moreover, since Ψ is smooth and compactly supported in $\mathbb{R}^3 \times (0, T)$ we have, for $|\alpha| \leq 2$

$$D^\alpha \widetilde{\varphi}(t) = -\int_0^t \Phi(t - s) * D^\alpha \widetilde{\Psi}(s) \, \mathrm{d}s,$$

and it follows from Lemma 6.1 that $D^\alpha \widetilde{\varphi} \in C([0,T], L^2)$ (note that $\Phi \in L^1_{loc}([0,\infty); L^1)$ due to the bound $\|\Phi(t)\|_1 \le C_0$). Therefore, we also obtain $\widetilde{\varphi}_t = \Delta \widetilde{\varphi} - \widetilde{\Psi} \in C([0,T]; L^2)$. Hence $\varphi(x,t) = \widetilde{\varphi}(x, T-t)$ is smooth, and it satisfies $\varphi(t) \equiv 0$ for $t \in [T', T]$, and

$$\sup_{t \in [0,T)} (\|\partial_t \varphi(t)\| + \|D^\alpha \varphi(t)\|) = \sup_{t \in [0,T]} (\|\partial_t \widetilde{\varphi}(t)\| + \|D^\alpha \widetilde{\varphi}(t)\|),$$

that is, φ satisfies (6.116). We can therefore use the first part and (6.118) to write

$$0 = \int_0^T \int (\varphi_t + \Delta \varphi) w \, dx \, dt = \int_0^T \int \Psi w \, dx \, dt. \qquad \square$$

6.5.2 The extension of Young's inequality for convolutions

We will prove the following lemma (Lemma 6.1).

Lemma 6.46 *Assume $p, q, r \ge 1$ are such that $1/q = 1/p + 1/r - 1$, $A \in L^1_{loc}([0,T); L^p)$ and $B \in C((0,T); L^r)$ with $\|B(t)\|_r$ bounded as $t \to 0^+$ then u defined by*

$$u(t) := \int_0^t A(t-s) * B(s) \, ds$$

belongs to $C([0,T); L^q)$ and

$$\|u(t)\|_q \le \int_0^t \|A(t-s)\|_p \|B(s)\|_r \, ds. \qquad (6.119)$$

Proof The bound (6.119) is clear from the integral version of Minkowski inequality (6.4) and from Young's inequality (6.6). This bound also implies the continuity $u(t) \to 0$ in L^q as $t \to 0^+$. It remains to show that $u \in C((0,T); L^q)$. For this reason let $T' < T$ and $\varepsilon > 0$. Let $M > 0$ be such that

$$\max_{s \in [0,T']} \|B(s)\|_r \le M, \quad \int_0^{T'} \|A(s)\|_p \, ds \le M.$$

From the assumption on A we see that there exist $\eta > 0$ such that

$$\int_0^\eta \|A(s)\|_p \, ds \le \frac{\varepsilon}{4M}.$$

Since B is uniformly continuous into L^r on $[\eta/2, T']$ then there exists $\delta \in (0, \eta/2)$ such that

$$\|B(s) - B(t)\|_r < \frac{\varepsilon}{4M}$$

whenever $s, t \in [\eta/2, T']$, and $|t - s| < \delta$. Letting $t_1, t_2 \in (0, T']$ be such that $t_1 < t_2$, $t_2 - t_1 < \delta$ we obtain

$$u(t_2) - u(t_1) = \int_0^{t_2} A(t_2 - s) * B(s) \, ds - \int_0^{t_1} A(t_1 - s) * B(s) \, ds$$

$$= \int_0^{t_2} A(t_2 - s) * B(s) \, ds - \int_{t_2-t_1}^{t_2} A(t_2 - s) * B(s - (t_2 - t_1)) \, ds$$

$$= \int_0^{t_2-t_1} A(t_2 - s) * B(s) \, ds$$

$$+ \int_{t_2-t_1}^{t_2} A(t_2 - s) * (B(s) - B(s - (t_2 - t_1))) \, ds.$$

$$(6.120)$$

Thus, using the integral Minkowski inequality (6.4) and Young's inequality for convolutions (6.6) we obtain

$$\|u(t_2) - u(t_1)\|_q \leq \int_0^{t_2-t_1} \|A(t_2 - s)\|_p \|B(s)\|_r \, ds$$

$$+ \int_{t_2-t_1}^{t_2} \|A(t_2 - s)\|_p \|B(s) - B(s - (t_2 - t_1))\|_r \, ds.$$

Since $t_2 - t_1 < \delta < \eta$ the first term on the right-hand side is bounded by $\varepsilon/4$. As for the second term, if $t_2 < \eta$ it can be similarly bounded by $\varepsilon/2$. If not, then we decompose it into $\int_{t_2-t_1}^{\eta} + \int_{\eta}^{t_2}$, we bound the first of the resulting integrals by $\varepsilon/2$. As for the second one, observe that $s, s - (t_2 - t_2) \in [\eta/2, T']$ whenever $s \in [\eta, T']$ to write

$$\int_{\eta}^{t_2} \|A(t_2 - s)\|_p \|B(s) - B(s - (t_2 - t_1))\|_r \, ds$$

$$\leq \frac{\varepsilon}{4M} \int_{\eta}^{t_2} \|A(t_2 - s)\|_p \, ds \leq \varepsilon/4.$$

Therefore in either case $\|u(t_2) - u(t_1)\|_q \leq \varepsilon$, which shows that $u \in C((0, T']; L^q)$. Since T' was chosen arbitrarily the claim follows. $\quad\square$

6.5.3 Decay estimates of $P(x, t)$

Let $P(x, t) := \frac{1}{|x|} \int_0^{|x|} \frac{e^{-\xi^2/4t}}{t^{1/2}} \, d\xi$ (see (6.12)). We have the following decay estimates.

Theorem 6.47 *For every $m \in \mathbb{N}$ we have*

$$|\nabla^m P(x, t)| \leq \frac{C_m}{(|x|^2 + t)^{(m+1)/2}} \qquad (6.121)$$

for all $x \in \mathbb{R}^3$, $t > 0$.

Proof Let us first assume that $t = 1$ (the general case will follow from the rescaling $P(x,t) = \frac{1}{\sqrt{t}}P(x/\sqrt{t},1)$). We claim that any partial derivative $D^\alpha P(x,1)$ of order $|\alpha| = m$ is of the form

$$D^\alpha P(x,1) = \frac{Q_\alpha(x)}{|x|^{2m}}P(x,1) + E_\alpha(x) \qquad (6.122)$$

for $m \geq 0$ and $|x| > 0$, where $Q_\alpha(x)$ denotes a polynomial of degree less than or equal to $|\alpha| = m$ and $E_\alpha(x)$ denotes a function that is smooth in $|x| > 0$ and whose derivatives of all orders decay exponentially when $|x| \to \infty$. The case $m = 0$ follows trivially with $E_0 \equiv 0$. For the inductive step, assume (6.122) holds for all partial derivatives $D^\alpha P(x,t)$ with $|\alpha| = m \geq 0$ and write, for $i = 1,2,3$,

$$
\begin{aligned}
\partial_{x_i} D^\alpha P(x,1) &= \partial_{x_i}\left(\frac{Q_\alpha(x)}{|x|^{2m}}P(x,1) + E_\alpha(x)\right) \\
&= \frac{\widetilde{Q}_{m-1}(x)}{|x|^{2m}}P(x,1) - \frac{2mQ_\alpha(x)\,x_i}{|x|^{2m+2}}P(x,1) \\
&\quad + \frac{Q_\alpha(x)}{|x|^{2m}}\left(\frac{x_i}{|x|^2}P(x,1) - \frac{x_i}{2|x|}e^{-|x|^2/4}\right) + \partial_{x_i}E_\alpha(x) \\
&= \frac{P(x,1)}{|x|^{2m+2}}\left(|x|^2\widetilde{Q}_{m-1}(x) - (2m-1)Q_\alpha(x)x_i\right) \\
&\quad - \frac{Q_\alpha(x)x_i}{2|x|^{2m+1}}e^{-|x|^2/4} + \partial_{x_i}E_\alpha(x),
\end{aligned}
$$

where $\widetilde{Q}_{m-1}(x)$ denotes some polynomial of degree less than or equal to $m - 1$. Clearly, the last bracket is some polynomial of degree less than or equal to $m + 1$ and the remaining two terms are smooth in $|x| > 0$ and decay exponentially (with all derivatives) as $|x| \to \infty$. Hence the induction follows. Because $P(x,1)$ decays like $|x|^{-1}$ as $|x| \to \infty$ we see from (6.122) that the estimate

$$|\nabla^m P(x,1)| \leq \frac{C_m}{|x|^{m+1}} \leq \frac{C_m}{(|x|^2 + 1)^{(m+1)/2}}$$

holds for $|x| \geq 1$. On the other hand, as $P(\cdot,1)$ is a smooth function (see Section 6.1.2) we have $|\nabla^m P(x,1)| \leq C_m \leq C_m(|x|^2 + 1)^{-(m+1)/2}$ for all $x \in B(0,1)$. Hence (6.121) follows in the case $t = 1$. Finally, the rescaling $P(x,t) = \frac{1}{\sqrt{t}} P(x/\sqrt{t}, 1)$ yields

$$|\nabla^m P(x,t)| = \left| t^{-(m+1)/2} \left[\nabla^m P(y,1) \right]_{y=x/\sqrt{t}} \right|$$

$$\leq t^{-(m+1)/2} \frac{C_m}{\left(\left| x/\sqrt{t} \right|^2 + 1 \right)^{(m+1)/2}} = \frac{C_m}{(|x|^2 + t)^{(m+1)/2}}. \quad \square$$

6.5.4 Properties of the Stokes equations

Here we present proofs of some results from Section 6.2. Namely we complete the proof of Theorem 6.8, show property (iii) of the representation (6.22) and the continuity $\nabla u \in C((0,T); L^\infty)$ given the representation (6.33) with $Y \in \mathcal{H}^{1/2}((0,T))$ (as defined in (6.3)), as well as the uniqueness of distributional solutions of the Stokes equations (Theorem 6.11).

Proof of Theorem 6.8

Recall that it remains to verify that if for some $R > 0$

$$F \in C^\infty(\mathbb{R}^3 \times [0,T); \mathbb{R}^3) \quad \text{and} \quad \operatorname{supp} F(t) \subset B(0,R) \text{ for } t \in [0,T)$$

then the pair of functions u_2, p given by (6.22), (6.23) is a classical solution of the problem

$$\begin{cases} \partial_t u_2 - \Delta u_2 + \nabla p &= F, \\ \operatorname{div} u_2 &= 0, \\ u_2(0) &= 0, \end{cases} \tag{6.123}$$

and $u_2 \in C([0,T); L^2)$. Indeed, as remarked after the statement of Theorem 6.8, $u = u_1 + u_2$ is a classical solution of the Stokes equations (6.20), (6.21) with $u(0) = u_0$ and $u \in C([0,T); L^2)$.

Note that, since F is smooth and compactly supported in space and since $\mathcal{T} \in L^1_{\text{loc}}([0,T); L^2)$ (see (6.16)) we see from Lemma 6.1 that $u_2 \in C([0,T); L^2)$ and $\|u_2(t)\| \to 0$ as $t \to 0^+$ (which means that u_2 satisfies the initial condition $u_2(0) = 0$). Moreover, since $\mathcal{T} \in C((0,\infty); L^2)$ (see (6.17)), we deduce that the functions ∇u_2, Δu_2, $\partial_t u_2$ are continuous (an application of the Dominated Convergence Theorem) and belong to $C((0,T); L^2)$ (a consequence of Lemma 6.1). Similarly ∇p is continuous and $\nabla p \in C([0,T); L^2)$ (by an application of Lemma 6.4). Therefore, since the Fourier transform is an isometry from L^2 into L^2, we see that u_2 and p satisfy (6.123) if and only if

$$\partial_t \widehat{u_2}(\xi,t) + 4\pi^2 |\xi|^2 \widehat{u_2}(\xi,t) + 2\pi i\, \xi\, \widehat{p}(\xi,t) = \widehat{F}(\xi,t) \tag{6.124}$$

and

$$\xi \cdot \widehat{u_2}(\xi, t) = 0 \qquad (6.125)$$

hold for $t \in (0, T)$ and almost every $\xi \in \mathbb{R}^3$. Here $\widehat{u_2}$, $\widehat{p}$, $\widehat{F}$ denote the Fourier transform (see (6.7), which is also denoted below by $\mathcal{F}$) of u_2, p, F, respectively. Since p satisfies $\Delta p = \operatorname{div} F$ we obtain

$$2\pi \mathrm{i}\, \xi\, \widehat{p}(\xi, t) = \frac{\xi \otimes \xi}{|\xi|^2} \widehat{F}(\xi, t),$$

where $\xi \otimes \xi$ denotes the 3×3 matrix with components $\xi_i \xi_j$. Therefore (6.124) is equivalent to

$$\left(\partial_t + 4\pi^2 |\xi|^2 \right) \widehat{u_2}(\xi, t) = \left(I - \frac{\xi \otimes \xi}{|\xi|^2} \right) \widehat{F}(\xi, t). \qquad (6.126)$$

Now since

$$\mathcal{F}\left[\Phi(\cdot, t) \right] = \mathrm{e}^{-4\pi^2 t |\xi|^2} \qquad (6.127)$$

(see e.g. Theorem 1.13 in Chapter 1 of Stein & Weiss (1971) for a proof of this fact) and $-\Delta P = \Phi$ (see (6.13)), we obtain

$$\mathcal{F}\left[P(\cdot, t) \right] = \frac{1}{4\pi^2 |\xi|^2} \mathrm{e}^{4\pi^2 t |\xi|^2},$$

and so

$$\mathcal{F}\left[\partial_i \partial_j P(\cdot, t) \right] = \frac{-\xi_i \xi_j}{|\xi|^2} \mathrm{e}^{-4\pi^2 t |\xi|^2}.$$

Hence for all $t > 0$

$$\mathcal{F}[\mathcal{T}(t)] = \left(I - \frac{\xi \otimes \xi}{|\xi|^2} \right) \mathrm{e}^{-4\pi^2 t |\xi|^2}. \qquad (6.128)$$

Since $u_2(t) = \int_0^t \mathcal{T}(t - s) * F(s)\, \mathrm{d}s$,

$$\widehat{u_2}(\xi, t) = \int_0^t \left(I - \frac{\xi \otimes \xi}{|\xi|^2} \right) \widehat{F}(\xi, s) \mathrm{e}^{-4\pi^2 (t - s) |\xi|^2}\, \mathrm{d}s, \qquad \xi \neq 0, \quad (6.129)$$

and so (6.125) holds for $\xi \neq 0$ and

$$\left(\partial_t + 4\pi^2 |\xi|^2 \right) \widehat{u_2}(\xi, t)$$

$$= \left(\partial_t + 4\pi^2 |\xi|^2 \right) \int_0^t \left(I - \frac{\xi \otimes \xi}{|\xi|^2} \right) \mathrm{e}^{-4\pi^2 (t - s) |\xi|^2} \widehat{F}(\xi, s)\, \mathrm{d}s$$

$$= \left(I - \frac{\xi \otimes \xi}{|\xi|^2} \right) \widehat{F}(\xi, t)$$

for $\xi \neq 0$, that is (6.124), as claimed.

Property (iii) of the representation (6.22)

We need to show that if $F \in C([0,T]; L^2)$, $u_0 \in L^2$ and u is given by (6.22), that is,

$$u(t) = u_1(t) + u_2(t) = \Phi(t) * u_0 + \int_0^t \mathcal{T}(t - s) * F(s) \, ds,$$

then $u \in C([0,T]; L^2)$ with

$$\|u(t)\| \le \int_0^t \|F(s)\| \, ds + \|u_0\| \qquad \text{for } t \in (0,T). \tag{6.130}$$

Moreover u satisfies the energy dissipation equality

$$\|u(t)\|^2 - \|u_0\|^2 + 2 \int_0^t \|\nabla u(s)\|^2 \, ds = 2 \int_0^t \int u \cdot F \, dx \, ds \tag{6.131}$$

for $t \in (0,T)$.

We prove these claims by considering two cases.

Case 1. F is regular, that is, satisfies (6.28) for some $R > 0$. For such F, due to Theorem 6.8 (or rather to the proof above), $u = u_1 + u_2$ satisfies $u(0) = u_0$, and

$$\partial_t u_1 - \Delta u_1 = 0, \quad \partial_t u_2 - \Delta u_2 + \nabla p = F \qquad \text{in } \mathbb{R}^3 \times (0,T) \tag{6.132}$$

and $u_1, u_2 \in C([0,T]; L^2)$, where p is given by (6.23).

Note that for $y \in B(0,R)$ and $|x| > 2R$

$$|x|/2 < |x| - R < |x - y| < |x| + R$$

and using (6.15) we can write, for such x, and for all t

$$|u_2(x,t)| \le \int_0^t \int_{B(0,R)} \frac{C_0}{(|x-y|^2 + t - s)^{3/2}} |F(y,s)| \, dy \, ds$$

$$\le \frac{C|B(0,R)|^{1/2} t}{|x|^3} \max_{s \in [0,t]} \|F(s)\|,$$

which gives a decay $\sim |x|^{-3}$ of $|u_2(x,t)|$ as $|x| \to \infty$ that is uniform on any compact time interval $[0, T']$, where $T' < T$. Similarly one can derive a decay $\sim |x|^{-5}$ of $|\Delta u_2(x,t)|$ and decay $\sim |x|^{-3}$ of $|\nabla p(x,t)|$. Thus (6.132) gives the decay $\sim |x|^{-3}$ of $\partial_t u_2(x,t)$. Letting $\delta, t \in (0,T)$, $t > \delta$ and using this decay as well as the fact that u_1 and all its derivatives belong to $C((0,T); L^2)$ (see (ii) in Appendix 6.5.1) we can integrate the equality

$$u \cdot \partial_t u - u \cdot \Delta u + u \cdot \nabla p = u \cdot F$$

on $\mathbb{R}^3 \times (\delta, t)$ to obtain

$$\|u(t)\|^2 - \|u(\delta)\|^2 + 2\int_\delta^t \|\nabla u(s)\|^2 \, ds = 2\int_\delta^t \int u \cdot F \, dx \, ds. \quad (6.133)$$

Since $u \in C([0,T]; L^2)$ we can take the limit as $\delta \to 0^+$ (and apply the Monotone Convergence Theorem) to obtain (6.131). As for the bound (6.130) note that the above equality and the Cauchy-Schwarz inequality give

$$\frac{d}{dt}\|u(t)\|^2 \le 2\int u(t) \cdot F(t) \, dx \le 2\|u(t)\| \, \|F(t)\|, \qquad t \in (0,T),$$

that is

$$\frac{d}{dt}\|u(t)\| \le \|F(t)\|, \qquad t \in (0,T).$$

Integrating this inequality in t gives (6.130).

Case 2. $F \in C([0,T]; L^2)$. Consider a compact time interval $[0,T'] \subset [0,T)$ and a sequence $\{F_R\} \subset C([0,T']; L^2)$ such that $F_R \in C^\infty(\mathbb{R}^3 \times \mathbb{R})$, supp $F_R(t) \subset B(0,R)$ for all t and

$$F_R \to F \qquad \text{in } C([0,T']; L^2) \text{ as } R \to \infty.$$

(See Lemma 6.55 for a proof of the existence of such a sequence.) Let u_R denote the velocity field corresponding to F_R (i.e. given by (6.22) with F replaced by F_R). Since the representation formula (6.22) is linear we can apply (6.130) to the difference $u_{R_1} - u_{R_2}$ for $R_1, R_2 > 0$ to see that $\{u_R\}$ is Cauchy in $C([0,T']; L^2)$ as $R \to \infty$. Thus $u_R \to u'$ in $C([0,T']; L^2)$ as $R \to \infty$ for some $u' \in C([0,T']; L^2)$ with $u'(0) = u_0$. However, Lemma 6.7 (i) applied to the difference $u_R - u$ gives $\|u(t) - u_R(t)\|_\infty \to 0$ for every $t \in (0,T']$, and thus $u = u'$ (since convergence in L^2 gives convergence almost everywhere on a subsequence). Hence $u \in C([0,T']; L^2)$ and (6.130) follows for $t \in [0,T']$ by taking $R \to \infty$ in the corresponding inequality for u_R. As for the energy dissipation equality (6.131), let $\delta \in (0,T')$ and note that Lemma 6.7 (ii) applied to the difference $u_R - u$ gives

$$\nabla u_R \to \nabla u \qquad \text{in } C([\delta, T']; L^2) \text{ as } R \to \infty.$$

(Note that the convergence is not in $C([0,T']; L^2)$ since each ∇u_R need not belong to this space.) Since Case 1 gives, for each $R > 0$, $t \in [0,T']$,

$$\frac{1}{2}\|u_R(t)\|^2 - \frac{1}{2}\|u_R(\delta)\|^2 + \int_\delta^t \|\nabla u_R(s)\|^2 \, ds = \int_\delta^t \int u_R \cdot F_R \, dx \, ds,$$

(see (6.133)) we can take the limit as $R \to \infty$ and then take the limit as $\delta \to 0^+$ (and apply the Monotone Convergence Theorem again) to obtain (6.131) for all $t \in [0, T']$.

The continuity of $\nabla u \in C((0, T); L^\infty)$ for u given by (6.33)
We need to show that if $Y \in C((0, T); L^\infty)$ is weakly divergence free, $\|Y(t)\|_\infty$ remains bounded as $t \to 0^+$ and $Y \in \mathcal{H}^{1/2}((0, T))$ with the corresponding constant $C_0(t)$ bounded as $t \to 0^+$, and if u_2 is given by

$$u_2(t) := \int_0^t \nabla \mathcal{T}(t - s) * [Y(s)Y(s)] \, ds$$

(recall the notation (6.35)) then $\nabla u \in C((0, T); L^\infty)$.

In order to prove it, fix $T' < T$ and let $M > 0$ be such that

$$\|Y(t)\|_\infty, C_0(t) \leq M \qquad \text{for } t \in [0, T'].$$

Fix $\varepsilon > 0$. Let $\eta > 0$ be such that

$$\int_0^t \frac{1}{(t - s)^{3/4}} \, ds \leq \frac{\varepsilon}{M^2} \qquad \text{for } t \leq 2\eta.$$

Since for each pair i, k we have $Y_i Y_k \in C((0, T); L^\infty)$, $Y_i Y_k$ is uniformly continuous into L^∞ on time interval $[\eta/2, T']$. Thus there exists $\delta \in (0, \eta/2)$ such that for all i, k

$$\|Y_i(t)Y_k(t) - Y_i(s)Y_k(s)\|_\infty \leq \frac{\varepsilon \eta}{T'}$$

whenever $s, t \in [\eta/2, T']$ and $|t - s| < \delta$. Now let $t_1, t_2 \in [0, T']$ be such that $t_1 < t_2$ and $t_2 - t_1 < \delta$, and calculate

$$\partial_l u_{2,j}(t_1) - \partial_l u_{2,j}(t_2) = \int_0^{t_2 - t_1} \partial_{li} \mathcal{T}_{jk}(t_2 - s) * [Y_i(s)Y_k(s)] \, ds$$

$$+ \int_{t_2 - t_1}^{t_2} \partial_{li} \mathcal{T}_{jk}(t_2 - s) * [Y_i(s)Y_k(s)$$

$$- Y_i(s - (t_2 - t_1))Y_k(s - (t_2 - t_1))] \, ds$$

(cf. calculation (6.120)) and denote the two integrals on the right-hand side by I_1, I_2 respectively. As for I_1, using the same trick of employing the Hölder continuity of $Y_k(s)$ as indicated in (6.38), we obtain

$$|I_1| \leq \int_0^{t_2-t_1} \int |\partial_{li}\mathcal{T}_{jk}(x-y,t_2-s)Y_i(y,s)\left[Y_k(y,s)-Y_k(x,s)\right]| \, dy \, ds$$

$$\leq M^2 \int_0^{t_2-t_1} \int \frac{C|x-y|^{1/2}}{(|x-y|^2+(t_2-s))^{5/2}} \, dy \, ds$$

$$= M^2 \int_0^{t_2-t_1} \frac{C}{(t_2-s)^{3/4}} \, ds \leq C\varepsilon.$$

If $t_2 \leq 2\eta$ then one can bound $|I_2|$ in a similar way to obtain

$$|I_2| \leq 2C\varepsilon.$$

Otherwise, we write $\int_{t_2-t_1}^{t_2} = \int_{t_2-t_1}^{\eta} + \int_{\eta}^{t_2-\eta} + \int_{t_2-\eta}^{t_2}$ and denote the resulting three integrals by $I_{2,1}$, $I_{2,2}$, $I_{2,3}$, respectively. Since the length of the intervals of integration in $I_{2,1}$, $I_{2,3}$ is less than 2η we obtain, as above,

$$|I_{2,1}|, |I_{2,3}| \leq 2C\varepsilon.$$

For $I_{2,2}$ note that $s, s-(t_2-t_1) \in [\eta/2, T']$ for each s from the interval of integration to write

$$|I_{2,2}| \leq \int_\eta^{t_2-\eta} \|\partial_{li}\mathcal{T}_{jk}(t_2-s)\|_1 \|Y_i(s)Y_k(s)$$

$$- Y_i(s-(t_2-t_1))Y_k(s-(t_2-t_1))\|_\infty \, ds$$

$$\leq \frac{\varepsilon\eta}{T'} \int_\eta^{t_2-\eta} \frac{C}{t_2-s} \, ds \leq \frac{\varepsilon\eta}{T'} \int_\eta^{t_2-\eta} \frac{C}{\eta} \, ds \leq C\varepsilon,$$

where we used the Minkowski inequality (6.4), Young's inequality (6.6) and the bound $\|\nabla^2 \mathcal{T}(t)\|_1 \leq C/t$ (see (6.15)). Thus altogether

$$|\partial_l u_{2,j}(t_1) - \partial_l u_{2,j}(t_2)| \leq 5C\varepsilon \qquad l,j=1,2,3$$

and the continuity $\nabla u_2 \in C((0,T);L^\infty)$ follows.

Uniqueness of distributional solutions of the Stokes equations
We will show that if $u \in C([0,T];L^2)$ is weakly divergence free and $p \in L^1_{\text{loc}}(\mathbb{R}^3 \times [0,T))$ satisfy

$$\int_0^T \int \left((\varphi_t + \Delta\varphi) \cdot u + p \operatorname{div}\varphi\right) \, dx \, dt = 0 \qquad (6.134)$$

for all $\varphi \in C_0^\infty(\mathbb{R}^3 \times [0,T);\mathbb{R}^3)$, then $u \equiv 0$.

Proof Fix $\psi \in C_0^\infty(\mathbb{R}^n \times [0,T))$ let $\varphi := \nabla\psi$. Then $\operatorname{div}\varphi = \Delta\psi$ and so (6.134) gives

$$\int_0^T \int p\,\Delta\psi\,\mathrm{d}x\,\mathrm{d}t = \int_0^T \int \nabla(\psi_t + \Delta\psi)\cdot u\,\mathrm{d}x\,\mathrm{d}t = 0 \qquad (6.135)$$

since u is weakly divergence free.

For $\varepsilon > 0$ let

$$v(x,t) := \int_0^t (J_\varepsilon u)(x,s)\,\mathrm{d}s, \qquad q(x,t) := \int_0^t (J_\varepsilon p)(x,s)\,\mathrm{d}s.$$

We first show that v, q also satisfy (6.134),

$$\int_0^T \int ((\varphi_t + \Delta\varphi)\cdot v + q\operatorname{div}\varphi)\,\mathrm{d}x\,\mathrm{d}t = 0 \qquad (6.136)$$

for all $\varphi \in C_0^\infty(\mathbb{R}^3 \times [0,T);\mathbb{R}^3)$, or equivalently

$$\begin{aligned}
0 &= \int_0^T \int \int_0^t \int (\eta_\varepsilon(y)u(x-y,s)\cdot(\varphi_t(x,t) + \Delta\varphi(x,t)) \\
&\qquad\qquad + \eta_\varepsilon(y)p(x-y,s)\operatorname{div}\varphi(x,t))\,\mathrm{d}y\,\mathrm{d}s\,\mathrm{d}x\,\mathrm{d}t \\
&= \int \eta_\varepsilon(y) \int \int_0^T \left(\int_0^t u(x,s)\,\mathrm{d}s \cdot (\varphi_t(x+y,t) + \Delta\varphi(x+y,t)) \right. \\
&\qquad\qquad \left. + \int_0^t p(x,s)\,\mathrm{d}s \operatorname{div}\varphi(x+y,t) \right) \mathrm{d}t\,\mathrm{d}x\,\mathrm{d}y.
\end{aligned}$$

We will show that the expression under the y integral vanishes. In fact, for fixed $y \in \mathbb{R}^3$ let

$$\Psi(x,t) := -\int_t^{T'} \varphi(x+y,s)\,\mathrm{d}s,$$

where $T' < T$ is such that $\varphi(t) \equiv 0$ for $t \geq T'$. Clearly we obtain $\Psi \in C_0^\infty(\mathbb{R}^3 \times [0,T);\mathbb{R}^3)$ and so (6.134) gives

$$\begin{aligned}
0 &= \int \int_0^T u(x,t)\cdot(\Psi_t(x,t) + \Delta\Psi(x,t))\,\mathrm{d}t\,\mathrm{d}x \\
&\quad + \int \int_0^T q(x,t)\operatorname{div}\Psi(x,t)\,\mathrm{d}t\,\mathrm{d}x.
\end{aligned}$$

Integration by parts in t and the identity

$$\partial_t(\Psi_t(x,t) + \Delta\Psi(x,t)) = \varphi_t(x+y,t) + \Delta\varphi(x+y,t)$$

give

$$
\int \int_0^T \left(\int_0^t u(x,s)\,ds \cdot (\varphi_t(x+y,t) + \Delta\varphi(x+y,t)) \right.
$$
$$
\left. + \int_0^t p(x,s)\,ds \; \mathrm{div}\,\varphi(x+y,t) \right) dt\,dx = 0,
$$

as required. Therefore v, q indeed satisfy (6.136). Moreover, letting

$$
\Psi(x,t) := - \int_t^{T'} \psi(x+y,s)\,ds,
$$

this time for a scalar test function ψ we obtain from (6.135) that for each $y \in \mathbb{R}^3$

$$
0 = \int_0^T \int p\Delta\Psi\,dx\,dt = - \int_0^T \int p(x,t) \int_t^{T'} \Delta\psi(x+y,s)\,ds\,dx\,dt
$$
$$
= - \int_0^T \int \left(\int_0^t p(x,s)\,ds \right) \Delta\psi(x+y,t)\,dx\,dt.
$$

Hence Fubini's theorem gives

$$
\int_0^T \int q\Delta\psi\,dx\,dt = \int_0^T \int \int_0^t \int \eta_\varepsilon(y)p(x-y,s)\Delta\psi(x,t)\,dy\,ds\,dx\,dt
$$
$$
= \int \eta_\varepsilon(y) \int_0^T \int \left(\int_0^t p(x,s)\,ds \right) \Delta\psi(x+y,t)\,dx\,dt\,dy = 0,
$$

that is, like p, q satisfies (6.135). Therefore (6.136) applied with $\Delta\varphi$ in place of φ gives

$$
0 = \int_0^T \int \Delta(\varphi_t + \Delta\varphi) \cdot v\,dx\,dt = \int_0^T \int (\varphi_t + \Delta\varphi) \cdot \Delta v\,dx\,dt.
$$

The uniqueness of weak solutions to the heat equation (see Lemma 6.45) now implies $\Delta v \equiv 0$ almost everywhere in $\mathbb{R}^3 \times [0,T)$, hence $\Delta v \equiv 0$ everywhere by continuity of Δv. Thus Liouville's theorem implies that $v(t)$ is constant for each t. Therefore $v(t) \equiv 0$ at each t due to the fact $v \in C([0,T); L^2)$. Hence, differentiating the definition of v (in t), we see that $J_\varepsilon u(t) \equiv 0$ for each t and ε. The almost everywhere convergence of the mollification (see Lemma 6.2, (v)) gives $u(t) \equiv 0$ for each t, as required. □

6.5.5 Integral inequalities

Lemma 6.48 *Let $g > 0$ be a continuous function on $(0, T)$ that is locally integrable on $[0, T)$. Let $f, \varphi : (0, T) \to \mathbb{R}^+$ be such that f^2 and φ^2 are integrable near zero, with φ continuous. Assume that*

$$f(t) \leq \int_0^t g(t - s) f(s)^2 \, ds + a(t), \qquad (6.137)$$

$$\varphi(t) \geq \int_0^t g(t - s) \varphi(s)^2 \, ds + b(t) \qquad (6.138)$$

for all $t \in (0, T)$, where a, b are continuous functions satisfying $a \leq b$. Then $f \leq \varphi$ on $(0, T)$.

Proof The proof proceeds in three steps.

Step 1. The case $a(t) \leq b(t) - \delta$ for $t \in (0, \tau)$ for some $\delta, \tau > 0$.
    Let

$$I := \left\{ t' : \int_0^t g(t - s) f(s)^2 \, ds + a(t) < \varphi(t) \text{ for all } t \in (0, t'] \right\}.$$

Note that (6.137) gives $f(t) < \varphi(t)$ for $t \in I$. Let $t_0 \in (0, \tau)$ be such that

$$\left| \int_0^t g(t - s) f(s)^2 \, ds \right| < \frac{\delta}{2}, \quad \left| \int_0^t g(t - s) \varphi(s)^2 \, ds \right| < \frac{\delta}{2}$$

for $t \in (0, t_0]$. Then, for $t \in (0, t_0]$, (6.138) gives

$$\int_0^t g(t - s) f(s)^2 \, ds + a(t) < \delta/2 + b(t) - \delta$$

$$\leq \varphi(t) - \int_0^t g(t - s) \varphi(s)^2 \, ds - \delta/2 < \varphi(t),$$

that is $t_0 \in I$. Now let $T' := \sup I$. We need to show that $T' = T$. Suppose otherwise that $T' < T$. Then

$$\int_0^{T'} g(T' - s) f(s)^2 \, ds + a(T') < \int_0^{T'} g(T' - s) \varphi(s)^2 \, ds + b(T') \leq \varphi(T')$$

by (6.138). By continuity we obtain $\int_0^t g(t - s) f(s)^2 \, ds + a(t) < \varphi(t)$ for $t \in [T', T'']$ for some $T'' > T'$. Hence $T'' \in I$, which contradicts the definition of T'. Therefore indeed $T' = T$ and the lemma follows in this case.

Step 2. The case $\liminf_{t \to 0^+} (b(t) - a(t)) = 0$: there exists $t_0 > 0$ such that $f(t) \leq \varphi(t)$ for $t \in (0, t_0)$.

Let $t_0 > 0$ be small enough such that

$$\left| \int_0^t g(t-s) \, ds \right| < \frac{1}{4}, \quad \left| \int_0^t g(t-s) f(s) \, ds \right| < \frac{1}{4} \qquad \text{for } t \in [0, t_0].$$

Let $\varepsilon \in (0,1)$ and set $f_\varepsilon(t) := f(t) - \varepsilon$. Then $\varepsilon/2 + 3\varepsilon^2/4 - \varepsilon \le -\varepsilon/4$ and so f_ε satisfies the inequality

$$\begin{aligned}
f_\varepsilon(t) &\le \int_0^t g(t-s)(f_\varepsilon(s) + \varepsilon)^2 \, ds + a(t) - \varepsilon \\
&= \int_0^t g(t-s) f_\varepsilon(s)^2 \, ds + 2\varepsilon \int_0^t g(t-s)(f(s) - \varepsilon) \, ds \\
&\quad + \varepsilon^2 \int_0^t g(t-s) \, ds + a(t) - \varepsilon \\
&\le \int_0^t g(t-s) f_\varepsilon(s)^2 \, ds + \varepsilon/2 + \varepsilon^2/2 + \varepsilon^2/4 + a(t) - \varepsilon \\
&\le \int_0^t g(t-s) f_\varepsilon(s)^2 \, ds + a(t) - \varepsilon/4
\end{aligned}$$

for $t \in (0, t_0]$. Because $a(t) - \varepsilon/4 \le b(t) - \varepsilon/4$ on $(0, t_0)$, similarly as in Step 1 we obtain $f_\varepsilon \le \varphi$ on $(0, t_0)$. The claim follows by taking the limit $\varepsilon \to 0^+$.

Step 3. The case $\liminf_{t \to 0^+}(b(t) - a(t)) = 0$: $f(t) \le \varphi(t)$ *for every* $t \in (0, T)$.
  Let

$$I_1 := \{t \in (0, T) \,:\, f(s) \le \varphi(s) \text{ for } s \in (0, t)\}.$$

Let $t_1 := \sup I_1$. Note that Step 2 gives $t_1 \ge t_0 > 0$. Suppose that $t_1 < T$. Let $F(t) := f(t_1 + t)$, $\Phi(t) := \varphi(t_1 + t)$. Then F, Φ satisfy

$$F(t) \le \int_0^t g(t-s) F(s)^2 \, ds + A(t),$$

$$\Phi(t) \ge \int_0^t g(t-s) \Phi(s)^2 \, ds + B(t)$$

for $t \in (T - t_1)$, where

$$A(t) := a(t_1 + t) + \int_0^{t_1} g(t_1 + t - s) f(s)^2 \, ds,$$

$$B(t) := b(t_1 + t) + \int_0^{t_1} g(t_1 + t - s)\varphi(s)^2 \, \mathrm{d}s$$

$$\geq A(t) + b(t_1 + t) - a(t_1 + t).$$

Noting that A, B are continuous (by the Dominated Convergence Theorem) and that $A(t) \leq B(t)$ for all $t \in (0, T - t_1)$, we can apply Step 1 and 2 to the functions F, Φ, to conclude that $F(t) \leq \Phi(t)$ for all $t \in (0, t_2]$ for some $t_2 > 0$. Thus $f(t) \leq \varphi(t)$ for all $t \in [0, t_1 + t_2)$, which contradicts the definition of t_1. $\square$

The above lemma can be modified to fit several other settings, for example we have the following results.

Corollary 6.49 *Let g, a, b be as in Lemma 6.5 and let h satisfy the same conditions as g. Let functions $f, \varphi \colon (0, T) \to \mathbb{R}$ be such that $f(t)$, $\varphi(t)$ are bounded as $t \to 0^+$, φ is continuous and*

$$\begin{cases} f(t) & \leq \int_0^t g(t-s)h(s)f(s)\, \mathrm{d}s + a(t), \\ \varphi(t) & \geq \int_0^t g(t-s)h(s)\varphi(s)\, \mathrm{d}s + b(t) \end{cases}$$

for all $t \in (0, T)$. Then $f \leq \varphi$ on $(0, T)$.

Corollary 6.50 *Let $a, b > 0$ be such that $a \leq b$ and let $h, g \colon (0, \infty) \to \mathbb{R}^+$ be continuous functions such that g is locally integrable on $[0, \infty)$ and h is integrable on $(1, \infty)$. Suppose also that there exist $\tau > 0$, $C' > 0$ such that*

$$h(s) \geq (C')^2 g(s) \quad \textit{for } s \in (0, \tau),$$
$$h(s) \leq (C')^2 g(s) \quad \textit{for } s \in [\tau, \infty),$$

and for all $t > 0$

$$C' > \int_0^t \min\left((C')^2 g(t-s), h(t-s)\right) \mathrm{d}s + b.$$

If $T > 0$ and f is a positive function on $(0, T)$ that is bounded near 0 and

$$f(t) \leq \int_0^t \min\left(g(t-s)f(s)^2, h(t-s)\right) \mathrm{d}s + a$$

for $t \in (0, T)$, then $f \leq C'$ on $(0, T)$.

Proof If $t < \tau$ then $\min\left((C')^2 g(t-s), h(t-s)\right) = (C')^2 g(t-s)$ and so Lemma 6.48 gives $f(t) \leq C'$ for such t. Thus letting

$$t_0 := \sup\{t' > 0 \colon f(t) \leq C' \text{ for } t < t'\}$$

we see that $t_0 \geq \tau > 0$. If $t_0 < T$ we obtain

$$\int_0^{t_0} \min\left(g(t_0 - s)f(s)^2, h(t_0 - s)\right) \mathrm{d}s + a$$

$$\leq \int_0^{t_0} \min\left(g(t_0 - s)(C')^2, h(t_0 - s)\right) \mathrm{d}s + b < C'.$$

Thus, by continuity

$$\int_0^t \min\left(g(t - s)f(s)^2, h(t - s)\right) \mathrm{d}s + a < C'$$

for $t \in [t_0, t_0 + \delta)$ for some $\delta > 0$, which contradicts the definition of t_0. Thus $t_0 = T$, as required. □

6.5.6 The Volterra equation

In this section we show that the equation

$$\varphi(x) = C \int_0^x \frac{\varphi(y)}{\sqrt{x - y}} \, \mathrm{d}y + D, \qquad (6.139)$$

where $C, D > 0$, has a unique solution $\varphi \in C([0, \infty))$. This is equivalent to showing that (6.139) has a unique solution $\varphi \in C([0, T])$ for every $T > 0$. We can rewrite (6.139) in the form

$$\varphi - A\varphi = D, \qquad (6.140)$$

where

$$A\varphi(x) := \int_0^T K(x, y)\varphi(y) \, \mathrm{d}y \qquad (6.141)$$

with $K(x, y) := C \chi_{\{y < x\}}(x - y)^{-1/2}$. This is an example of the Volterra integral equation of the 2nd kind with a weakly singular kernel K, that is any K such that for all $x, y \in [0, T]$ with $x \neq y$ K is continuous at (x, y) and $|K(x, y)| \leq M|x - y|^{\alpha - 1}$ for some $\alpha \in (0, 1]$, $M > 0$. In what follows we apply the theory of such equations to (6.140). We consider only the case $\alpha = 1/2$; other cases follow similarly. We follow the arguments from Kress (2014).

We first note that the set of compact operators is closed in the operator norm.

Lemma 6.51 *Let X be a Banach space and $A_n \in L(X)$ be a sequence of compact operators such that $\|A_n - A\| \to 0$ as $n \to \infty$ for some $A \in L(X)$. Then A is compact.*

This is elementary (see e.g. Kress (2014), p. 26, for the proof). We now show that A (defined by (6.141)) is a compact operator on $X := C([0,T])$ by cutting off the singularity and using the above lemma.

Lemma 6.52   *The operator $A : X \to X$ is compact.*

Proof   We see that A is continuous by writing

$$
|A\varphi(t)| = C \left| \int_0^t \frac{\varphi(s)}{\sqrt{t-s}} \, \mathrm{d}s \right|
$$

$$
\leq C\|\varphi\|_{\sup} \int_0^t (t-s)^{-1/2} \, \mathrm{d}s = 2C\|\varphi\|_{\sup} \, t^{1/2} \leq 2CT^{1/2}\|\varphi\|_{\sup}.
$$

For $n \in \mathbb{N}$ we define a cut-off K_n of the kernel K by

$$
K_n(t,s) := \begin{cases} h(n|t-s|)K(t,s) & t \neq s \\ 0 & t = s, \end{cases}
$$

where $h : [0,\infty) \to [0,1]$ is a continuous function such that $h(t) = 0$ for $t \in [0,1/2]$ and $h(t) = 1$ for $t \geq 1$. Because $K_n \in C([0,T]^2)$ for every n, the corresponding integral operators A_n are compact by the Arzelà-Ascoli theorem. Moreover

$$
|A\varphi(t) - A_n\varphi(t)| = \left| \int_{t-1/n}^t (1 - h(n|t-s|))K(t,s)\varphi(s) \, \mathrm{d}s \right|
$$

$$
\leq \|\varphi\|_{\sup} \int_{t-1/n}^t \frac{C}{\sqrt{t-s}} \, \mathrm{d}s = C\|\varphi\|_{\sup} \, n^{-1/2} \to 0
$$

as $n \to \infty$ uniformly in t. Hence $\|A_n - A\| \to 0$ and Lemma 6.51 gives compactness of A. $\square$

We now show the existence and uniqueness of a solution to the original Volterra equation by applying the Fredholm Alternative Theorem.

Theorem 6.53   *Equation* (6.140) *has a unique solution $\varphi \in C([0,T])$.*

Proof   Because X is a Banach space and $A : X \to X$ is compact we can apply the Fredholm Alternative Theorem to conclude that (6.140) has a unique solution if the equation

$$
\varphi - A\varphi = 0 \tag{6.142}
$$

has no non-zero solutions. We will use induction to show that any solution φ to this homogeneous problem satisfies

$$
|\varphi(t)| \leq \|\varphi\|_{\sup} \frac{M^k t^k}{k!} \tag{6.143}
$$

for all $t \in [0,T]$, $k \in \mathbb{N}$, where $M > 0$ is a constant. The base case $k = 0$ is trivial. For the inductive step we first note that for any $t, s \in [0,T]$ with $0 \le s < t \le T$ we have

$$\int_s^t \frac{d\tau}{\sqrt{t-\tau}\sqrt{\tau-s}} = \int_0^1 \frac{dz}{\sqrt{z(1-z)}} =: I,$$

by the change of variable $z := \frac{\tau-s}{t-s}$. Now assume that a solution φ to (6.142) satisfies $|\varphi(t)| \le \|\varphi\|_{\sup} \frac{M^k t^k}{k!}$ for some k. Then, because $\varphi = A\varphi = A^2\varphi$, we have for all $t \in [0,T]$

$$|\varphi(t)| = |A^2\varphi(t)| = C^2 \left| \int_0^t \int_0^\tau \frac{\varphi(s)}{\sqrt{\tau-s}\sqrt{t-\tau}} \, ds \, d\tau \right|$$

$$\le C^2 \int_0^t \int_0^\tau \frac{|\varphi(s)|}{\sqrt{\tau-s}\sqrt{t-\tau}} \, ds \, d\tau = C^2 \int_0^t \int_s^t \frac{|\varphi(s)|}{\sqrt{\tau-s}\sqrt{t-\tau}} \, d\tau \, ds$$

$$= IC^2 \int_0^t |\varphi(s)| \, ds \le \|\varphi\|_{\sup} \frac{IC^2 M^k}{k!} \int_0^t s^k \, ds = \|\varphi\|_{\sup} \frac{IC^2 M^k t^{k+1}}{(k+1)!},$$

where we also used Fubini's theorem. The bound (6.143) now follows with $M := IC^2$. Taking the limit $k \to \infty$ in (6.143) gives $\varphi \equiv 0$.   □

6.5.7 A proof of (6.85) without the use of the Lemma 6.4

Here we give an elementary proof of (6.85),

$$\|p\|^2 \le C\|\nabla u\|^3 \|u\|.$$

Leray (1934b) discusses this inequality on pp. 233-234, but gives few details, particularly about the equation following (5.4), which we discuss below (see (6.145)).

First note that the representation formula for p (see (6.79)) can be rewritten in the form

$$p = \frac{-1}{4\pi} \int \nabla \left(\frac{1}{|x-y|} \right) \cdot g(y) \, dy,$$

where $g := (J_\epsilon u \cdot \nabla)u$. Therefore

$$\|p\|^2 = \frac{1}{(4\pi)^2} \iiint \left[\nabla \frac{1}{|x-y|} \cdot g(y) \right] \left[\nabla \frac{1}{|x-z|} \cdot g(z) \right] dx \, dy \, dz$$

$$= \frac{-1}{(4\pi)^2} \iiint \left[\frac{1}{|x-y|} \operatorname{div} g(y) \right] \left[\nabla \frac{1}{|x-z|} \cdot g(z) \right] dx \, dy \, dz.$$

Since $\nabla_x |x-y|^{-1} = \nabla_y |x-y|^{-1}$, integration by parts in x and then in

y gives

$$\|p\|^2 = \frac{-1}{(4\pi)^2} \iiint \left[\frac{1}{|x-y|} \nabla(\operatorname{div} g(y)) \right] \cdot \frac{g(z)}{|x-z|} \, dx \, dy \, dz.$$

Now the calculus identity

$$\nabla(\operatorname{div} g) = \Delta g + \operatorname{curl}(\operatorname{curl} g)$$

gives

$$\|p\|^2 = \frac{1}{4\pi} \iint \frac{g(x) \cdot g(z)}{|x-z|} \, dx \, dz$$

$$- \frac{1}{(4\pi)^2} \iiint \left[\frac{1}{|x-y|} \operatorname{curl}(\operatorname{curl} g(y)) \right] \cdot \frac{g(z)}{|x-z|} \, dx \, dy \, dz.$$
$$(6.144)$$

Since the ith component of $\operatorname{curl} g$ is given by $(\operatorname{curl} g)_i = \varepsilon_{ijk} \partial_j g_k$, where the coefficients

$$\varepsilon_{ijk} := \begin{cases} 1 & \text{if } ijk \text{ is an even permutation of 123,} \\ -1 & \text{if } ijk \text{ is an odd permutation of 123,} \\ 0 & \text{otherwise} \end{cases}$$

satisfy $\varepsilon_{ijk} = -\varepsilon_{kji}$, we can write the last triple integral in (6.144) as

$$\iiint \frac{\varepsilon_{ijk} \partial_j \left(\operatorname{curl} g(y)\right)_k g_i(z)}{|x-y| \, |x-z|} \, dx \, dy \, dz,$$

which integrated by parts three times (first in y then in x and z) gives

$$- \iiint \frac{\varepsilon_{ijk} \left(\operatorname{curl} g(y)\right)_k \partial_j g_i(z)}{|x-y| \, |x-z|} \, dx \, dy \, dz$$

$$= \iiint \frac{\left(\operatorname{curl} g(y)\right)_k \varepsilon_{kji} \partial_j g_i(z)}{|x-y| \, |x-z|} \, dx \, dy \, dz$$

$$= \iiint \frac{\left(\operatorname{curl} g(y)\right) \cdot \left(\operatorname{curl} g(z)\right)}{|x-y| \, |x-z|} \, dx \, dy \, dz = \int |F(x)|^2 \, dx \geq 0,$$

where $F(x) := \int \operatorname{curl} g(y) / |x-y| \, dy$. Therefore (6.144) gives

$$\|p\|^2 \leq \frac{1}{4\pi} \iint \frac{g(x) \cdot g(z)}{|x-z|} \, dx \, dz$$

$$= \frac{1}{4\pi} \iint \frac{(J_\varepsilon u_k(x)) \partial_k u_i(x)(J_\varepsilon u_j(z)) \partial_j u_i(z)}{|x-z|} \, dx \, dz$$
$$(6.145)$$

(this inequality appears in Leray (1934b) on p. 233 with "$\leq$" wrongly replaced by "$=$"), from where the Cauchy–Schwarz inequality and Lemma

6.3 give

$$\|p\|^2 \le \frac{1}{4\pi} \int \|\nabla u\| \left(\int \frac{|J_\varepsilon u_k(x)|^2}{|x-z|^2} \, \mathrm{d}x \right)^{1/2} |J_\varepsilon u_j(z)| \, |\nabla u(z)| \, \mathrm{d}z$$

$$\le C\|\nabla u\|^2 \int |J_\varepsilon u_j(z)| \, |\nabla u(z)| \, \mathrm{d}z \le C\|\nabla u\|^3 \|u\|,$$

where we also used $\|J_\varepsilon(\nabla u)\| \le \|\nabla u\|$ and that $\|J_\varepsilon u\| \le \|u\|$ (see Lemma 6.2 (i)).

6.5.8 Smooth approximation of the forcing

Lemma 6.54 (Dini's lemma) *Let I be a compact interval and let $f_n, f \in C(I; \mathbb{R})$ be continuous functions such that $f_n(t) \to f(t)$ as $n \to \infty$ and $f_{n+1}(t) \le f_n(t)$ for each $t \in I$. Then $\|f_n - f\|_{C(I)} \to 0$.*

Proof This is elementary. □

Lemma 6.55 *Let $p \in [1,\infty)$ and $F \in C([0,T), L^p)$. Then for any $T' \in (0,T)$ and any $\varepsilon > 0$ there exists $\widetilde{F} \in C^\infty(\mathbb{R}^3 \times \mathbb{R})$ and $R > 0$ such that $\mathrm{supp}\, \widetilde{F}(t) \subset B(0,R)$ for all $t \in \mathbb{R}$ and*

$$\|F - \widetilde{F}\|_{C([0,T'],L^2)} < \varepsilon. \tag{6.146}$$

Moreover $\max_{t\in[0,T']} \|\widetilde{F}(t)\|_\infty \lesssim \max_{t\in[0,T']} \|F(t)\|_\infty$.

Proof It suffices to consider $T < \infty$. First extend $F|_{[0,T']}$ in time from $[0,T']$ to the whole line by taking

$$F^1(\cdot,t) := \begin{cases} F(\cdot,0) & t < 0 \\ F(\cdot,t) & t \in [0,T'] \\ F(\cdot,T') & t > T' \end{cases}.$$

For $R > 0$ let

$$F_R(x,t) := \chi_{B(0,R)}(x) F^1(x,t).$$

Clearly $F_R \in C(\mathbb{R}; L^p)$ (as a product of two such functions) and so $\|F_R(t) - F^1(t)\|_p$ is continuous in t for each R. Therefore, noting that $\|F_R(t) - F^1(t)\|_p$ is a non-increasing function of R converging to zero as $R \to \infty$ for each $t \in [0,T']$, we can use Dini's Lemma 6.54 to fix $R > 0$ such that

$$\|F_R - F\|_{C([0,T'];L^p)} = \|F_R - F^1\|_{C([0,T'];L^p)} < \varepsilon/3. \tag{6.147}$$

We will now mollify F_R to obtain $\widetilde{F}$. Let η_δ, ξ_δ be mollifiers in $\mathbb{R}^3$ and

$\mathbb{R}$ respectively, that is let $\xi(t) := C \exp(1/(|t|^2 - 1))$ for $t \in (-1, 1)$ and $\xi(t) := 0$ if $t \notin (0, 1)$, where C is chosen such that $\int_{\mathbb{R}} \xi = 1$, and let $\xi_\delta(t) := \xi(t/\delta)/\delta$, $\eta(x) := \xi(|x|)$, $\eta_\delta(x) := \eta(x/\delta)/\delta^3$. Define the mollification F_R^δ of F_R by

$$F_R^\delta(x, t) := \int_{\mathbb{R}} \xi_\delta(s) \int \eta_\delta(y) F_R(x - y, t - s) \, dy \, ds$$

$$= \int_{\mathbb{R}} \xi(s) \int \eta(y) F_R(x - \delta y, t - \delta s) \, dy \, ds.$$

Clearly $\widetilde{F} := F_R^\delta(x, t)$ has the required regularity for each $\delta > 0$ (in particular $\max_{t \in [0, T']} \|\widetilde{F}(t)\|_\infty \leq \max_{t \in [0, T']} \|F(t)\|_\infty$ holds by the property of mollifiers (see (i) in Lemma 6.2)). Therefore the proof will be complete if we show the approximation property (6.146) for some $\delta > 0$. Since

$$F_R(x, t) - F_R^\delta(x, t)$$

$$= \int_{\mathbb{R}} \xi(s) \int \eta(y) \left(F_R(x, t) - F_R(x - \delta y, t - \delta s) \right) \, dy \, ds,$$

we can use the Minkowski inequality (see 6.5) and the triangle inequality

$$\|F_R(x - \delta y, t - \delta s) - F_R(x, t)\|_p \leq \|F_R(x - \delta y, t - \delta s) - F_R(x - \delta y, t)\|_p$$
$$+ \|F_R(x - \delta y, t) - F_R(x, t)\|_p$$

to write

$$\|F_R(t) - F_R^\delta(t)\|_p \leq \int_{\mathbb{R}} \xi(s) \int \eta(y) \|F_R(\cdot - \delta y, t - \delta s) - F_R(\cdot, t)\|_p \, dy \, ds$$

$$\leq \int_{\mathbb{R}} \xi(s) \|F_R(t - \delta s) - F_R(t)\|_p \, ds$$

$$+ \int \eta(y) \|F_R(\cdot - \delta y, t) - F_R(\cdot, t)\|_p \, dy.$$

$$(6.148)$$

From the definition of F^1, F_R is uniformly continuous on $[-1, T' + 1]$ into L^p, so there exists a sufficiently small $\delta_1 > 0$ such that for $\delta \in (0, \delta_1)$

$$\|F_R(t - \delta s) - F_R(t)\|_p < \varepsilon/3$$

for all $t \in [0, T']$, and so the first term on the right-hand side of (6.148) is less than $\varepsilon/3$ for all $t \in [0, T']$. As for the second term, we will show that there exists $\delta \in (0, \delta_1)$ such that

$$\|F_R(\cdot - z, t) - F_R(\cdot, t)\|_p < \varepsilon/3,$$

whenever $|z| < \delta$ and $t \in [0, T']$. Indeed, by the continuity of translation in space of L^p functions for each $t_0 \in [0, T']$ there exists a δ_{t_0} such that $\|F_R(\cdot - z, t_0) - F_R(\cdot, t_0)\|_p < \varepsilon/3$ whenever $|z| < \delta_{t_0}$. Moreover, the continuity of F_R in time into L^p and triangle inequality gives that $\|F_R(\cdot - z, t) - F_R(\cdot, t)\|_p < \varepsilon/3$ whenever $|z| < \delta_{t_0}$ and t belongs to some open set J_{t_0} containing t_0. By compactness of $[0, T']$ we obtain a finite cover $\{J_{t_i}\}_{i=1,\dots,m}$ of $[0, T']$ consisting of such open sets and δ is obtained by taking the minimum of the corresponding δ_{t_i}, $i = 1, \dots, m$. This means in particular that

$$\|F_R(\cdot - \delta y, t) - F_R(\cdot, t)\|_p < \varepsilon/3$$

whenever $|y| < 1$, that is for all $y \in \operatorname{supp} \eta$. Therefore, for such δ the second term on the right-hand side of (6.148) is bounded by $\varepsilon/3$ uniformly in $t \in [0, T']$. The required approximation property (6.146) therefore follows directly from (6.147) and (6.148). □

References

Adams, R.A. and Fournier, J.J.F. (2003) *Sobolev spaces* Elsevier/Academic Press, Amsterdam

Beale, J.T., Kato, T., & Majda, A. (1984) Remarks on the breakdown of smooth solutions for the 3-D Euler equation. *Comm. Math. Phys.* **94**, 61–66.

Besicovitch, A.S. (1935) On the sum of digits of real numbers represented in the dyadic system. *Math. Ann.* **110**, no. 1, 321–330.

Besicovitch, A.S. & Taylor, S.J. (1954) On the complementary intervals of a linear closed set of zero Lebesgue measure. *J. London Math. Soc.* **29**, 449–459.

Bouligand, G. (1928) Ensembles impropres et nombre dimensionnel. *Bull. Sciences Mathématiques* **52**, (September), 320–334, 361–376.

Caffarelli, L., Kohn, R., & Nirenberg, L. (1982) Partial regularity of suitable weak solutions of the Navier-Stokes equations. *Comm. Pure Appl. Math.* **35**, no. 6, 771–831.

Carothers, N.L. (2000) *Real analysis.* Cambridge University Press, Cambridge.

Constantin, P. (2008) Euler and Navier–Stokes Equations. *Publ. Mat* **52**, no. 2, 235–265.

Constantin, P. & Foias, C. (1988) *Navier–Stokes Equations.* The University of Chicago Press, Chicago.

Escauriaza, L., Seregin, G.A., & Šverák, V. (2003) $L_{3,\infty}$-solutions of Navier–Stokes equations and backwards uniqueness. *Russian Math. Surveys* **58**, 211–250.

Evans, L.C. (2010) *Partial differential equations.* American Mathematical Society. Providence R.I.

Evans, L.C. & Gariepy, R.F. (2015) *Measure theory and fine properties of functions*. Revised edn. Textbooks in Mathematics. CRC Press, Boca Raton, FL.

Falconer, K. (2014) *Fractal geometry, Mathematical foundations and applications*. John Wiley & Sons, Ltd., Chichester.

Fefferman, C. (2006) Existence and Smoothness of the Navier–Stokes Equation. In: Carlson, J.A., Jaffe, A., & Wiles, A. (eds). *The Millennium Prize Problems*. AMS.

Fujita, H. & Kato, T. (1964) On the Navier–Stokes initial value problem. I. *Arch. Ration. Mech. Anal.* **16**, 269–315.

Giga, M.H., Giga, Y., & Saal, J. (2010) *Nonlinear partial differential equations, Asymptotic behavior of solutions and self-similar solutions*. Progress in Nonlinear Differential Equations and their Applications **79**. Birkhäuser Boston, Inc., Boston, MA.

Hausdorff, F. (1918) Dimension und äußeres Maß. *Math. Ann.* **79**, no. 1–2, 157–179.

Helly, E. (1912) Über lineare Funktionaloperationen. *Sitzungsberichte der Kaiserlichen Akademie der Wissenschaften zu Wien. Mathematisch-Naturwissenschaftlichen Klasse,* **121**, II A1, 265–297.

Hopf, E. (1951) Über die Anfangswertaufgabe für die hydrodynamischen Grundgleichungen. *Math. Nachr.* **4**, 213–231. (An English translation due to Andreas Klöckner is available at *http://www.dam.brown.edu/people/menon/publications/notes/hopf-NS.pdf*).

Kato, T. (1984) Strong L^p-solutions of the Navier–Stokes equations in $\mathbb{R}^m$ with applications to weak solutions. *Math. Zeit.* **187**, 471–480.

Kiselev, A. & Ladyzhenskaya, O.A. (1957) On the existence and uniqueness of the solution of the nonstationary problem for a viscous, incompressible fluid. *Izv. Akad. Nauk SSSR. Ser. Mat.* **21**, 655–680.

Kress, R. (2014) *Linear integral equations*. Applied Mathematical Sciences **82**. Springer, New York.

Ladyženskaja, O.A. (1959) Solution "in the large" of the nonstationary boundary value problem for the Navier–Stokes system in two space variables. *Comm. Pure Appl. Math.* **12**, 427–433.

Ladyženskaja, O.A. (1967) Uniqueness and smoothness of generalized solutions of Navier–Stokes equations. *Zap. Naučn. Sem. Leningrad. Otdel. Mat. Inst. Steklov. (LOMI)* **5**, 169–185.

Lapidus, M.L. & van Frankenhuijsen, M. (2006) *Fractal geometry, complex dimensions and zeta functions*. Springer Monographs in Mathematics. Springer, New York.

Lax, P. (1998) *Jean Leray and Partial Differential Equations*. Springer-Verlag, Berlin; Société Mathématique de France, Paris. Introduction to *Jean Leray, Selected papers. Œuvres scientifiques. Vol. II: Fluid dynamics and real partial differential equations/Équations aux dérivées partielles réelles et mécanique des fluides*, Springer.

Lemarié-Rieusset, P.G. (2002) *Recent developments in the Navier-Stokes problem.* Chapman & Hall/CRC Research Notes in Mathematics **431**. Chapman & Hall/CRC, Boca Raton, FL.

Lemarié-Rieusset, P.G. (2016) *The Navier-Stokes problem in the 21st century.* CRC Press, Boca Raton, FL.

Leray, J. (1933) Étude de diverses équations intégrales non linéaires et de quelques problèmes que pose l'hydrodynamique. *J. Math. Pures Appl. (9)* **12**, 1–82.

Leray, J. (1934a) Essai sur les mouvements plans d'un liquide visquex que limitent des parois. *J. Math. Pures Appl. (9)* **13**, 331–418. (Available in *Jean Leray, Selected papers. Œuvres scientifiques. Vol. II: Fluid dynamics and real partial differential equations/Équations aux dérivées partielles réelles et mécanique des fluides*, Springer, 1998).

Leray, J. (1934b) Sur le mouvement d'un liquide visqueux emplissant l'espace. *Acta Math.* **63**, 193–248. (An English translation due to Robert Terrell is available at *http://www.math.cornell.edu/~bterrell/leray.pdf* and *https://arxiv.org/abs/1604.02484*.).

Majda, A.J. & Bertozzi, A.L. (2002) *Vorticity and incompressible flow.* Cambridge university Press, Cambridge.

Navier, C.L.M.H. (1822) Mémoire sur les lois du mouvement des fluides. *Mém. Ac. R. Sc. de l'Institut de France* **6**, 389–440.

Nečas, J., Růžička, M., & Šverák, V. (1996) On Leray's self-similar solutions of the Navier–Stokes equations. *Acta Math.* **176**, no. 2, 283–294.

Oseen, C.W. (1911) Sur les formules de green généralisées qui se présentent dans l'hydrodynamique et sur quelquesunes de leurs applications. *Acta. Math.* **34**, 205–288.

Pontrjagin, L. & Schnirelmann, L. (1932) Sur une propriété métrique de la dimension. *Ann. of Math.* **33**, no. 1, 156–162.

Prodi, G. (1959) Un teorema di unicità per le equazioni di Navier–Stokes. *Ann. Mat. Pura Appl.* **48**, 173–182.

Robinson, J.C. (2006) Regularity and singularity in the three-dimensional Navier-Stokes equations. *Bol. Soc. Esp. Mat. Apl. SeMA* **35**, 43–71.

Robinson, J.C., Rodrigo, J.L., & Sadowski, W. (2016) *The three-dimensional Navier–Stokes equations. Classical Theory.* Cambridge University Press, Cambridge.

Roubíček, T. (2013) *Nonlinear partial differential equations with applications*, Birkhäuser/Springer Basel AG, Basel

Scheffer, V. (1976) Turbulence and Hausdorff dimension. In *Turbulence and Navier–Stokes equations* (Proc. Conf., Univ. Paris-Sud, Orsay, 1975), Springer Lecture Notes in Mathematics **565**. Springer Verlag, Berlin, 174–183.

Scheffer, V. (1977) Hausdorff measure and the Navier–Stokes equations. *Comm. Math. Phys.* **55**, 97–112.

Schwartz, L. (1950) *Théorie des distributions. Tome I.* Actualités Sci. Ind., no. 1091, Publ. Inst. Math. Univ. Strasbourg 9. Hermann & Cie., Paris.

Serrin, J. (1962) On the interior regularity of weak solutions of the Navier-Stokes equations. *Arch. Rational Mech. Anal.* **9**, 187–195.

Sobolev, S. (1936) Méthode nouvelle à résoudre le problème de Cauchy pour les équations linéaires hyperboliques normales. *Rec. Math. [Mat. Sbornik]* *N.S.* **1(43)**, no. 1, 39–72.

Sohr, H. (2001) *The Navier–Stokes equations, An elementary functional analytic approach.* Birkhäuser Advanced Texts. Birkhäuser Verlag, Basel.

Stein, E.M. & Weiss, G. (1971) *Introduction to Fourier analysis on Euclidean spaces.* Princeton Mathematical Series, **32**. Princeton University Press, Princeton, N.J.

Stokes, G.G. (1845) On the theories of the internal friction of fluids in motion, and of the equilibrium and motion of elastic solids. *Trans. Cam. Phil. Soc.* **8**, 287–319.

Strichartz, R. S. (1994) *A guide to distribution theory and Fourier transforms,* CRC Press, Boca Raton, FL

Temam, R. (2001) *Navier–Stokes equations.* AMS Chelsea Publishing, Providence, RI.

7

Stable mild Navier–Stokes solutions by iteration of linear singular Volterra integral equations

Reimund Rautmann

Institut für Mathematik der Universität Paderborn,
D33095 Paderborn. Germany.
`rautmann@math.uni-paderborn.de`

Dedicated to Professor Solonnikov on the occasion of his 85th birthday

Abstract

By their use of mild solutions, Fujita-Kato and later on Giga-Miyakawa opened the way to solving the initial-boundary value problem for the Navier–Stokes equations with the help of the contracting mapping principle in suitable Banach spaces, on any smoothly bounded domain $\Omega \subset \mathbb{R}^n, n \geq 2$, globally in time in case of sufficiently small data. We will consider a variant of these classical approximation schemes: by iterative solution of linear singular Volterra integral equations, on any compact time interval J, again we find the existence of a unique mild Navier–Stokes solution under smallness conditions, but moreover we get the stability of each (possibly large) mild solution, inside a scale of Banach spaces which are imbedded in some $C^0(J, L^r(\Omega))$, $1 < r < \infty$.

7.1 The initial-boundary value problem of the Navier–Stokes equations

The initial-boundary value problem of the Navier–Stokes equations

$$\frac{\partial}{\partial t}v - \Delta v + v \cdot \nabla v + \nabla p = f, \tag{7.1}$$

$$\operatorname{div} v = 0, \tag{7.2}$$

$$v_{|\partial\Omega} = 0, \tag{7.3}$$

$$v_{|t=0} = v(0, \cdot) \tag{7.4}$$

Published as part of *Partial Differential Equations in Fluid Mechanics*, edited by C.L. Fefferman, J.C. Robinson, & J.L. Rodrigo. © Cambridge University Press 2018.

describes the evolution at times $t \in [0, a]$ of the velocity $v(t, x) \in \mathbb{R}^n$ and of the pressure function $p(t, x) \in \mathbb{R}$ in a viscous incompressible fluid flow at points x of some bounded open set $\Omega \subset \mathbb{R}^n$, $n \geq 2$, with smooth boundary $\partial\Omega$. For simplicity we assume constant mass density $\rho = 1$ and viscosity constant $\nu = 1$.

We often omit the domain in the notation for Sobolev spaces, writing $H^{m,r} = H^{m,r}(\Omega)$ and $H^{0,r} = L^r(\Omega)$, $m \in \mathbb{N}$, $1 < r < \infty$.

Equation (7.2) (understood in the generalized sense) holds in the space

$$L^r_\sigma := \overline{C^\infty_{c,\sigma}(\Omega)}^{\|\cdot\|_{L^r}},$$

the L^r-closure of the divergence-free C^∞-vector functions that have compact support in Ω; similarly the boundary condition (7.3) holds in the space

$$\overset{\circ}{H}{}^{1,r} := \overline{C^\infty_c(\Omega)}^{\|\cdot\|_{H^{1,r}}},$$

the $H^{1,r}$-closure of the C^∞-vector functions having compact support in Ω. The Helmholtz–Weyl projection $P : L^r(\Omega) \to L^r_\sigma(\Omega)$ and the Laplacian $\Delta = \sum_{j=1}^n \frac{\partial^2}{\partial x_j^2}$ define the Stokes operator $A := -P\Delta$ with its domain

$$D_A = H^{2,r} \cap \overset{\circ}{H}{}^{1,r} \cap L^r_\sigma(\Omega),$$

the operator A being the generator of the analytic semigroup e^{-tA}, $t \geq 0$. Therefore the fractional powers A^α with their domains D_{A^α}, $\alpha \in \mathbb{R}$, are well defined, D_{A^α} being continuously imbedded in the fractional order Sobolev space $H^{2\alpha,r}(\Omega) \cap L^r_\sigma(\Omega)$ if $\alpha > 0$, and A^α being bounded on L^r_σ if $\alpha \leq 0$, (Fujita & Kato, 1964; Giga, 1981, 1982, 1985; Pazy, 1983; Sohr, 2001; Solonnikov, 1977; von Wahl, 1985).

Application of P on both sides of (7.1) leads to the evolution equation

$$\frac{\partial}{\partial t}v + Av + P(v \cdot \nabla v) = Pf, t > 0, v_{|t=0} = v(0). \tag{7.5}$$

The equation

$$Av + P(v \cdot \nabla v) = Pf, \quad x \in \Omega, \tag{7.6}$$

models the stationary case $\partial v / \partial t \equiv 0$.

7.2 Results on stability of Navier–Stokes solutions

For a short review of known results on stability of Navier–Stokes solutions, besides (7.5) or (7.6) we consider neighbouring equations

$$\frac{\partial}{\partial t}\widetilde{v} + A\widetilde{v} + P(\widetilde{v}\cdot\nabla\widetilde{v}) = P\widetilde{f}, \widetilde{v}_{|t=0} = \widetilde{v}(0), \qquad (7.7)$$

with data $\widetilde{v}(0), P\widetilde{f}$. Subtracting (7.5) or (7.6) from (7.7), for the differences $u = \widetilde{v} - v, Pg = P\widetilde{f} - Pf$ we find

$$\frac{\partial}{\partial t}u + Au + P[u\cdot\nabla u + v\cdot\nabla u + u\cdot\nabla v] = Pg, t > 0, u_{|t=0} = u(0). \quad (7.8)$$

The (uniform) stability of any solution v to (7.5) on $(0,\infty)$ (or to (7.6)) requires that with this v in (7.8)

(i) for any sufficiently small data $(u(0), Pg)$, the solution u to (7.8) exists on $(0,\infty)$, and

(ii) some norm $\|u(t)\|_*$ can be controlled (uniformly in t) by a suitable norm of the data of u.

Any solution v to (7.5) on $(0,\infty)$ (or to (7.6)) is asymptotically stable, if (i) holds, but instead of (ii)

(iii) some norm $\|u(t)\|_*$ tends to zero as $t \to \infty$.

On domains $\Omega \subset \mathbb{R}^3$ with uniformly smooth boundary $\partial\Omega$, Ponce et al. (1994) showed that if v is any strong solution to (7.5) on $(0,\infty)$ with data

$$v(0) \in \overset{\circ}{H}{}^{1,2} \cap L^2_\sigma, \quad \text{and} \quad Pf \in L^1([0,\infty), L^2_\sigma(\Omega)) \cap L^2([0,\infty), L^2_\sigma(\Omega))$$

then if $\nabla v \in L^4([0,\infty), L^2(\Omega))$ there exists a $\delta > 0$ such that for any

$$u(0) \in \overset{\circ}{H}{}^{1,2} \cap L^2_\sigma \quad \text{and} \quad Pg \in L^1([0,\infty), L^2_\sigma(\Omega)) \cap L^2([0,\infty), L^2_\sigma(\Omega))$$

with

$$\|u(0)\|_{H^1} + \int_0^\infty \|Pg(t)\|_{L^2}\, \mathrm{d}t \leq \delta$$

there exists a unique global strong solution u to (7.8). This solution u satisfies

$$\sup_{t\geq 0} \|u(t)\|_{H^1(\Omega)} \leq M(\delta),$$

where $M(\delta) \to 0$ for $\delta \to +0$.

On domains $\Omega \subset \mathbb{R}^3$ with uniform C^3-boundary $\partial\Omega$, Heywood (1980) has proved the existence of global strong solutions v to (7.5) with $f = 0$ in case of sufficiently small $\|v(0)\|$, v having the decay rate

$$\sup_{x \in \Omega} |v(t,x)| \le \text{const} \cdot t^{-\frac{1}{2}}$$

as $t \to \infty$.

On smoothly bounded domains $\Omega \subset \mathbb{R}^n, n \ge 3$, Kozono & Ozawa (1990) have found the existence of a unique strong solution u to (7.8) on $(0, \infty)$ under smallness assumptions on $v \in D_A$ from (7.6), for all sufficiently regular and small data $u(0), Pg$. They have established decay rates of $\|A^\alpha u(t)\|_{L^r(\Omega)}$ with $t \to \infty$, $\gamma \le \alpha < 1 - \delta$ with some restrictions to $r > \max\left(\frac{n}{3}, 1\right), \alpha, \gamma, \delta$ in dependence on n. The main tool in their proof is the analysis of the perturbed Stokes operator

$$Lu = Au + P[u \cdot \nabla v + v \cdot \nabla u]$$

and of the semigroup e^{-Lt} generated by L under suitable restrictions.

Kozono & Ogawa (1994) consider smooth solutions v to (7.6) on exterior domains $\Omega \subset \mathbb{R}^n$, $n \ge 3$, Ω having a compact complement $\mathcal{C}\Omega \subset \mathbb{R}^n$ and smooth boundary $\partial\Omega$. In the case of sufficiently small $\|v\|_{L^n(\Omega)} + \|\nabla v\|_{L^{\frac{n}{2}}(\Omega)}$ and small $\|u(0)\|_{L^n(\Omega)}$ they established unique global strong solutions u to (7.8) with $Pg = 0$ which decay such that $t^{\frac{1}{4}} \|u(t)\|_{L^{2n}(\Omega)}$ tends to zero as $t \to \infty$. In addition they give explicit decay rates for $u(t)$ and $\nabla u(t)$ in a scale of $L^p(\Omega)$ norms. Their proof uses the analysis of the perturbed Stokes operator L and $L^q - L^r$-estimates of the semigroup e^{-Lt} mentioned above.

On exterior domains $\Omega \subset \mathbb{R}^3$ that are the complement of a finite number of isolated smoothly-bounded compact sets, Masuda (1975) treated weak solutions u to (7.8) with some stationary solution $v \in C^1(\bar{\Omega})$ of (7.6), $\nabla v \in L^3(\Omega)$, v obeying $\sup_{x \in \Omega} |x| \cdot |v(x) - v^\infty| < \frac{1}{2}$, where $v^\infty \in \mathbb{R}^3$ is some constant vector. Under restrictions on g he proved that u becomes a strong solution to (7.8) for sufficiently large t, and he has shown decay rates for $\|\nabla u(t)\|_{L^2(\Omega)}, \|u(t)\|_{L^2(\Omega)}$ as $t \to \infty$.

On domains $\Omega \subset \mathbb{R}^3$ with uniform C^3-boundary, Kozono (2000) has proved the asymptotic stability of weak solutions v to (7.5) which belong to Serrin's class $L^p((0,\infty), L^q(\Omega))$, $\frac{2}{p} + \frac{3}{q} = 1$, $q > 3$. When $\partial\Omega$ is compact he gives explicit decay rates for solutions $u(t)$ to (7.8), and for $\nabla u(t)$ as $t \to \infty$ in a scale of $L^r(\Omega)$-norms.

Considering equation (7.5) with $f = 0$ in the whole $\mathbb{R}^3$, Beirão da Veiga & Secchi (1987) show the asymptotic stability in $L^p(\mathbb{R}^3), p > 3$, of any strong solution $v \in L^\infty((0,\infty), L^{p+2}(\mathbb{R}^3)), v(0) \in L^1 \cap L^{p+2}(\mathbb{R}^3)$

to (7.5), and they find explicit decay rates of $\|u(t)\|_{L^p(\mathbb{R}^3)}$ for small $\|u(0)\|_{L^p(\mathbb{R}^3)}$ in (7.8).

On the whole of $\mathbb{R}^n$, $n \geq 3$, Wiegner (1990) has proved a general result on asymptotic stability and decay rates of strong solutions v to (7.5) in Serrin's class $L^s((0,\infty), L^p(\mathbb{R}^n))$, $\frac{2}{s} + \frac{n}{p} = 1, p > n$.

In the whole of $\mathbb{R}^n$, $n \geq 2$, Kawanago (1998) shows the stability of strong solutions $v \in C^0([0,\infty), L_\sigma^n(\mathbb{R}^n))$ to (7.5) with $Pf = 0$, that fulfil

$$\lim_{t\to\infty} \|v(t)\|_{L^n(\mathbb{R}^n)} = 0 \quad \text{and} \quad \int_0^\infty \|v(t)\|_{L^{n+2}(\mathbb{R}^n)}^{n+2}\, dt < \infty.$$

In the general frame of Besov spaces, stability theorems for Navier–Stokes solutions on the whole $\mathbb{R}^3$ have been proved in Auscher, Dubois, & Tchamitchian (2004), Gallagher et al. (2003), and Lemarié-Rieusset (2016).

7.3 Bounds on $P(u \cdot \nabla v)$ and on e^{-tA}

The key for the construction of solutions to (7.5) are norm estimates for the nonlinear term.

Lemma 7.1 (Fujita & Kato, 1964; Giga & Miyakawa, 1985)
Let $0 \leq \delta < \frac{1}{2}[1 + n(1 - \frac{1}{r})]$, $u \in D_{A^\theta}$, and $v \in D_{A^\rho}$. Then

$$\|A^{-\delta}P(u \cdot \nabla v)\|_{L^r(\Omega)} \leq M \|A^\theta u\|_{L^r(\Omega)} \|A^\rho v\|_{L^r(\Omega)}, \qquad (7.9)$$

for some constant $M = M(\delta, \theta, \rho, r)$, provided that

$$\frac{1}{2}\left[1 + \frac{n}{r}\right] \leq \delta + \theta + \rho, \quad 0 < \theta, \quad 0 < \rho, \quad \frac{1}{2} < \delta + \rho, \quad 1 < r < \infty.$$

A special case of (7.9) is given by Sobolevskii's inequality

$$\|A^{-\frac{1}{4}}P(u\nabla v)\|_{L^2(\Omega)} \leq M\|A^{\frac{1}{2}}u\|_{L^2(\Omega)}\|A^{\frac{1}{2}}v\|_{L^2(\Omega)} \qquad (7.10)$$

when $n = 3$, see Sobolevskii (1959).

Let J be any real interval and X any Banach space with norm $\|\cdot\|_X$. As usual, by $C^0(J, X)$ we denote the Banach space of continuous maps $F: J \to X$, equipped with the supremum norm

$$\|F\| := \sup_{t \in J} \|F(t)\|_X.$$

We will need the following lemma in what follows, which collects results from Fujita & Kato (1964), Giga (1981), Giga (1982), Miyakawa (1981), Pazy (1983), Sohr (2001), and von Wahl (1985).

Lemma 7.2 *Let $u \in L^r_\sigma(\Omega)$ and take $t > 0$, $\alpha \geq 0$. Then*

(i) $\|A^\alpha e^{-tA} u\|_{L^r(\Omega)} \leq C_\alpha\, t^{-\alpha}\, \|u\|_{L^r(\Omega)}$, *where $C_\alpha > 0$;*

(ii) *for $\alpha > 0$*

$$\|t^\alpha\, A^\alpha\, e^{-tA} u\|_{L^r(\Omega)} \to 0 \qquad as \qquad t \to 0^+$$

and

$$\|(1 - e^{-tA}) u\|_{L^r(\Omega)} \to 0 \qquad as \qquad t \to 0^+;$$

(iii) *A^α and e^{-tA} commute on D_{A^α}; and*

(iv) *if $u \in C^0([0,a], L^r_\sigma)$, $\alpha \in [0,1)$, and $a \in (0,\infty)$, then U defined by*

$$U(t) = A^\alpha \int_0^t e^{-(t-s)A} u(s)\, ds = \int_0^t A^\alpha e^{-(t-s)A} u(s)\, ds$$

satisfies $U \in C^0([0,a], L^r_\sigma)$, with

$$\|U\| \leq T_{a,\alpha}\, C_\alpha\, \|u\|,$$

where

$$T_{t,\alpha} := \frac{t^{1-\alpha}}{1-\alpha} = \int_0^t \frac{ds}{(t-s)^\alpha}.$$

7.4 The approximation schemes of Fujita–Kato and Giga–Miyakawa: sketch of the proof in Fujita & Kato (1964) for the uniform bound of the approximations

For solving (7.5), Fujita & Kato (1964) and Giga & Miyakawa (1985) used the approximation scheme

$$A^\alpha v_{m+1}(t) = A^\alpha v_0(t) - \int_0^t A^{\alpha+\delta} e^{-(t-s)A} A^{-\delta} P(v_m \cdot \nabla v_m)(s)\, ds,$$

$$\tag{7.11}$$

$$A^\alpha v_0(t) = A^\alpha e^{-tA} v(0) + \int_0^t A^{\alpha+\delta} e^{-(t-s)A} A^{-\delta} P f(s)\, ds, \qquad (7.12)$$

$m \in \mathbb{N}, t \in (0, a], v(0) \in D_{A^\gamma}$, with suitable restrictions on Pf and on the exponents $n, r, \alpha, \gamma, \delta$, (Giga & Miyakawa, 1985, pp. 271–275).

In Fujita & Kato (1964), considering the case $n = 3$, $r = 2$, $\gamma = \frac{1}{4}$,

$\alpha \geq \frac{1}{2}$, the authors proved existence and convergence of the sequence $(A^{\frac{1}{2}} v_m(t))_{m \in \mathbb{N}}$ from (7.11) with respect to the weighted norm

$$K_m(t) = \sup_{0 < \tau \leq t} \tau^{\frac{1}{4}} \| A^{\frac{1}{2}} v_m(\tau) \|_{L^2(\Omega)} \tag{7.13}$$

$t \in (0, a]$, under smallness assumptions on $K_0(t)$.

Sobolevskii's inequality (7.10) and Lemma 7.2 lead from (7.11) to

$$K_{m+1}(t) \leq K_0(t) + C\, K_m^2(t) := Q(K_m) \tag{7.14}$$

with $C = M\, B(\frac{1}{4}, \frac{1}{2})$, where $B(\frac{1}{4}, \frac{1}{2})$ is the beta function. When

$$4C K_0(t) < 1 \tag{7.15}$$

there exists a positive lower root

$$\chi_t = \frac{1}{2C} \left[1 - \sqrt{1 - 4C K_0(t)} \right]$$

of the equation

$$Q(\chi) = \chi \tag{7.16}$$

which is quadratic in the real variable χ. Equation (7.16) implies that $K_0(t) \leq \chi_t$, and, as seen by induction from (7.14),

$$K_m(t) \leq \chi_t \quad \text{for all } m \in \mathbb{N}. \tag{7.17}$$

Therefore the monotonicity of $K_m(t)$ in t implies that (7.17) holds uniformly in $t \in (0, a]$, if $K_0(a) < \frac{1}{4C}$. Similarly the convergence of the sequence $(A^{\frac{1}{2}} v(t))_{m \in \mathbb{N}}, t \in (0, a]$ results under smallness conditions on $K_0(t)$ (even on $(0, \infty)$), see pp. 284–290 in Fujita & Kato (1964).

7.5 Stable mild Navier–Stokes solutions Theorems 7.4–7.6

Mild solutions of (7.5) are the solutions to its integral representation

$$v(t) = v_0(t) - \int_0^t e^{-(t-s)A} P(v \cdot \nabla v)(s)\, ds,$$
$$v_0(t) = e^{-tA} v(0) + \int_0^t e^{-(t-s)A} P f(s)\, ds, \tag{7.18}$$

$t \in (0, a]$. The right-hand sides in (7.18) are well defined under suitable regularity assumptions according to the two lemmas above.

In order to prove the stability of mild Navier–Stokes solutions with

$v \in C^0([0,a], D_{A^\alpha})$, when $\|\|A^\alpha v\|\|$ and $a \in (0, \infty)$ may be large, we will restrict ourselves to exponents $\alpha = \theta = \rho$, for which Lemma 7.1 holds provided that

$$0 \leq \delta < \frac{1}{2}\left[1 + n(1 - \frac{1}{r})\right], \quad \frac{1}{2}\left[1 + \frac{n}{r}\right] \leq \delta + 2\alpha,$$
$$\frac{1}{2} < \alpha + \delta, \quad 0 < \alpha, \quad 1 < r < \infty, \quad 2 \leq n \in \mathbb{N}. \tag{7.19}$$

In the following let the fixed values n, r, α, δ always obey (7.19) and additionally

$$\alpha + \delta < 1. \tag{7.20}$$

For simplicity we will write

$$\|v\| = \|v\|_{L^r(\Omega)}, \quad \beta = \alpha + \delta, \quad J = [0, a].$$

Note that

(i) when $n = 3, r = 2$, the requirements (7.19), (7.20) hold with

$$\alpha \in \left(\frac{1}{4}, \frac{5}{8}\right], \quad \frac{5}{4} - 2\alpha < \delta < 1 - \alpha \quad \text{or} \quad \alpha \in \left(\frac{5}{8}, 1\right), \quad 0 \leq \delta < 1 - \alpha, \tag{7.21}$$

(ii) when $n = 3 = r$, these requirements are fulfilled with

$$\alpha \in \left(0, \frac{1}{2}\right], \quad 1 - 2\alpha < \delta < 1 - \alpha \quad \text{or} \quad \alpha \in (\frac{1}{2}, 1), \quad 0 \leq \delta < 1 - \alpha. \tag{7.22}$$

As we will see, mild solutions $v \in C^0(J, D_{A^\alpha})$ to (7.18) result from constructing solutions $A^\alpha v \in C^0(J, L^r_\sigma(\Omega))$ to the nonlinear integral equation

$$A^\alpha v(t) = A^\alpha v_0(t) - \int_0^t A^\beta e^{-(t-s)A} A^{-\delta} P(v \cdot \nabla v)(s)\, ds,$$
$$A^\alpha v_0(t) = A^\alpha e^{-tA} v(0) + \int_0^t A^\beta e^{-(t-s)A} A^{-\delta} Pf(s)\, ds, \tag{7.23}$$

$t \in J$, with prescribed $v(0) \in D_{A^\alpha}$, $A^{-\delta} Pf \in C^0(J, L^r_\sigma)$, and $\beta = \alpha + \delta$.
    Problems with nearby data satisfy

$$A^\alpha \widetilde{v}(t) = A^\alpha \widetilde{v}_0(t) - \int_0^t A^\beta e^{-(t-s)A} A^{-\delta} P(\widetilde{v} \cdot \nabla \widetilde{v})(s)\, ds,$$
$$A^\alpha \widetilde{v}_0(t) = A^\alpha e^{-tA} \widetilde{v}(0) + \int_0^t A^\beta e^{-(t-s)A} A^{-\delta} P\widetilde{f}(s)\, ds, \tag{7.24}$$

$t \in J$, with given $\widetilde{v}(0) \in D_{A^\alpha}, A^{-\delta} P \widetilde{f} \in C^0(J, L_\sigma^r)$,

$$\|A^\alpha(\widetilde{v}(0) - v(0))\| \leq \zeta, \qquad \left\|\left|A^{-\delta} P(\widetilde{f} - f)\right|\right\| \leq \zeta$$

and any (possibly small) constant $\zeta > 0$.

Definition 7.3   Any solution $v \in C^0(J, D_{A^\alpha})$ to (7.23) will be called *stable* if for each prescribed $\varepsilon > 0$ there exists some bound $\zeta > 0$, where $\zeta = \zeta(\varepsilon)$ depends on ε, such that each neighbouring problem (7.24) with the bound $\zeta = \zeta(\varepsilon)$ admits some solution $\widetilde{v} \in C^0(J, D_{A^\alpha})$ that obeys $\|A^\alpha(\widetilde{v} - v)\| \leq \varepsilon$.

Subtracting equation (7.23) from (7.24), we obtain coupled integral equations for the differences $u = \widetilde{v} - v$ and $g = \widetilde{f} - f$:

$$A^\alpha u(t) = A^\alpha u_0(t)$$
$$- \int_0^t A^\beta e^{-(t-s)A} A^{-\delta} P\left[u \cdot \nabla u + \kappa(v \cdot \nabla u + u \cdot \nabla v)\right](s)\, \mathrm{d}s,$$

$$A^\alpha u_0(t) = A^\alpha e^{-tA} u(0) + \int_0^t A^\beta e^{-(t-s)A} A^{-\delta} P g(s)\, \mathrm{d}s,$$

$$(7.25)$$

where $\kappa = 1, t \in J$.

The stability of any solution $v \in C^0(J, D_{A^\alpha})$ to (7.23) is guaranteed by the following theorem.

Theorem 7.4   *Let* $v \in C^0(J, D_{A^\alpha})$ *and* $\kappa = 0$ *or* $\kappa = 1$. *Then*

(i) *for each* $\varepsilon > 0$ *there exists some* $\zeta > 0$, $\zeta = \zeta(\varepsilon)$ *depending on* ε, *such that for* $u(0) \in D_{A^\alpha}$ *and* $A^{-\delta} P g \in C^0(J, L_\sigma^r)$ *with* $\|A^\alpha u_0\| \leq \zeta$, *the integral equation (7.25) admits a solution* $u \in C^0(J, D_{A^\alpha})$ *with* $\|A^\alpha u\| \leq \varepsilon$,

(ii) *any solution* $u \in C^0(J, D_{A^\alpha})$ *of (7.25) is uniquely determined by its data* $A^\alpha u_0 \in C^0(J, L_\sigma^r)$.

Setting $\kappa = 0$ in this theorem yields the following result.

Theorem 7.5   *For any prescribed* $v(0) \in D_{A^\alpha}, A^{-\delta} P f \in C^0(J, L_\sigma^r)$, *for which the norms* $\|A^\alpha v(0)\|$ *and* $\left\|\left|A^{-\delta} P f\right|\right\|$ *are sufficiently small there exists a unique solution* $v \in C^0(J, D_{A^\alpha})$ *to (7.23)*.

Theorem 7.5 is special case of the results in Fujita & Kato (1964) and Giga & Miyakawa (1985), which have been proved there in another way.

Setting $\kappa = 1$ in Theorem 7.4 we obtain the following result.

Theorem 7.6   *Each solution* $v \in C^0(J, D_{A^\alpha})$ *to (7.23) is stable*.

Note that Theorem 7.4 implies that the set of all data $v(0) \in D_{A^\alpha}$, $A^{-\delta} P f \in C^0(J, L^r_\sigma)$ that admit solutions $v \in C^0(J, D_{A^\alpha})$ to (7.23) is open with respect to the supremum norm of $A^\alpha v_0$. For this norm, from Lemma 7.1 and 7.2 we find the estimate

$$\||A^\alpha v_0\|| \le C_0 \, \||A^\alpha v(0)\|| + T_{a,\beta} C_\beta \||A^{-\delta} P f\||.$$

Theorems 7.4–7.6 will be proved in Section 7.7 below. As we will see, iterative schemes that converge on the whole of J to the unique solution of (7.25) (even in case of large $\||A^\alpha v\||$) can be constructed by iterative solution of linear singular Volterra integral equations.

7.6 Basic results on linear singular Volterra integral equations

For any Banach space X with norm $\| \cdot \|_X$ let $B = B(X)$ denote the Banach space of bounded linear operators $S \colon X \to X$, S with norm $\|S\|_B$. On the triangle

$$\mathcal{T} := \{(t, s) \in \mathbb{R}^2 : 0 \le s \le t \le a\}, \qquad 0 < a < \infty,$$

let $H \colon \mathcal{T} \to B(X)$ with $\|H(t, s)\|_B \le N$ denote any uniformly bounded, strongly continuous function.

Setting $J = [0, a]$ we obtain the following result.

Proposition 7.7 *To any given* $g \in C^0(J, X), \lambda \in \mathbb{R}, \alpha \in (0, 1)$, *the singular Volterra integral equation*

$$u(t) = g(t) + \lambda \int_0^t \frac{H(t, s)}{(t - s)^\alpha} u(s) \, ds \qquad (7.26)$$

admits a unique solution $u \in C^0(J, X)$.

Proposition 7.8 *The solution* u *of* (7.26) *from Proposition 7.7 has the representation*

$$u(t) = g(t) + \lambda \int_0^t \frac{\widetilde{H}(t, s, \lambda)}{(t - s)^\alpha} g(s) \, ds \qquad (7.27)$$

by means of the resolvent kernel

$$\widetilde{H}(t, s, \lambda) \in B(X), \qquad \|\widetilde{H}(t, s, \lambda)\|_B \le \widetilde{N}(N, \lambda, a).$$

The weakly singular kernel

$$K(t,s,\lambda) := \frac{\widetilde{H}(t,s,\lambda)}{(t-s)^\alpha} \qquad (7.28)$$

is given by the following power expansion, which is strongly convergent in $B(X)$:

$$K(t,s,\lambda) = \sum_{m=1}^{\infty} \lambda^{m-1} \cdot K_m(t,s), \qquad (7.29)$$

where

$$K_1(t,s) := \frac{H(t,s)}{(t-s)^\alpha}, \quad K_m(t,s) := \int_s^t K_1(t,\sigma)\, K_{m-1}(\sigma,s)\, d\sigma, \ m \geq 2, \qquad (7.30)$$

and

$$\widetilde{N}(N,\lambda,a) := \sum_{m=1}^{\infty} |\lambda|^{m-1} a^{(m-1)(1-\alpha)} N^m \gamma_m, \qquad (7.31)$$

with

$$\gamma_m := (\Gamma(1-\alpha))^m [\Gamma(m \cdot (1-\alpha))]^{-1}, \qquad (7.32)$$

$\widetilde{N}$ being monotone increasing in N, a, and $\lambda > 0$. Here Γ denotes the Gamma function.

Note that

$$\gamma_m^{1/m} = \left[0\left(\frac{1}{m(1-\alpha)} \right) \right]^{1-\alpha-\frac{1}{2m}} \to 0$$

as $m \to \infty$.

The proofs of both propositions above, which are given on pages 17–18 of Mikhlin (1957) for the case of a real-valued continuous function $u \in C^0(J, \mathbb{R})$, extend immediately to our abstract case, if we recall the abstract Riemann integration and the abstract Cauchy–Hadamard Theorem to be found on pp. 59, 62–64, 66, and 96 of Hille & Phillips (1957).

In the special case $X = \mathbb{R}$ let $H \colon \mathcal{T} \to \mathbb{R}_+ := [0,\infty)$ denote some uniformly bounded, continuous function,

$$0 \leq H(t,s) \leq N, \qquad (7.33)$$

with constant $N > 0$ for all $(t,s) \in \mathcal{T}$.

Due to the Propositions 7.7 and 7.8 above, for any given $g \in C^0(J, \mathbb{R})$,

$\lambda > 0$, $0 < \alpha < 1$, the unique solution $u \in C^0(J, \mathbb{R})$ of the Volterra integral equation

$$u(t) = g(t) + \lambda \int_0^t \frac{H(t,s)}{(t-s)^\alpha} \, u(s) \, ds \qquad (7.34)$$

can be written as

$$u(t) = g(t) + \int_0^t \frac{\widetilde{H}(t,s,\lambda)}{(t-s)^\alpha} \, g(s) \, ds, \qquad t \in J. \qquad (7.35)$$

The following result (see Amann, 1995; Rautmann, 2016) will therefore prove to be extremely useful.

Proposition 7.9 (The singular Gronwall inequality) *Suppose that* (7.33) *holds,* $\lambda > 0$, *and* $g \in C^0(J, \mathbb{R})$. *If* $v \in C^0(J, \mathbb{R})$ *satisfies the inequality*

$$v(t) \le g(t) + \lambda \int_0^t \frac{H(t,s)}{(t-s)^\alpha} \, v(s) \, ds, \qquad t \in J, \qquad 0 < \alpha < 1, \quad (7.36)$$

then v *is bounded from above by the solution* u *from* (7.35).

7.7 Proof of the theorems

We will obtain solutions $u \in C^0(J, D_{A^\alpha})$ to (7.25) by means of the iteration scheme

$$A^\alpha u_{m+1}(t) = A^\alpha u_0(t) - \int_0^t A^\beta e^{-(t-s)A} A^{-\delta} P \big[u_m \cdot \nabla u_{m+1} \qquad (7.37)$$

$$+ \kappa \cdot (u_{m+1} \cdot \nabla v + v \cdot \nabla u_{m+1}) \big](s) \, ds,$$

$$A^\alpha u_0(t) = A^\alpha e^{-tA} u(0) + \int_0^t A^\beta e^{-(t-s)A} A^{-\delta} \, Pg(s) \, ds,$$

where $m \in \mathbb{N}$, $\kappa = 0$ or $\kappa = 1$, $\beta = \alpha + \delta$, $t \in J$, for given $u(0) \in D_{A^\alpha}$, $A^{-\delta} Pg \in C^0(J, L^r_\sigma)$, and $v \in C^0(J, D_{A^\alpha})$.

To point out the Volterra type of the problem in (7.37), with some given $u_m \in C^0(J, D_{A^\alpha})$ we set $z(s) := (u_m(s), v(s)) \in D_{A^\alpha} \times D_{A^\alpha}$ and define the kernel

$$H(t, s, z(s)) = H(t, s) F(z(s)),$$

$$H(t, s) = (t-s)^\beta A^\beta e^{-(t-s)A},$$

$$F(z(s))\widetilde{u} = -A^{-\delta} P \big[u_m \cdot \nabla A^{-\alpha} \widetilde{u} + \kappa ((A^{-\alpha} \widetilde{u}) \cdot \nabla v + v \cdot \nabla A^{-\alpha} \widetilde{u}) \big](s),$$

where

$$\tilde{u} \in L_\sigma^r, \quad (t,s) \in \mathcal{T} = \{(t,s) \in \mathbb{R}^2 : 0 \le s \le t \le a\}, \quad [0,a] = J.$$

In this notation, with any $\tilde{u} := A^\alpha u_{m+1}(s) \in L_\sigma^r$, (7.37) reads

$$A^\alpha u_{m+1}(t) = A^\alpha u_0(t) + \int_0^t \frac{H(t,s,z(s))}{(t-s)^\beta} A^\alpha u_{m+1}(s)\,ds, \quad (7.38)$$

with $A^\alpha u_0(t)$ from (7.37). We will see that this integral equation fulfils the requirements of Proposition 7.7.

Proposition 7.10 *Take $u(0) \in D_{A^\alpha}$, $A^{-\delta} Pg \in C^0(J, L_\sigma^r)$, and choose $v \in C^0(J, D_{A^\alpha})$ with $\|A^\alpha v\| \le c_v$. Then*

(i) $A^\alpha u_0 \in C^0(J, L_\sigma^r)$ *with*

$$\|A^\alpha u_0\| \le C_0 \|A^\alpha u(0)\| + T_{a,\beta} C_\beta \|A^{-\delta} Pg\| =: c_0.$$

In addition assume that $u_m \in C^0(J, D_{A^\alpha})$, $\|A^\alpha u_m\| \le c_m$, and that $z = (u_m, v)$ for some $m \in \mathbb{N}$. Then

(ii) $H(t,s,z(s)) \in B(L_\sigma^r), (t,s) \in \mathcal{T}$;
(iii) *there is a monotone increasing continuous function $N: \mathbb{R}_+ \to \mathbb{R}_+$ such that*

$$\|H(t,s,z(s))\|_{B(L_\sigma^r)} \le N(c_m) = N_{c_m} = C_\beta M(c_m + 2\kappa c_v);$$

N_{c_m} *depends on c_v in a monotone way, too; and*
(iv) $H(t,s,z(s))$ *is strongly L_σ^r-continuous with respect to $(t,s) \in \mathcal{T}$.*

Proof With $0 < \beta = \alpha + \delta < 1$, parts (i)–(iii) follow immediately from Lemmas 7.1 and 7.2. To prove (iv), we show that

$$D := \|[H(t,s,z(s)) - H(\tau,\sigma,z(\sigma))]\tilde{u}\| \to 0$$

as $(t,s) \to (\tau,\sigma)$ within $\mathcal{T}$.

Without loss of generality we may assume that

$$0 \le s \le t \le \tau \le a, \quad 0 \le \sigma \le \tau, \quad (7.39)$$

for any fixed $(\tau,\sigma) \in \mathcal{T}$. Since $(A^\alpha u_m) \in C^0(J, L_\sigma^r)$, $(A^\alpha v) \in C^0(J, L_\sigma^r)$, and $\tilde{u} \in L_\sigma^r$, in the inequality

$$D \le \|H(t,s)[F(z(s)) - F(z(\sigma))]\tilde{u}\| + \|[H(t,s) - H(\tau,\sigma)]F(z(\sigma))\tilde{u}\|$$
$$=: D_1 + D_2,$$

the first term D_1 on the right-hand side tends to zero as $|s - \sigma| \to 0$, as we see from Lemmas 7.1 and 7.2.

To show that the second term D_2 tends to zero, too, firstly in addition to (7.39) we suppose that

$$0 \le t - s \le \tau - \sigma. \tag{7.40}$$

Then $\sigma = \tau$ implies that $s = t$, and thus $D_2 = 0$. Setting $w := F(z(s))\tilde{u}$, we find that

$$D_2 \le \|A^\beta e^{-(t-s)A}(t-s)^\beta [1 - e^{-[(\tau-t)-(\sigma-s)]A}]w\|$$
$$+ \|A^\beta e^{-(\tau-\sigma)A}[(t-s)^\beta - (\tau-\sigma)^\beta]w\|$$
$$=: D_{21} + D_{22}. \tag{7.41}$$

Observing that $0 \le \eta := [\tau - t] - [\sigma - s]$ because of (7.40), from Lemmas 7.1 and 7.2 we find that $D_{21} \to 0$ if $(t, s) \to (\tau, \sigma)$. Furthermore by our assumption $0 \le \sigma < \tau$ with $(\tau, \sigma) \in \mathcal{T}$ being fixed, the convergence $(t, s) \to (\tau, \sigma)$ implies that $(t-s)^\beta = (\tau-\sigma)^\beta(1+\varepsilon)$ with values $|\varepsilon| \to 0$. Therefore recalling Lemmas 7.1 and 7.2 again we get

$$D_{22} = \|A^\beta e^{-(\tau-\sigma)A} (\tau-\sigma)^\beta w\| \, |\varepsilon| \to 0$$

as $|\varepsilon| \to 0$. We conclude similarly in the remaining case (7.39) when we have $0 \le \tau - \sigma < t - s$. □

Corollary 7.11 *With the assumptions of Proposition 7.10, the integral equation (7.37) admits a unique solution $u_{m+1} \in C^0(J, D_{A^\alpha})$.*

Proof Because of Proposition 7.10, Proposition 7.7 applies to (7.37), giving the existence of the unique solution $A^\alpha u_{m+1} \in C^0(J, L^r_\sigma)$, thus there exists $\|\|A^\alpha u_{m+1}\|\| = c_{m+1} < \infty$. Then the boundedness of $A^{-\alpha}$ on L^r_σ implies $u_{m+1} \in C^0(J, D_{A^\alpha})$. □

This corollary shows that the sequence of solutions $u_m \in C^0(J, D_{A^\alpha})$ to the scheme (7.37) is well defined for all $m \in \mathbb{N}$.

Recalling the bound $N_{c_m} \ge \|H(t, s, z(s))\|_{B(L^r_\sigma)}$ in part (iii) of Proposition 7.10, from the linear integral equations (7.37) in the Banach space $C^0(J, L^r_\sigma)$ we will get linear integral inequalities for the continuous real functions $\|A^\alpha u_m(t)\|$ and $\|A^\alpha(u_{m+1}(t) - u_m(t))\|$. As we will see by means of the singular Gronwall inequality (Proposition 7.9 above), these integral inequalities imply boundedness and convergence of the $A^\alpha u_m \in C^0(J, L^r_\sigma), m \in \mathbb{N}$, uniformly in $t \in J$ under smallness assumptions on $\|\|A^\alpha u_0\|\|$ only.

With the assumptions of Proposition 7.10 using Lemmas 7.1 and 7.2, from the integral equation (7.37) we find the integral inequality

$$\|A^\alpha u_{m+1}(t)\| \le c_0 + \int_0^t \frac{N_{c_m}}{(t-s)^\beta} \|A^\alpha u_{m+1}(s)\| \, ds, \qquad t \in J, \quad (7.42)$$

where

$$N_{c_m} = C_\beta \, M \, (c_m + 2\kappa c_v), \tag{7.43}$$

$$c_0 = C_0 \, \|A^\alpha u(0)\| + T_{a,\beta} \, C_\beta \, \|\|A^{-\delta} Pg\|\|. \tag{7.44}$$

Consequently Proposition 7.9 implies

$$\|\|A^\alpha u_{m+1}\|\| \le c_0 \left[1 + \int_0^a \frac{\widetilde{N}(N_{c_m})}{(t-s)^\beta} \, ds \right] \le c_0 [1 + T_{a,\beta} \, \widetilde{N}(N_{c_m})], \quad (7.45)$$

$\widetilde{N}(N_{c_m}) \ge \widetilde{H}(t,s)$ denoting the bound (7.31) on the resolvent kernel $\widetilde{H}$ of the constant kernel N_{c_m} in (7.42).

Proposition 7.12 *Suppose that $u(0) \in D_{A^\alpha}$, $A^{-\delta} Pg \in C^0(J, L_\sigma^r)$, and $v \in C^0(J, D_{A^\alpha})$ with $\|\|A^\alpha v\|\| \le c_v$. Denote by $A^\alpha u_m \in C^0(J, L_\sigma^r)$ the solutions to (7.37), $m \in \mathbb{N}$, with $\|\|A^\alpha u_m\|\| = c_m$ for some constants c_m, and let c be any constant such that*

$$\|\|A^\alpha u_0\|\| \le c_0 \le \frac{c}{1 + \widetilde{N}(N_c) T_{a,\beta}}. \tag{7.46}$$

Then

$$\|\|A^\alpha u_m\|\| \le c \quad \text{for all } m \in \mathbb{N}. \tag{7.47}$$

Note that since $\widetilde{N}(N_c)$ is monotone increasing in $c > 0$, for all $c \le \bar{c} < \infty$, it follows that

$$\frac{c}{1 + \widetilde{N}(N_c) T_{a,\beta}} \ge \frac{c}{1 + \widetilde{N}(N_{\bar{c}}) T_{a,\beta}} > 0.$$

Proof We conclude by induction: (7.47) holds true with $m = 0$ because of (7.46). Assuming (7.46), and additionally (7.47) for $u_0, \dots, u_m$, from (7.45) we see $\|\|A^\alpha u_{m+1}\|\| \le c$, which proves our claim for all $m \in \mathbb{N}$. □

To show the convergence of the $A^\alpha u_m(t)$, we will find suitable bounds to the differences

$$d_m(t) := u_{m+1}(t) - u_m(t).$$

Proposition 7.13 *Under the assumptions of Proposition 7.12, additionally let*

$$c \le \frac{q}{T_{a,\beta} \, C_\beta \, M \, [1 + T_{a,\beta} \, \widetilde{N}(N_c)]} \tag{7.48}$$

with some $q \in (0,1)$. Then the $A^{\alpha}u_m \in C^0(J, L^r_\sigma)$ are uniformly convergent:

$$\||A^{\alpha}d_{m+1}\|| \leq q\,\||A^{\alpha}d_m\||, \quad m \in \mathbb{N}, \tag{7.49}$$

and thus

$$\||A^{\alpha}d_m\|| \leq q^m\,\||A^{\alpha}d_0\||. \tag{7.50}$$

Proof The equations (7.37) for $A^{\alpha}u_m(t)$, $m \in \mathbb{N}$, imply that the differences $d_m(t) = u_{m+1}(t) - u_m(t)$ satisfy

$$A^{\alpha}d_{m+1}(t) = -\int_0^t A^{\beta} e^{-(t-s)A} A^{-\delta} P(d_m \nabla u_{m+2})(s)\,\mathrm{d}s$$

$$-\int_0^t A^{\beta} e^{-(t-s)A} A^{-\delta} P\big[u_m \nabla d_{m+1} + \kappa\,(d_{m+1}\nabla v + v \cdot \nabla d_{m+1})\big](s)\,\mathrm{d}s.$$

Recalling the uniform bounds $\||A^{\alpha}u_m\|| \leq c$, $m \in \mathbb{N}$, in Proposition 7.12, and our requirement that $\||A^{\alpha}v\|| \leq c_v$, from Lemmas 7.1, 7.2 we obtain the integral inequality

$$\|A^{\alpha}d_{m+1}(t)\| \leq c\,\bar{g}\,\||A^{\alpha}d_m\|| + \int_0^t \frac{N_c}{(t-s)^{\beta}}\|A^{\alpha}d_{m+1}(s)\|\,\mathrm{d}s \tag{7.51}$$

with $\bar{g} = T_{a,\beta}\,C_{\beta}\,M$, $N_c = C_{\beta}\,M\,[c + 2\kappa c_v]$ for the continuous real functions $\|A^{\alpha}d_{m+1}(t)\|$, $t \in J$, $m \in \mathbb{N}$.

Because of Proposition 7.9, from (7.51) we get

$$\||A^{\alpha}d_{m+1}\|| \leq c\,\bar{g}\,\||A^{\alpha}d_m\|| \left[1 + \int_0^a \frac{\widetilde{N}(N_c)}{(t-s)^{\beta}}\,\mathrm{d}s\right] \tag{7.52}$$

$$\leq c\,\bar{g}\,\||A^{\alpha}d_m\||\,[1 + T_{a,\beta}\widetilde{N}(N_c)]$$

where $\widetilde{N}(N_c)$ denotes the bound (7.31) to the kernel $\widetilde{H}(t,s)$ in the resolvent to the constant kernel N_c in (7.51).

Evidently (7.52) implies (7.49) and (7.50) because of our requirement (7.48). This proves our claim. $\qquad\square$

Proposition 7.14 *Let $u(0) \in D_{A^{\alpha}}$, $A^{-\delta}Pg \in C^0(J, L^r_\sigma)$, and $A^{\alpha}v \in C^0(J, L^r_\sigma)$, such that*

$$\||A^{\alpha}u_0\|| \leq c_0 \leq \frac{c}{1 + \widetilde{N}(N_c)\,T_{a,\beta}},$$

where

$$c \leq \frac{q}{T_{a,\beta}\,C_{\beta}\,M\,[1 + \widetilde{N}(N_c)\,T_{a,\beta}]},$$

for some $q \in (0,1)$. Then

(i) *The Cauchy sequence $A^\alpha u_m \in C^0(J, L^r_\sigma)$ defines the Cauchy sequence $u_m \in C^0(J, L^r_\sigma)$;*
(ii) $\lim_{m\to\infty} u_m = u \in C^0(J, L^r_\sigma)$; $\lim_{m\to\infty} A^\alpha u_m = \tilde{u} \in C^0(J, L^r_\sigma)$;
(iii) *$u(t) \in D_{A^\alpha}$ and $\tilde{u}(t) = A^\alpha u(t)$ for all $t \in J$;*
(iv) *$A^\alpha u(t)$ represents a solution of the integral equation (7.25); and*
(v) *the error estimate*

$$\||A^\alpha(u - u_m)\|| \leq \frac{q^m}{1-q}\||A^\alpha(u_1 - u_0)\||, \quad m \in \mathbb{N}$$

holds.

Proof Parts (i) and (ii) clearly hold because of Proposition 7.13 and the boundedness of the linear operator $A^{-\alpha}$ on L^r_σ. Therefore for each $t \in J$, (iii) results from the closedness of the linear operator A^α. Thus by Lemmas 7.1 and 7.2, the right-hand side of (7.25) is well defined. We prove the statement (iv) by estimating the difference D^* of the right-hand sides in (7.25) and (7.37): by Lemmas 7.1, 7.2 and Proposition 7.12 we find

$$\|D^*\| = \left\| \int_0^t A^\beta e^{-(t-s)A} A^{-\delta} P\big[(u - u_m) \cdot \nabla u + u_m \cdot \nabla(u - u_{m+1}) \right.$$

$$\left. + \kappa((u - u_{m+1}) \cdot \nabla v + v \cdot \nabla(u - u_{m+1})\big](s)\, ds \right\|$$

$$\leq T_{a,\beta}\, C_\beta\, M \big[(\||A^\alpha(u - u_m)\|| + \||A^\alpha(u - u_{m+1})\||)c$$

$$+ 2\kappa\||A^\alpha(u - u_{m+1})\||c_v\big],$$

which $\to 0$ as $m \to \infty$ because of parts (ii) and (iii).

Finally, to prove (v), from (7.50) we find

$$\||A^\alpha(u_{m+k} - u_m)\|| \leq \frac{q^m}{1-q}\||A^\alpha(u_1 - u_0)\||, \tag{7.53}$$

$m, k \in \mathbb{N}$, which in the limit $k \to \infty$ gives the inequality in (v).

Thus far we have proved Theorem 7.4 (i) under the restriction (7.48) on $\varepsilon = c$. For arbitrary $\varepsilon \in (0,\infty)$, recalling the continuity and monotonicity of the function $N(c) = N_c$ in Proposition 7.10 (iii) and of $\tilde{N}(N_c)$ in (7.31), we conclude that there exist some $\varepsilon_1 > 0$ and some constant $\tilde{N}_1 > 0, \tilde{N}_1$ dependent on ε_1, such that $T_{a,\beta} \cdot \tilde{N}(N_c) \leq \tilde{N}_1$ for all $c \in (0, \varepsilon_1]$. Therefore (7.48) holds true especially for all $c > 0$,

$$c := \varepsilon_2 \leq \min\left\{ \frac{q}{T_{a,\beta}\, C_\beta\, M(1 + \tilde{N}_1)}, \varepsilon_1, \varepsilon \right\}.$$

Consequently Propositions 7.12–7.14 above show that for any choice of data $A^\alpha u_0 \in C^0(J, L^r_\sigma)$ with $|||A^\alpha u_0||| \le \zeta(\varepsilon) := c_0, c_0$ from (7.46), there exists a solution $u \in C^0(J, D_{A^\alpha})$ to (7.25) with $|||A^\alpha u||| \le c = \varepsilon_2 \le \varepsilon$. □

We can now complete the proof of Theorem 7.4 by showing the uniqueness of solutions $u \in C^0(J, D_{A^\alpha})$ to (7.25). Let $u(0) = \tilde{u}(0) \in D_{A^\alpha}$, $A^{-\delta}Pg \in C^0(J, L^r_\sigma)$, $v \in C^0(J, D_{A^\alpha})$ with $|||A^\alpha v||| \le c_v$, and suppose that $\tilde{u} \in C^0(J, D_{A^\alpha})$ is another solution of (7.25).

Using (7.25) the difference $d(t) = \tilde{u}(t) - u(t)$ satisfies

$$A^\alpha d(t) = -\int_0^t A^\beta e^{-(t-s)A} A^{-\delta} P[d \cdot \nabla \tilde{u} + u \cdot \nabla d + \kappa(v \cdot \nabla d + d \cdot \nabla v)](s)\,\mathrm{d}s.$$

(7.54)

Writing $c_u = |||A^\alpha u|||$ and $c_{\tilde{u}} = |||A^\alpha \tilde{u}|||$, Lemmas 7.1 and 7.2 give

$$\|A^\alpha d(t)\| \le \int_0^t \frac{C_\beta M}{(t-s)^\beta} [c_u + c_{\tilde{u}} + 2\kappa c_v] \|A^\alpha d(s)\|\,\mathrm{d}s. \qquad (7.55)$$

From (7.55), applying Proposition 7.9 with $g = 0$ we find that $d(t) = 0$ for all $t \in J$, which completes the proof of Theorem 7.4.

Theorem 7.5 results from Theorem 7.4 with $\kappa = 0$. Finally, Theorem 7.4 with $\kappa = 1$ gives Theorem 7.6: since equation (7.23) for $v \in C^0(J, D_{A^\alpha})$ together with equation (7.25) for $u \in C^0(J, D_{A^\alpha})$ implies equation (7.24) for $\tilde{v} = v + u \in C^0(J, D_{A^\alpha})$, to each $\varepsilon > 0$ and to any prescribed data $\tilde{v}(0) \in D_{A^\alpha}, A^{-\delta}P\tilde{f} \in C^0(J, L^r_\sigma)$ fulfilling

$$|||A^\alpha(\tilde{v}_0 - v_0)||| \le \zeta(\varepsilon),$$

Theorem 7.4 with $\kappa = 1$ guarantees the existence of the solution $\tilde{v} = (v + u) \in C^0(J, D_{A^\alpha})$ to (7.24) that obeys

$$|||A^\alpha(\tilde{v} - v)||| = |||A^\alpha u||| \le \varepsilon.$$

References

Amann, H. (1995) *Linear and Quasilinear Parabolic Problems*. Birkhäuser Verlag, Basel.

Auscher, P., Dubois, S., & Tchamitchian, P. (2004) On the stability of global solutions to Navier-Stokes equations in the space. *J. Math. Pures Appl.* **83**, 673–697.

Beirão da Veiga, H., & Secchi, P. (1987) L^p-stability for strong solutions of the Navier-Stokes equations in the whole space. *Arch. Rat. Mech. Anal.* **98**, 65–70.

Fujita, H., & Kato, T. (1964) On the Navier-Stokes initial value problem I. *Arch. Rat. Mech. Anal* **16**, 269–315.

Gallagher, I., Iftimie, D., & Planchon, F. (2003) Asymptotics and stability for global solutions to the Navier-Stokes equations. *Ann. Inst. Fourier* **53**, 1387–1424.

Giga, Y. (1981) Analyticity of the semigroup generated by the Stokes operator in L_r spaces. *Math. Z.* **178**, 297–329.

Giga, Y. (1982) The Navier-Stokes initial value problem in L^p and related problems. *Lecture Notes in Num. Appl. Anal.* **5**, 37–54.

Giga, Y. (1985) Domains of fractional powers of the Stokes operator in L_r spaces. *Arch. Rat. Mech. Anal* **89**, 251–265.

Giga, Y., & Miyakawa, T. (1985) Solutions in L_r of the Navier-Stokes initial value problem. *Arch. Rat. Mech. Anal.* **89**, 267–281.

Heywood, J. G. (1980) The Navier-Stokes Equations: On the Existence, Regularity and Decay of Solutions. *Indiana Univ. Math. J.* **29**, 639–681.

Hille, E., & Phillips, R.S. (1957) *Funtional Analysis and Semigroups*. American Math. Society, Providence, Rhode Island.

Kawanago, T. (1998) Stability estimate of strong solutions for the Navier-Stokes system and its applications. *Electron J. Differential Equations (electronic)* **15**, 1–23.

Kozono, H. (2000) Asymptotic stability of large solutions with large perturbations to the Navier-Stokes equations. *J. Functional Anal.* **176**, 153–197.

Kozono, H., & Ogawa, T. (1994) On stability of Navier-Stokes flow in exterior domains. *Arch. Rat. Mech. Anal.* **128**, 1–31.

Kozono, H., & Ozawa, T. (1990) Stability in L^r for the Navier-Stokes flow in an n-dimensional bounded domain. *J. Math. Anal. Appl.* **152**, 35–45.

Lemarié-Rieusset, P.G. (2016) *The Navier–Stokes Problem in the 21st Century*. CRC Press, Taylor & Francis Group, New York.

Masuda, K. (1975) On the stability of incompressible viscous fluid motions past objects. *J. Math. Soc. Japan* **27**, 294–327.

Mikhlin, S.G. (1957) *Integral Equations*. Pergamon Press, London.

Miyakawa, T. (1981) On the initial value problem for the Navier Stokes equations in L^p spaces. *Hiroshima Math. J.* **11**, 9–20.

Pazy, A. (1983) *Semigroups of Linear Operators and Applications to Partial Differential Equations*. Springer Verlag, New York.

Ponce, G., Racke, R., Sideris, T.C., & Titi, E.S. (1994) Global stability of large solutions to the 3D Navier-Stokes equations. *Comm. Math. Phys.* **159**, 329–341.

Rautmann, R. (2016) Note on monotonicity in singular Volterra Integral equations. *Dynamic Systems and Appl.* **25**, 531–542.

Sobolevskii, P.E. (1959) On non-stationary equations of hydrodynamics for viscous fluid. *Doklady Akad. Nauk USSR* **128**, 45–48 (in Russian).

Sohr, H. (2001) *The Navier–Stokes Equations*. Birkhäuser Verlag, Basel.

Solonnikov, V.A. (1977) Estimates for solutions of nonstationary Navier–Stokes equations. *J. Soviet Math.* **8**, 467–529.

von Wahl, W. (1985) *The Equations of Navier–Stokes and Abstract Parabolic Equations*. Vieweg Verlag, Braunschweig.

Wiegner, M. (1990) Decay and stability in L^p for strong solutions of the Cauchy problem for the Navier–Stokes equations. Pages 95–99 of: Heywood, J.G., Masuda, K., Rautmann, R., & Solonnikov, V.A. (eds), *The Navier–Stokes equations*. Springer Lecture Notes in Math., vol. 1431.

8

Energy conservation in the 3D Euler equations on $\mathbb{T}^2 \times \mathbb{R}_+$

James C. Robinson

Mathematics Institute, University of Warwick,
Coventry, CV4 7AL. UK.
j.c.robinson@warwick.ac.uk

José L. Rodrigo
Mathematics Institute, University of Warwick,
Coventry, CV4 7AL. UK.
j.rodrigo@warwick.ac.uk

Jack W.D. Skipper
Institute of Applied Mathematics, Leibniz University Hannover,
Welfengarten 1, 30167 Hannover. Germany.
skipper@ifam.uni-hannover.de

Abstract

The aim of this paper is to prove energy conservation for the incompressible Euler equations in a domain with boundary. We work in the domain $\mathbb{T}^2 \times \mathbb{R}_+$, where the boundary is both flat and has finite measure; in this geometry we do not require any estimates on the pressure, unlike the proof in general bounded domains due to Bardos & Titi (2018).

However, first we study the equations on domains without boundary (the whole space $\mathbb{R}^3$, the torus $\mathbb{T}^3$, and the hybrid space $\mathbb{T}^2 \times \mathbb{R}$). We make use of some arguments due to Duchon & Robert (2000) to prove energy conservation under the assumption that $u \in L^3(0,T; L^3(\mathbb{R}^3))$ and

$$\lim_{|y| \to 0} \frac{1}{|y|} \int_0^T \int_{\mathbb{R}^3} |u(x+y) - u(x)|^3 \, dx \, dt = 0$$

or

$$\int_0^T \int_{\mathbb{R}^3} \int_{\mathbb{R}^3} \frac{|u(x) - u(y)|^3}{|x-y|^{4+\delta}} \, dx \, dy \, dt < \infty, \qquad \delta > 0,$$

the second of which is equivalent to $u \in L^3(0,T; W^{\alpha,3}(\mathbb{R}^3))$, $\alpha > 1/3$.

We use the first of these two conditions to prove energy conservation for a weak solution u on $D_+ := \mathbb{T}^2 \times \mathbb{R}_+$. We extend u to a solution defined on $\mathbb{T}^2 \times \mathbb{R}$ and then use the condition on this domain to prove energy conservation for a weak solution $u \in L^3(0,T; L^3(D_+))$ that sat-

Published as part of *Partial Differential Equations in Fluid Mechanics*, edited by C.L. Fefferman, J.C. Robinson, & J.L. Rodrigo. © Cambridge University Press 2018.

isfies

$$\lim_{|y| \to 0} \frac{1}{|y|} \int_0^T \iint_{\mathbb{T}^2} \int_{|y|}^{\infty} |u(t, x + y) - u(t, x)|^3 \, \mathrm{d}x_3 \, \mathrm{d}x_1 \, \mathrm{d}x_2 \, \mathrm{d}t = 0,$$

and $u \in L^3(0, T; C^0(\mathbb{T}^2 \times [0, \delta])$ for some $\delta > 0$.

8.1 Introduction

Energy conservation for solutions of the incompressible Euler equations

$$\partial_t u + (u \cdot \nabla)u + \nabla p = 0 \qquad \nabla \cdot u = 0$$

has long been a topic of interest. While for sufficiently smooth solutions u a standard integration-by-parts argument shows that energy is conserved ($\|u(t)\|_{L^2} = \|u(0)\|_{L^2}$ for every $t \geq 0$) for weak solutions $u \in L^\infty(0, T; L^2) \cap L^3(0, T; L^3)$ we do not have the regularity needed to perform these operations. Onsager (1949) conjectured that weak solutions to the Euler equations satisfying a Hölder continuity condition of order greater than one third should conserve energy.

Prior to this work the study of energy conservation for this system has primarily been carried out on domains without boundary, either the whole space $\mathbb{R}^3$ or the torus $\mathbb{T}^3$. In this paper we aim to treat the question on the domain $\mathbb{T}^2 \times \mathbb{R}_+$, which involves a flat boundary with finite measure.

The first proof of energy conservation for weak solutions was given by Eyink (1994) on the torus, assuming that $u(\cdot, t) \in C_*^\alpha$ for $\alpha > 1/3$ with a uniform bound for $t \in [0, T]$. A definition of the space C_*^α equivalent to that of Eyink's is as follows: expand u as the Fourier series

$$u = \sum_{k \in \mathbb{Z}^3} \hat{u}_k \mathrm{e}^{\mathrm{i}k \cdot x},$$

imposing conditions to ensure that u is real ($\hat{u}_k = \overline{\hat{u}_{-k}}$) and is divergence free ($k \cdot \hat{u}_k = 0$); then $u \in C_*^\alpha(\mathbb{T}^3)$ if

$$\sum_{k \in \mathbb{Z}^3} |k|^\alpha |\hat{u}_k| < \infty.$$

Requiring $u \in C_*^\alpha$ with $\alpha > 1/3$ is a stronger condition than the one-third Hölder continuity conjectured by Onsager.

Subsequently Constantin, E, & Titi (1994) gave a short proof of energy conservation, in the framework of Besov spaces (but still on the torus), under the weaker assumption that

$$u \in L^3(0, T; B^{\alpha}_{3,\infty}) \quad \text{with} \quad \alpha > 1/3. \qquad (8.1)$$

As $C^{\alpha} \subset B^{\alpha}_{3,\infty}$ this proves Onsager's Conjecture. Here $B^s_{p,r}$ denotes a Besov space as defined in Bahouri, Chemin, & Danchin (2011) and Lemarié-Rieusset (2002).

Duchon & Robert (2000) showed that solutions satisfying a weaker regularity condition still conserve energy. They derived a local energy equation that contains a term $D(u)$ representing the dissipation or production of energy caused by the lack of smoothness of u; this term can be seen as a local version of Onsager's original statistically averaged description of energy dissipation. They showed that if u satisfies

$$\int |u(t, x + \xi) - u(t, x)|^3 \, \mathrm{d}x \le C(t) |\xi| \sigma(|\xi|), \qquad (8.2)$$

with $\sigma(a) \to 0$ as $a \to 0$, $C \in L^1(0, T)$, then $\|D(u)\|_{L^1(0,T,L^1(\mathbb{T}^3))} = 0$ and hence the kinetic energy is conserved. The condition in (8.2) is weaker than (8.1). A detailed review examining this and further work relating to Onsager's conjecture is given by Eyink & Sreenivasan (2006).

More recently energy conservation was shown by Cheskidov et al. (2008) for $u \in L^3(0, T; B^{1/3}_{3,c(\mathbb{N})})$, where $B^{1/3}_{3,c(\mathbb{N})}$ is a subspace of $B^{1/3}_{3,\infty}$. In fact Cheskidov et al. (2008) showed that energy conservation holds for solutions satisfying the still weaker condition

$$\lim_{q \to \infty} \int_0^T 2^q \|\Delta_q u\|^3_{L^3} \, \mathrm{d}t = 0,$$

where Δ_q performs a smooth restriction of u into Fourier modes of order 2^q. In a follow-up paper Shvydkoy (2009) (see also Shvydkoy, 2010) states that this condition is equivalent to

$$\lim_{|y| \to 0} \frac{1}{|y|} \int_0^T \int |u(x + y) - u(x)|^3 \, \mathrm{d}x \, \mathrm{d}t = 0 \qquad (8.3)$$

(see Skipper (2018) for a proof) and proves a local energy balance under this condition. Observe that this condition is less restrictive than (8.2).

In this paper we use an approach similar to that of Shvydkoy (2009), but rather than basing our argument on the approach of Constantin et al. (1994) we adopt some of the ideas from Duchon & Robert (2000) and give a direct proof that energy conservation follows on the whole domain (this simplifies matters since the pressure no longer plays a role) under the condition that

$$\int_{\mathbb{R}^3} \int_{\mathbb{R}^3} \nabla \varphi_\varepsilon(\xi) \cdot (u(x+\xi) - u(x)) |u(x+\xi) - u(x)|^2 \, d\xi \, dx \to 0$$

as $\varepsilon \to 0$, where φ is a radially symmetric mollifier.

Given this condition it is relatively simple to show energy conservation under the assumption (8.3), which we do in Theorem 8.9, and under the alternative condition

$$\int_0^T \int_{\mathbb{R}^3} \int_{\mathbb{R}^3} \frac{|u(x) - u(y)|^3}{|x-y|^{4+\delta}} \, dx \, dy < \infty, \qquad \delta > 0, \qquad (8.4)$$

which is equivalent to requiring $u \in L^3(0, T; W^{\alpha, 3}(\mathbb{R}^3))$ for some $\alpha > 1/3$ (Theorem 8.10).

For the most significant new contribution of this paper we use condition (8.3) to analyse energy conservation in the domain $D_+ := \mathbb{T}^2 \times \mathbb{R}_+$. We show that if (u, p) is a weak solution on D_+ then (u_R, p) is a weak solution on D_-, where u_R is an appropriately 'reflected' version of u, and that $u + u_R$ is a weak solution on $D := \mathbb{T}^2 \times \mathbb{R}$. It follows that energy is conserved for u_E under condition (8.3); from here we deduce energy conservation for u under the conditions

$$\lim_{|y| \to 0} \frac{1}{|y|} \int_{t_1}^{t_2} \iint_{\mathbb{T}^2} \int_{|y|}^\infty |u(t, x+y) - u(t, x)|^3 \, dx_3 \, dx_1 \, dx_2 \, dt = 0$$

and $u \in L^3(0, T; C^0(\mathbb{T}^2 \times [0, \delta))$ for some $\delta > 0$, see Theorem 8.15.

Our method makes significant use of the symmetry of the domain, and it is not clear how to extend it to a more general bounded domain. Since we completed this paper, the problem on such a general bounded domain has been treated by Bardos & Titi (2018); their work treats the equations in distributional form and relies on estimates for the pressure.

8.2 Energy conservation without boundaries

In this section we treat the incompressible Euler equations on a domain without boundaries: $\mathbb{R}^3$, $\mathbb{T}^3$, or one of the hybrid domains $\mathbb{T} \times \mathbb{R}^2$ or $\mathbb{T}^2 \times \mathbb{R}$. We write D in what follows to denote any one of these domains, being careful to highlight any differences required in the definitions/arguments required to deal with the periodic or hybrid cases.

8.2.1 Weak solutions of the Euler equations

For vector-valued functions f, g and matrix-valued functions F, G we use the notation

$$\langle f, g \rangle = \int_D f_i(x) g_i(x) \, dx \quad \text{and} \quad \langle F : G \rangle = \int_D F_{ij}(x) G_{ij}(x) \, dx,$$

employing Einstein's summation convention (sum over repeated indices).

We use the notation $\mathcal{D}(D)$ to denote the collection of C^∞ functions with compact support in D, and $\mathcal{S}(D)$ for the collection of all C^∞ functions with Schwartz-like decay in the unbounded directions of D, e.g. for $\mathbb{T}^2 \times \mathbb{R}$ we require

$$\sup_{x \in \mathbb{T}^2 \times \mathbb{R}} |\partial^\alpha \varphi| |x_3|^k < \infty,$$

for all $\alpha, k \geq 0$ where α is a multi-index over all the spatial variables (x_1, x_2, x_3). Note that in periodic directions the requirement of 'compact support' is trivially satisfied. The spaces $\mathcal{D}_\sigma(D)$ and $\mathcal{S}_\sigma(D)$ consist of all divergence-free elements of $\mathcal{D}(D)$ or $\mathcal{S}(D)$.

We denote by $H_\sigma(D)$ the closure of $\mathcal{D}_\sigma(D)$ in the norm of $L^2(D)$; this coincides with the closure of $\mathcal{S}_\sigma(D)$ in the same norm.

Elements of $H_\sigma(D)$ are divergence free in the sense of distributions, i.e.

$$\langle u, \nabla \varphi \rangle = 0 \qquad \text{for all} \quad \varphi \in \mathcal{D}(D); \tag{8.5}$$

in fact this equality holds for all $\varphi \in \mathcal{S}(D)$, and even for every $\varphi \in H^1(D)$: indeed, since $\mathcal{S}_\sigma(D)$ is dense in $H_\sigma(D)$, for any $u \in H_\sigma(D)$ we can find $(u_n) \in \mathcal{S}_\sigma(D)$ such that $u_n \to u$ in $H^1(D)$, and then for any $\varphi \in H^1(D)$ we have

$$\langle u, \nabla \varphi \rangle = \lim_{n \to \infty} \langle u_n, \nabla \varphi \rangle = \lim_{n \to \infty} \langle \nabla \cdot u_n, \varphi \rangle = 0$$

(cf. Lemma 2.11 in Robinson et al., 2016, for example).

In a slight abuse of notation we denote by $C_w([0,T]; H_\sigma)$ the collection of all functions $u \colon [0,T] \to H_\sigma(D)$ that are weakly continuous into L^2, i.e.

$$t \mapsto \langle u(t), \varphi \rangle \tag{8.6}$$

is continuous for every $\varphi \in L^2(D)$. Note if $u \in C_w([0,T]; H_\sigma)$ then $u(t) \in H_\sigma$ for every $t \in [0,T]$ (not only almost every t), and that $C_w([0,T]; H_\sigma) \subset L^\infty(0,T; H_\sigma)$.

We take as our space-time test functions the elements of

$$\mathcal{S}_\sigma^T := \{ \psi \in C^\infty(D \times [0,T]) : \ \psi(\cdot, t) \in \mathcal{S}_\sigma(D) \text{ for all } t \in [0,T] \}.$$

We choose these functions to take values in $\mathcal{S}_\sigma$ (rather than in $\mathcal{D}_\sigma$) since the property of compact support is not preserved by the Helmholtz decomposition, whereas such a decomposition respects Schwartz-like decay.

Lemma 8.1   *Any* $\psi \in \mathcal{S}$ *can be decomposed as* $\psi = \varphi + \nabla\chi$, *where* $\varphi \in \mathcal{S}_\sigma$ *and* $\chi \in \mathcal{S}$, *and*

$$\|\varphi\|_{H^s} + \|\nabla\chi\|_{H^s} \le C_s\|\psi\|_{H^s} \tag{8.7}$$

for each $s \ge 0$.

Proof (Cf. Theorem 2.6 and Exercise 5.2 in Robinson, Rodrigo, & Sadowksi, 2016.) Since $\psi \in \mathcal{S}$ we can write ψ as a hybrid Fourier series/inverse Fourier transform, using Fourier series in the periodic directions and the Fourier transform in the unbounded directions. Since u is real valued we have $\hat{u}(k) = \overline{\hat{u}(-k)}$ for the components of k in $\mathbb{Z}$. Further, if we are considering the case $D = \mathbb{T}^3$ then $\hat{u}(0) = 0$ so that u has zero mean. For example, in the case $D = \mathbb{T}^2 \times \mathbb{R}$ we have

$$\psi(x) = \int_{-\infty}^{\infty} \sum_{(k_1,k_2)\in\mathbb{Z}^2} \hat{u}(k)\mathrm{e}^{\mathrm{i}k\cdot x}\,\mathrm{d}k_3,$$

and we can set

$$\varphi(x) = \int_{-\infty}^{\infty} \sum_{(k_1,k_2)\in\mathbb{Z}^2} \left(I - \frac{k \otimes k}{|k|^2}\right)\hat{u}(k)\mathrm{e}^{\mathrm{i}k\cdot x}\,\mathrm{d}k_3,$$

and

$$\chi(x) = \int_{-\infty}^{\infty} \sum_{(k_1,k_2)\in\mathbb{Z}^2} \frac{k \cdot \hat{u}(k)}{|k|^2}\mathrm{e}^{\mathrm{i}k\cdot x}\,\mathrm{d}k_3;$$

in the fully periodic case we omit the $k \otimes k/|k|^2$ term when $k = 0$. It is easy to check that these functions have the stated properties.   □

Assuming that u is a smooth solution of the Euler equations

$$\partial_t u + (u \cdot \nabla)u + \nabla p = 0 \qquad \nabla \cdot u = 0$$

if we multiply by an element of $\mathcal{S}_\sigma^T$ and integrate by parts in space and time then we obtain (8.8) below; the pressure term vanishes since there are no boundaries and ψ is divergence free. Requiring only (8.8) to hold we obtain our definition of a weak solution.

Definition 8.2 (Weak Solution)   We say that $u \in C_{\mathrm{w}}([0,T];H_\sigma)$ is a weak solution of the Euler equations on $[0, T]$, arising from the initial condition $u(0) \in H_\sigma$, if

$$\langle u(t), \psi(t) \rangle - \langle u(0), \psi(0) \rangle - \int_0^t \langle u(\tau), \partial_t \psi(\tau) \rangle \, d\tau$$

$$= \int_0^t \langle u(\tau) \otimes u(\tau) : \nabla \psi(\tau) \rangle \, d\tau \quad (8.8)$$

for every $t \in [0, T]$ and any $\psi \in \mathcal{S}_\sigma^T$.

We note here that replacing $\mathcal{S}_\sigma^T$ by $\mathcal{D}_\sigma^T$ leads to an equivalent definition (via a simpler version of the argument of Lemma 8.3, below).

Throughout the paper we let φ be a radial function in $C_c^\infty(B(0,1))$ with $\int_{\mathbb{R}^3} \varphi = 1$ and for any $\varepsilon > 0$ we set $\varphi_\varepsilon(x) = \varepsilon^{-3} \varphi(x/\varepsilon)$. Then for any function f we define the mollification of f as $f_\varepsilon := \varphi_\varepsilon \star f$ where $\star$ denotes convolution. Thus

$$f_\varepsilon(x) = \varphi_\varepsilon \star f(x) := \int_{\mathbb{R}^3} \varphi_\varepsilon(x-y) f(y) \, dy = \int_{B(0,\varepsilon)} \varphi_\varepsilon(y) f(x-y) \, dy.$$

In the periodic directions we extend f by periodicity in this integration. We insist that φ is radially symmetric since this ensures that the operation of mollification satisfies the 'symmetry property', that is, for $u \in L^p$ and $v \in L^q$ with $\frac{1}{p} + \frac{1}{q} = 1$ we have

$$\langle \varphi_\varepsilon \star u, v \rangle = \langle u, \varphi_\varepsilon \star v \rangle. \quad (8.9)$$

Our aim in the next section is to show the validity of the following two equalities that follow from the definition of a weak solution in (8.8). The first is

$$\langle u(t), u_\varepsilon(t) \rangle - \langle u(0), u_\varepsilon(0) \rangle - \int_0^t \langle u(\tau), \partial_t u_\varepsilon(\tau) \rangle \, d\tau$$

$$= \int_0^t \langle u(\tau) \otimes u(\tau) : \nabla u_\varepsilon(\tau) \rangle \, d\tau; \quad (8.10)$$

this amounts to using u_ε, a mollification (in space only) of the solution u, as a test function in (8.8): we need to show that there is sufficient time regularity to do this, which we do in Section 8.2.2. The second is

$$\int_0^t \langle \partial_t u_\varepsilon(\tau), u(\tau) \rangle \, d\tau = -\int_0^t \langle \nabla \cdot [u(\tau) \otimes u(\tau)]_\varepsilon, u(\tau) \rangle \, d\tau, \quad (8.11)$$

assuming that $u \in L^3(0, T; L^3)$. One could see this heuristically as a "mollification of the equation" tested with u; we will show that this can be done in a rigorous way in Section 8.2.3. We can then add these equations and take the limit as $\varepsilon \to 0$ to obtain the equation for conservation (or otherwise) of energy (Section 8.2.4).

8.2.2 Using u_ε as a test function

We will show that if u is a weak solution then in fact (8.8) holds for a larger class of test functions with less time regularity. We denote by $C^{0,1}([0,T]; H_\sigma)$ the space of Lipschitz functions from $[0,T]$ into H_σ.

Lemma 8.3 *If u is a weak solution of the Euler equations in the sense of Definition 8.2 then (8.8) holds for every $\psi \in \mathcal{L}_\sigma$, where*

$$\mathcal{L}_\sigma := L^1(0,T; H^3) \cap C^{0,1}([0,T]; H_\sigma).$$

Proof We will extend the class of test functions allowed in the definition to $\psi \in \mathcal{L}_\sigma$ using a standard density argument. We first note that for a fixed u we can write (8.8) as $E(\psi) = 0$ for every $\psi \in \mathcal{S}_\sigma^T$, where

$$E(\psi) := \langle u(t), \psi(t) \rangle - \langle u(0), \psi(0) \rangle - \int_0^t \langle u(\tau), \partial_t \psi(\tau) \rangle \, d\tau$$

$$- \int_0^t \langle u(\tau) \otimes u(\tau) : \nabla \psi(\tau) \rangle \, d\tau. \tag{8.12}$$

Since E is linear in ψ and $\mathcal{S}_\sigma^T$ is dense in $\mathcal{L}_\sigma$, provided that

$$|E(\psi)| \le C\|\psi\|_{\mathcal{L}_\sigma} \qquad \text{for every } \psi \in \mathcal{S}_\sigma^T$$

we can define $E(f)$ for $f \in \mathcal{L}_\sigma$ by setting

$$E(f) = \lim_{n \to \infty} E(\psi_n)$$

whenever $\psi_n \to f$ in $\mathcal{L}_\sigma$. It will then follow that $E(f) = 0$ for every $f \in \mathcal{L}_\sigma$, as required.

To show that $|E(\psi)| \le C\|\psi\|_{\mathcal{L}_\sigma}$ for every $\psi \in \mathcal{S}_\sigma^T$ we proceed term-by-term. For the first two terms of (8.12) we have

$$|\langle u(t), \psi(t) \rangle - \langle u(0), \psi(0) \rangle| \le 2\|u\|_{L^\infty(0,T;L^2)} \|\psi\|_{L^\infty(0,T;L^2)}$$

$$\le 2\|u\|_{L^\infty(0,T;L^2)} \|\psi\|_{C^{0,1}(0,T;L^2)},$$

using the fact that $u \in C_{\mathrm{w}}([0,T]; H_\sigma)$. For the last term of (8.12) we observe that

$$\left| \int_0^t \langle u(\tau) \otimes u(\tau) : \nabla f(\tau) \rangle \, d\tau \right| \le T\|u\|_{L^\infty(0,T;L^2)}^2 \|\nabla \psi\|_{L^1(0,T;L^\infty)}$$

$$\le T\|u\|_{L^\infty(0,T;L^2)}^2 \|\psi\|_{L^1(0,T;H^3)}$$

(as $\psi \in H^3$ implies that $\nabla\psi \in L^\infty$ on three-dimensional domains). Finally for the third term of (8.12)

$$\left| \int_0^t \langle u(\tau), \partial_\tau \psi(\tau) \rangle \, d\tau \right| \leq \|u\|_{L^\infty(0,T;L^2)} \|\partial_\tau \psi\|_{L^1(0,T;L^2)};$$

the proof is finished provided that $\|\partial_\tau f\|_{L^1(0,T;L^2)} \leq \|f\|_{C^{0,1}(0,T;L^2)}$.

To show this, noting that $\psi \in C^\infty(D \times [0,T])$, we can write

$$\|\partial_\tau \psi\|_{L^1(0,T;L^2)} = \int_0^T \left(\int |\partial_\tau \psi(x,t)|^2 \, dx \right)^{1/2} dt$$

$$= \int_0^T \left[\int \left| \lim_{h \to 0} \frac{\psi(x,t+h) - \psi(x,t)}{h} \right|^2 dx \right]^{1/2} dt$$

$$= \int_0^T \left[\int \lim_{h \to 0} \left| \frac{\psi(x,t+h) - \psi(x,t)}{h} \right|^2 dx \right]^{1/2} dt$$

$$= \int_0^T \lim_{h \to 0} \frac{1}{|h|} \left[\int (\psi(x,t+h) - \psi(x,t))^2 \, dx \right]^{1/2} dt$$

$$\leq T \sup_t \lim_{h \to 0} \frac{1}{|h|} \left[\int |\psi(x,t+h) - \psi(x,t)|^2 \, dx \right]^{1/2} dt$$

$$= T \sup_t \lim_{h \to 0} \frac{\|\psi(\cdot,t+h) - \psi(\cdot,t)\|_{L^2}}{|h|}$$

$$\leq T \|\psi\|_{C^{0,1}(0,T;L^2)},$$

as required.   □

We now study the time regularity of u when paired with a sufficiently smooth function that is not necessarily divergence free.

Lemma 8.4 *If u is a weak solution of the Euler equations (in the sense of Definition 8.2) then*

$$|\langle u(t) - u(s), \varphi \rangle| \leq C|t - s| \quad \text{for all} \quad \varphi \in H^3(D), \qquad (8.13)$$

where C depends only on $\|u\|_{L^\infty(0,T;L^2)}$ and $\|\varphi\|_{H^3}$.

Proof We use Lemma 8.1 to decompose $\varphi \in \mathcal{S}(D)$ as $\varphi = \eta + \nabla\sigma$, where $\sigma \in \mathcal{S}(D)$ and $\eta \in \mathcal{S}_\sigma(D)$ with

$$\|\nabla\eta\|_{L^\infty} \leq \|\nabla\eta\|_{H^2} \leq \|\eta\|_{H^3} \leq C\|\varphi\|_{H^3},$$

using (8.7) and the fact that $H^2(D) \subset L^\infty(D)$. Since $u(t)$ is incompressible for every $t \in [0,T]$, we have

$$\langle u(t) - u(s), \varphi \rangle = \langle u(t) - u(s), \eta + \nabla\sigma \rangle = \langle u(t) - u(s), \eta \rangle.$$

Since $\eta \in \mathcal{S}_\sigma$ and $\partial_t \eta = 0$ it follows from the definition of a weak solution at times t and s that

$$\langle u(t) - u(s), \varphi \rangle = \int_s^t \langle u(\tau) \otimes u(\tau) : \nabla \eta \rangle \, d\tau$$

and hence

$$\begin{aligned}
|\langle u(t) - u(s), \varphi \rangle| &\leq \|u\|^2_{L^\infty(0,T;L^2)} \|\nabla \eta\|_{L^\infty} |t - s| \\
&\leq C \|u\|^2_{L^\infty(0,T;L^2)} \|\varphi\|_{H^3} |t - s|, \quad (8.14)
\end{aligned}$$

which gives (8.13) for all $\varphi \in \mathcal{S}$. We now want to extend to $\varphi \in H^3(D)$. Let $\psi \in \mathcal{S}$ and $\varphi \in H^3$ such that, using density, there exists $\varepsilon > 0$ such that $\|\psi - \varphi\|_{H^3} \leq \varepsilon$ then we have

$$\begin{aligned}
|\langle u(t) - u(s), \varphi \rangle| &\leq |\langle u(t) - u(s), \varphi - \psi \rangle| + |\langle u(t) - u(s), \psi \rangle| \\
&\leq C \|u\|_{L^\infty(0,T;L^2)} \|\varphi - \psi\|_{H^3} + |\langle u(t) - u(s), \psi \rangle| \\
&\leq C\varepsilon + |\langle u(t) - u(s), \psi \rangle|.
\end{aligned}$$

We can now use (8.14) to see that

$$\begin{aligned}
|\langle u(t) - u(s), \varphi \rangle| &\leq C\varepsilon + C \|u\|^2_{L^\infty(0,T;L^2)} \|\psi\|_{H^3} |t - s| \\
&\leq C\varepsilon + C \|u\|^2_{L^\infty(0,T;L^2)} \|\psi - \varphi\|_{H^3} |t - s| \\
&\quad + C \|u\|^2_{L^\infty(0,T;L^2)} \|\varphi\|_{H^3} |t - s| \\
&\leq C\varepsilon + C \|u\|^2_{L^\infty(0,T;L^2)} \|\varphi\|_{H^3} |t - s|.
\end{aligned}$$

Since we can make ε arbitrarily small we are done. $\qquad \square$

A striking corollary of this weak continuity in time is that a mollification *in space alone* yields a function that is Lipschitz continuous *in time*.

Corollary 8.5 *If u is a weak solution of the Euler equations we have $u_\varepsilon \in \mathcal{L}_\sigma$ for any $\varepsilon > 0$; in particular the function $u_\varepsilon(x, \cdot)$ is Lipschitz continuous in t as a function into $L^2(D)$:*

$$\|u_\varepsilon(\cdot, t) - u_\varepsilon(\cdot, s)\|_{L^2} \leq C_\varepsilon \|u\|^2_{L^\infty(0,T;L^2)} |t - s|. \quad (8.15)$$

Proof Take $f \in L^2(D)$ with $\|f\|_{L^2(D)} = 1$, and let $\varphi = f_\varepsilon$. Then $\varphi \in H^3(D)$, and using the symmetry property (8.9) we have

$$\begin{aligned}
\langle u(t) - u(s), \varphi \rangle &= \langle u(t) - u(s), f_\varepsilon \rangle \\
&= \langle (u_\varepsilon(t) - u_\varepsilon(s)), f \rangle.
\end{aligned}$$

Since we have $\|\varphi\|_{H^3} \leq C_\varepsilon \|f\|_{L^2} = C_\varepsilon$ it follows from Lemma 8.4 that

$$|\langle u_\varepsilon(t) - u_\varepsilon(s), f\rangle| \leq C_\varepsilon \|u\|_{L^\infty(0,T;L^2)}^2 |t - s|.$$

Since this holds for every $f \in L^2(D)$ with $\|f\|_{L^2(D)} = 1$ we obtain the inequality (8.15) and $u_\varepsilon \in C^{0,1}([0,T];L^2)$.

As mollification commutes with differentiation it follows that u_ε is divergence free. Finally, since $u \in L^\infty(0,T;L^2)$, we observe that we have $u_\varepsilon \in L^\infty(0,T;H^3)$ and

$$\|u_\varepsilon\|_{L^1(0,T;H^3)} \leq T\|u_\varepsilon\|_{L^\infty(0,T;H^3)}$$

as $[0,T]$ is bounded.   □

Since $u_\varepsilon \in \mathcal{L}_\sigma$ it follows from Lemma 8.3 that we can use u_ε as a test function in the definition of a weak solution and obtain

$$\langle u(t), u_\varepsilon(t)\rangle - \langle u(0), u_\varepsilon(0)\rangle - \int_0^t \langle u(\tau), \partial_t u_\varepsilon(\tau)\rangle \, d\tau$$

$$= \int_0^t \langle u(\tau) \otimes u(\tau) : \nabla u_\varepsilon(\tau)\rangle \, d\tau;$$

we have validated equation (8.10), the first of the two equalities we need.

8.2.3 'Mollifying the equation'

We will now prove (8.11). The trick is to test with a mollified test function and move the mollification from the test function onto the terms involving u; all terms are then smooth enough to allow for an integration by parts.

Lemma 8.6   *If u is a weak solution then*

$$\int_0^t \langle \partial_t u_\varepsilon, \varphi\rangle \, d\tau = -\int_0^t \langle \nabla \cdot [u \otimes u]_\varepsilon, \varphi\rangle \, d\tau \qquad (8.16)$$

for every $t \in [0,T]$ and any $\varphi \in \mathcal{S}_\sigma^T$.

Proof   Take $\varphi \in \mathcal{S}_\sigma^T$, and use $\psi := \varphi_\varepsilon \star \varphi$ as the test function in the weak formulation (8.8). Then

$$\langle u(t), (\varphi_\varepsilon \star \varphi)(t)\rangle - \langle u(0), (\varphi_\varepsilon \star \varphi)(0)\rangle - \int_0^t \langle u(\tau), \partial_t[\varphi_\varepsilon \star \varphi](\tau)\rangle \, d\tau$$

$$= \int_0^t \langle u(\tau) \otimes u(\tau) : \nabla[\varphi_\varepsilon \star \varphi](\tau)\rangle \, d\tau.$$

Since we have chosen φ to be radial, $\langle \varphi_\varepsilon \star u, v \rangle = \langle u, \varphi_\varepsilon \star v \rangle$ (see (8.9)), and so we can move the derivatives and mollification onto the terms involving u. We will do this in detail for the term on the right-hand side, since it is the most complicated; the other terms follow similarly. We obtain

$$\int_0^t \langle u(\tau) \otimes u(\tau) : \nabla[\varphi_\varepsilon \star \varphi](\tau) \rangle \, \mathrm{d}\tau = \int_0^t \langle u(\tau) \otimes u(\tau) : \varphi_\varepsilon \star \nabla\varphi(\tau) \rangle \, \mathrm{d}\tau$$

$$= \int_0^t \langle [u(\tau) \otimes u(\tau)]_\varepsilon : \nabla\varphi(\tau) \rangle \, \mathrm{d}\tau = -\int_0^t \langle \nabla \cdot [u(\tau) \otimes u(\tau)]_\varepsilon, \varphi(\tau) \rangle \, \mathrm{d}\tau.$$

This implies that

$$\langle u_\varepsilon(t), \varphi(t) \rangle - \langle u_\varepsilon(0), \varphi(0) \rangle - \int_0^t \langle u_\varepsilon(\tau), \partial_t \varphi(\tau) \rangle \, \mathrm{d}\tau$$

$$= -\int_0^t \langle \nabla \cdot [u(\tau) \otimes u(\tau)]_\varepsilon : \varphi(\tau) \rangle \, \mathrm{d}\tau.$$

Since u_ε and φ are both absolutely continuous in time, the integration-by-parts formula

$$\langle u_\varepsilon(t), \varphi(t) \rangle - \langle u_\varepsilon(0), \varphi(0) \rangle - \int_0^t \langle u_\varepsilon(\tau), \partial_t \varphi(\tau) \rangle \, \mathrm{d}\tau$$

$$= \int_0^t \langle \partial_t u_\varepsilon(\tau), \varphi(\tau) \rangle \, \mathrm{d}\tau$$

finishes the proof. $\qquad\qquad\qquad\qquad\qquad\qquad\qquad\qquad\qquad\qquad$ $\square$

We now show that (8.16) holds for a much larger class of functions than $\varphi \in \mathcal{S}_\sigma^T$.

Lemma 8.7 *If u is a weak solution and in addition $u \in L^3(0,T;L^3)$ then (8.16) holds for any $\varphi \in L^3(0,T;L^3) \cap C_\mathrm{w}(0,T;H_\sigma)$.*

(Recall that we use $C_\mathrm{w}(0,T;H_\sigma)$ to denote H_σ-valued functions that are weakly continuous into L^2.)

Proof First we will obtain from (8.16) an equation that holds for all test functions ψ from the space $\mathcal{S}(D \times [0,T])$, not just for $\psi \in \mathcal{S}_\sigma^T$. For this we will use the Leray projection $\mathbb{P}$, the projection onto divergence-free vector fields (see Robinson et al., 2016, for example). Since for any $\psi \in \mathcal{S}(D \times [0,T])$ we have $\mathbb{P}\psi \in \mathcal{S}_\sigma$, it follows from (8.16) that

$$\int_0^t \langle \partial_t u_\varepsilon + \nabla \cdot [u \otimes u]_\varepsilon, \mathbb{P}\psi \rangle \, \mathrm{d}\tau = 0.$$

Since $\mathbb{P}$ is symmetric ($\langle \mathbb{P}g, f \rangle = \langle g, \mathbb{P}f \rangle$) and $\mathbb{P}\partial_t u_\varepsilon = \partial_t u_\varepsilon$ (since $\mathbb{P}$ commutes with derivatives and u_ε is incompressible) we obtain

$$\int_0^t \langle \partial_t u_\varepsilon + \mathbb{P}(\nabla \cdot [u \otimes u]_\varepsilon), \psi \rangle \, \mathrm{d}\tau = 0 \qquad \text{for every} \quad \psi \in \mathcal{S}(D \times [0, T]).$$

Since u_ε is Lipschitz in time (as a function from $[0, T]$ into H_σ) its time derivative $\partial_t u_\varepsilon$ exists almost everywhere (see Theorem 5.5.4 in Albiac & Kalton (2016), for example) and is integrable; we can therefore deduce using the Fundamental Lemma of the Calculus of Variations ($u \in L^2(\Omega)$ with $\int_\Omega u \cdot \psi = 0$ for all $\psi \in C_c^\infty(\Omega)$ implies that $u = 0$ almost everywhere in Ω, see e.g. Lemma 3.2.3 in Jost & Li-Jost, 1998) that for almost every $(x, t) \in D \times [0, T]$

$$\partial_t u_\varepsilon + \mathbb{P}(\nabla \cdot (u \otimes u)_\varepsilon) = 0.$$

Observing that $\mathbb{P}\nabla \cdot (u \otimes u)_\varepsilon \in L^{3/2}(0, T; L^{3/2})$ and that $\partial_t u_\varepsilon$ has the same integrability since $\partial_t u_\varepsilon = -\mathbb{P}\nabla \cdot (u \otimes u)_\varepsilon$, we can now multiply this equality by any choice of function $\varphi \in L^3(0, T; L^3) \cap C_w(0, T; H_\sigma)$ and integrate, obtaining

$$\int_0^t \langle \partial_t u_\varepsilon, \varphi \rangle \mathrm{d}\tau = -\int_0^t \langle \mathbb{P}\nabla \cdot [u \otimes u]_\varepsilon, \varphi \rangle \, \mathrm{d}\tau$$

$$= -\int_0^t \langle \nabla \cdot [u \otimes u]_\varepsilon, \mathbb{P}\varphi \rangle \, \mathrm{d}\tau = -\int_0^t \langle \nabla \cdot [u \otimes u]_\varepsilon, \varphi \rangle \, \mathrm{d}\tau,$$

where we have used the fact that $\mathbb{P}\varphi = \varphi$ since $\varphi(t) \in H_\sigma$ for every $t \in [0, T]$. $\square$

Note that the condition on $u \in L^3(0, T; L^3)$ is stronger than necessary for the proof but since Theorem 8.9 will need this condition the above result will suffice for our purposes.

We can now use u as a test function in (8.16) and thereby obtain equation (8.11), the second of the equalities we need.

8.2.4 Energy Conservation

We can now add equations (8.10) and (8.11) to obtain

$$\langle u(t), u_\varepsilon(t) \rangle - \langle u(0), u_\varepsilon(0) \rangle =$$

$$\int_0^t \langle u(\tau) \otimes u(\tau) : \nabla u_\varepsilon(\tau) \rangle - \langle \nabla \cdot [u(\tau) \otimes u(\tau)]_\varepsilon, u(\tau) \rangle \, \mathrm{d}\tau \quad (8.17)$$

valid for any $u \in L^3(0, T; L^3) \cap C_w(0, T; H_\sigma)$ that is a weak solution to the Euler equations.

In order to proceed we will need the following identity. We note that its validity is entirely independent of the Euler equations, but relies crucially on the fact that φ is radially symmetric.

Lemma 8.8   *Suppose that $v \in L^3 \cap H_\sigma$ and define*

$$J_\varepsilon(v) := \int_{\mathbb{R}^3} \int_{\mathbb{R}^3} \nabla\varphi_\varepsilon(\xi) \cdot (v(x+\xi) - v(x))|v(x+\xi) - v(x)|^2 \, \mathrm{d}\xi \, \mathrm{d}x.$$

Then

$$\tfrac{1}{2} J_\varepsilon(v) = \langle \nabla \cdot [v(\tau) \otimes v(\tau)]_\varepsilon, v(\tau) \rangle - \langle v(\tau) \otimes v(\tau) : \nabla v_\varepsilon(\tau) \rangle.$$

Proof   We have

$$J_\varepsilon(v) = \int_{\mathbb{R}^3} \int_{\mathbb{R}^3} \partial_i\varphi_\varepsilon(\xi)(v_i(x+\xi) - v_i(x))(v_j(x+\xi) - v_j(x))$$

$$\times (v_j(x+\xi) - v_j(x)) \, \mathrm{d}\xi \, \mathrm{d}x.$$

Expanding the expression for $J_\varepsilon(v)$ yields

$$\int_{\mathbb{R}^3} \left\{ \int_{\mathbb{R}^3} \partial_i\varphi_\varepsilon(\xi)v_i(x+\xi)v_j(x+\xi)v_j(x+\xi) \, \mathrm{d}\xi \right.$$

$$- \int_{\mathbb{R}^3} \partial_i\varphi_\varepsilon(\xi)v_i(x)v_j(x)v_j(x) \, \mathrm{d}\xi$$

$$+ \int_{\mathbb{R}^3} \partial_i\varphi_\varepsilon(\xi)v_i(x+\xi)v_j(x)v_j(x) \, \mathrm{d}\xi$$

$$- \int_{\mathbb{R}^3} \partial_i\varphi_\varepsilon(\xi)v_i(x)v_j(x+\xi)v_j(x+\xi) \, \mathrm{d}\xi$$

$$+ 2\int_{\mathbb{R}^3} \partial_i\varphi_\varepsilon v_i(x)v_j(x+\xi)v_j(x) \, \mathrm{d}\xi$$

$$\left. - 2\int_{\mathbb{R}^3} \partial_i\varphi_\varepsilon v_i(x+\xi)v_j(x+\xi)v_j(x) \, \mathrm{d}\xi \right\} \mathrm{d}x.$$

Note that the second term is zero since φ_ε has compact support, and the third term is zero since v is incompressible. For the fourth term we can change variables and set $\eta = x + \xi$ to obtain

$$- \int_{\mathbb{R}^3} \int_{\mathbb{R}^3} \partial_{\eta_i}\varphi_\varepsilon(\eta - x)v_i(x)v_j(\eta)v_j(\eta) \, \mathrm{d}\eta \, \mathrm{d}x.$$

As $\partial_i\varphi_\varepsilon$ is an odd function we have

$$\int_{\mathbb{R}^3} \int_{\mathbb{R}^3} \partial_i\varphi_\varepsilon(x - \eta)v_i(x)v_j(\eta)v_j(\eta) \, \mathrm{d}\eta \, \mathrm{d}x,$$

which becomes

$$\int_{\mathbb{R}^3} v_i(x)\partial_{x_i}\left[\int_{\mathbb{R}^3} \varphi_\varepsilon(x-\eta)v_j(\eta)v_j(\eta)\,\mathrm{d}\eta\right]\mathrm{d}x$$

$$= \int_{\mathbb{R}^3} v_i(x)\partial_i([v_jv_j]_\varepsilon)\,\mathrm{d}x = 0,$$

where again the term becomes zero as we use the incompressibility of v. A similar calculation for the first term gives

$$\int_{\mathbb{R}^3}\int_{\mathbb{R}^3} \partial_i\varphi_\varepsilon(\xi)v_i(x+\xi)v_j(x+\xi)v_j(x+\xi)\,\mathrm{d}\xi\,\mathrm{d}x$$

$$= \int_{\mathbb{R}^3} \partial_i([v_iv_jv_j]_\varepsilon)(x)\,\mathrm{d}x = 0,$$

using periodicity. For the final two terms similar calculations yield

$$2\int_{\mathbb{R}^3} [v_j\partial_i(v_jv_j)_\varepsilon - v_jv_i\partial_i(v_j)_\varepsilon]\,\mathrm{d}x = 2[\langle\nabla\cdot[v\otimes v]_\varepsilon, v\rangle - \langle v\otimes v : \nabla v_\varepsilon\rangle]$$

and the result follows.   □

Note that here again the assumption that $v \in L^3$ is stronger than needed but will hold when we use the result in Theorem 8.9.

We now want to look at the limit as $\varepsilon \to 0$ and see what condition on the solution is needed for the right hand side of (8.17) to converge to zero.

Take $t \in [0, T]$; then

$$\langle u(t), u_\varepsilon(t)\rangle - \langle u(0), u_\varepsilon(0)\rangle$$

$$= \int_0^t \langle u(\tau)\otimes u(\tau) : \nabla u_\varepsilon(\tau)\rangle - \langle\nabla\cdot[u(\tau)\otimes u(\tau)]_\varepsilon, u(\tau)\rangle\,\mathrm{d}\tau$$

$$= -\frac{1}{2}\int_0^t J_\varepsilon(u(\tau))\,\mathrm{d}\tau.$$

Therefore talking the limit as $\varepsilon \to 0$, since $u \in C_w([0, T]; H_\sigma)$ we obtain

$$\|u(t)\|_{L^2} - \|u(0)\|_{L^2} = -\frac{1}{2}\lim_{\varepsilon\to 0}\int_0^t J_\varepsilon(u(\tau))\,\mathrm{d}\tau.$$

Hence any condition on u that guarantees that

$$\lim_{\varepsilon\to 0}\int_0^t J_\varepsilon(u(\tau))\,\mathrm{d}\tau \to 0 \qquad \text{as} \qquad \varepsilon \to 0 \tag{8.18}$$

ensures energy conservation. We give two such conditions in the next section.

8.3 Two spatial conditions for energy conservation in the absence of boundaries

First we provide another proof (cf. Shvydkoy, 2009) of energy conservation under condition (8.3).

Theorem 8.9 *If* $u \in L^3(0,T;L^3(D))$ *is a weak solution of the Euler equations that satisfies*

$$\lim_{|y|\to 0} \frac{1}{|y|} \int_0^T \int_D |u(t,x+y) - u(t,x)|^3 \, dx \, dt = 0 \qquad (8.19)$$

then energy is conserved on $[0,T]$.

Proof We take t_1, t_2 with $0 \le t_1 \le t_2 \le T$, and consider the integral of $|J_\varepsilon(u)|$ over $[t_1, t_2]$; our aim is to show that this is zero in the limit as $\varepsilon \to 0$. We start by noticing that

$$\int_{t_1}^{t_2} |J_\varepsilon(u)| \, dt \le \int_{t_1}^{t_2} \int_D \int_{\mathbb{R}^3} \frac{1}{\varepsilon^4} \left| \nabla\varphi\left(\frac{\xi}{\varepsilon}\right) \right| |u(x+\xi) - u(x)|^3 \, d\xi \, dx \, dt.$$

We can then change variables $\xi = \eta\varepsilon$ and obtain,

$$\int_{t_1}^{t_2} |J_\varepsilon(u)| \, dt \le \int_{t_1}^{t_2} \int_D \int_{\mathbb{R}^3} \frac{1}{\varepsilon} |\nabla\varphi(\eta)| \, |u(x+\varepsilon\eta) - u(x)|^3 \, d\eta \, dx \, dt.$$

Using Fubini's Theorem we can exchange the order of the integrals to yield

$$\int_{t_1}^{t_2} |J_\varepsilon(u)| \, dt \le \int_{\mathbb{R}^3} \int_{t_1}^{t_2} \int_D \frac{|u(x+\varepsilon\eta) - u(x)|^3}{|\varepsilon\eta|} \, dx \, dt \, |\eta| \, |\nabla\varphi(\eta)| \, d\eta.$$

Taking limits as ε goes to zero

$$\lim_{\varepsilon\to 0} \int_{t_1}^{t_2} |J_\varepsilon(u)| \, dt$$

$$\le \lim_{\varepsilon\to 0} \int_{\mathbb{R}^3} \int_{t_1}^{t_2} \int_D \frac{|u(x+\varepsilon\eta) - u(x)|^3}{|\varepsilon\eta|} \, dx \, dt \, |\eta| \, |\nabla\varphi(\eta)| \, d\eta.$$

We are finished if we can exchange the outer integral and limit. This can be done using the Dominated Convergence Theorem. To do this we define the non-negative function,

$$f(y) = \frac{1}{|y|} \int_{t_1}^{t_2} \int_D |u(x+y) - u(x)|^3 \, dx \, dt.$$

By assumption $\limsup_{|y|\to 0} f(y) = 0$, thus for any $\varepsilon > 0$, we have

$\sup_{y \in B_0(\varepsilon)} f(y) \leq K$ for some $K = K(\varepsilon)$. Further, $\text{supp}(\varphi)$ is compact. Combining these facts we obtain a dominating integrable function

$$g(\eta) := K|\eta| \, |\nabla \varphi(\eta)|,$$

and the result follows.   □

We now show how the general condition in (8.18) allows for a simple proof of energy conservation when $u \in L^3(0, T; W^{\alpha,3}(\mathbb{R}^3))$ for any $\alpha > 1/3$. The use of condition (8.20) to characterise this space is due independently to Aronszajn, Gagliardo, and Slobodeckij, see Di Nezza, Palatucci, & Valdinoci (2012), for example.

Theorem 8.10 *If u is a weak solution of the Euler equations on the whole space that satisfies $u \in L^3(0, T; W^{\alpha,3}(\mathbb{R}^3))$ for some $\alpha > 1/3$, i.e. if $u \in L^3(0, T; L^3(\mathbb{R}^3))$ and*

$$\int_{\mathbb{R}^3} \int_{\mathbb{R}^3} \frac{|u(x) - u(y)|^3}{|x - y|^{3+3\alpha}} \, dx \, dy < \infty, \tag{8.20}$$

then energy is conserved.

Proof First observe that for $\alpha > 1/3$ the space $W^{\alpha,3}$ has a factor $|x - y|^{4+\delta}$ in the denominator of (8.20), where $\delta = 3\alpha - 1 > 0$.

As in the previous proof, our starting point is that

$$\int_{t_1}^{t_2} |J_\varepsilon(u)| \, dt \leq \int_{t_1}^{t_2} \int_{\mathbb{R}^3} \int_{\mathbb{R}^3} \frac{1}{\varepsilon^4} \left| \nabla \varphi \left(\frac{\xi}{\varepsilon} \right) \right| |u(x + \xi) - u(x)|^3 \, d\xi \, dx \, dt.$$

We can write

$$\int_{t_1}^{t_2} \int_{\mathbb{R}^3} \int_{\mathbb{R}^3} \frac{1}{\varepsilon^4} \left| \nabla \varphi \left(\frac{y - x}{\varepsilon} \right) \right| |u(y) - u(x)|^3 \, dy \, dx \, dt$$

$$= \int_{t_1}^{t_2} \int_{\mathbb{R}^3} \int_{\mathbb{R}^3} \frac{1}{\varepsilon^4} \left| \nabla \varphi \left(\frac{y - x}{\varepsilon} \right) \right| |u(y) - u(x)|^3 \, dy \, dx \, dt$$

$$= \int_{t_1}^{t_2} \int_{\mathbb{R}^3} \int_{\mathbb{R}^3} \frac{|y - x|^{4+\delta}}{\varepsilon^4} \left| \nabla \varphi \left(\frac{y - x}{\varepsilon} \right) \right| \frac{|u(y) - u(x)|^3}{|y - x|^{4+\delta}} \, dy \, dx \, dt$$

$$\leq c K_\varphi \varepsilon^\delta \int_{t_1}^{t_2} \int_{\mathbb{R}^3} \int_{\mathbb{R}^3} \frac{|u(y) - u(x)|^3}{|y - x|^{4+\delta}} \, dy \, dx \, dt = c\varepsilon^\delta,$$

since $\|\nabla \varphi\|_{L^\infty} \leq K_\varphi$ and the integrand is only non-zero within the support of φ, i.e. where $|y - x| \leq 2\varepsilon$. Energy conservation now follows.

□

8.4 Energy Balance on $\mathbb{T}^2 \times \mathbb{R}_+$

In this section we first give a definition of a weak solution on the domain $D_+ := \mathbb{T}^2 \times \mathbb{R}_+$, where $\mathbb{R}_+ = [0, \infty)$, that only involves the pressure on the boundary. We then show that such a solution u can be extended to a weak solution u_E on the boundary-free domain $D := \mathbb{T}^2 \times \mathbb{R}$. (Note that in this section we reserve D for this particular domain.)

8.4.1 Weak solutions of the Euler equations on $D_+ := \mathbb{T}^2 \times \mathbb{R}_+$

We define $\mathcal{S}(D_+)$ and $\mathcal{S}_\sigma(D_+)$ by restricting functions in $\mathcal{S}(\mathbb{T}^2 \times \mathbb{R})$ and $\mathcal{S}_\sigma(\mathbb{T}^2 \times \mathbb{R})$ to D_+; this means that we have Schwartz-like decay in the unbounded direction, and that the functions have a smooth restriction to the boundary.

We let

$$\mathcal{S}_{n,\sigma}(D_+) := \{\varphi \in \mathcal{S}(D_+) : \nabla \cdot \varphi = 0 \text{ and } \varphi_3 = 0 \text{ on } \partial D_+\} \qquad (8.21)$$

and we define $H_\sigma(D_+)$ to be the completion of $\mathcal{S}_{n,\sigma}(D_+)$ in the norm of $L^2(D_+)$. Functions in $H_\sigma(D_+)$ are weakly divergence free in the sense that they satisfy

$$\langle u, \nabla\varphi \rangle = 0 \qquad \text{for every} \quad \varphi \in H^1(D_+); \qquad (8.22)$$

that this holds for every $\varphi \in H^1(D_+)$ and not only for $\varphi \in \mathcal{D}(D_+)$ (proved exactly as in Section 8.2.1) will be useful in what follows.

As before, in a slight abuse of notation we let $C_w([0,T]; H_\sigma(D_+))$ denote the collection of all functions $u\colon [0,T] \to H_\sigma(D_+)$ that are weakly continuous into $L^2(D_+)$ i.e.

$$t \mapsto \langle u(t), \varphi \rangle_{D_+}$$

is continuous for every $\varphi \in L^2(D_+)$.

The space of functions $C_w([0,T]; H_\sigma(D_+))$ is incompressible in the sense of distributions for almost every $t \in [0,T]$ and satisfies $u \cdot n = 0$ on ∂D_+ in the sense of the Gauss formula: for $w \in H^{1/2}(\partial D_+)$, $\widetilde{w} \in H^1(D_+)$ and for $u \in H_\sigma(D_+)$ we have the functional

$$F_u(w) = \int_{\partial D_+} w(u \cdot n) \, \mathrm{d}S_x := \int_{D_+} (\nabla \cdot u)\widetilde{w} \, \mathrm{d}x + \int_{D_+} u \cdot \nabla\widetilde{w} \, \mathrm{d}x = 0.$$
$$(8.23)$$

We define the space $\mathcal{S}_\sigma^T(D_+)$ to be

$$\{\psi \in C^\infty(D_+ \times [0,T]) : \psi(\cdot, t) \in \mathcal{S}_\sigma(D_+) \text{ for every } t \in [0,T]\},$$

which will be our space of test functions; note that these functions are smooth and incompressible, but there is no restriction on their values on ∂D_+.

To obtain a weak formulation of the equations on D_+ we consider first a smooth solution u with pressure p that satisfies the Euler equations

$$\begin{cases} \partial_t u + \nabla \cdot (u \otimes u) + \nabla p = 0 & \text{in } D_+ \\ \nabla \cdot u = 0 & \text{in } D_+ \\ u \cdot n = 0 & \text{on } \partial D_+, \end{cases}$$

where n is the normal to ∂D_+, so that the third equation is in fact $u_3 = 0$ on ∂D_+. We can now take the scalar product of the first line with a test function $\varphi \in \mathcal{S}_\sigma^T$ and integrate over space and time to give

$$\int_0^t \langle \partial_t u + \nabla \cdot (u \otimes u) + \nabla p, \varphi \rangle_{D_+} \, d\tau = 0.$$

We can now integrate by parts and obtain

$$\langle u(t), \varphi(t) \rangle_{D_+} - \langle u(0), \varphi(0) \rangle_{D_+} - \int_0^t \langle u, \partial_t \varphi \rangle_{D_+} \, d\tau$$

$$- \int_0^t \langle u \otimes u : \nabla \varphi \rangle_{D_+} \, d\tau - \langle u_3, u \cdot \varphi \rangle_{\partial D_+ \times [0,t]}$$

$$- \int_0^t \langle p, \nabla \cdot \varphi \rangle_{D_+} \, d\tau + \langle p, \varphi \cdot n \rangle_{\partial D_+ \times [0,t]} = 0.$$

We notice that as $u_3 = 0$ on ∂D_+ and $\nabla \cdot \varphi = 0$ in D_+ the two terms involving these expressions vanish and we have

$$\langle u(t), \varphi(t) \rangle_{D_+} - \langle u(0), \varphi(0) \rangle_{D_+} - \int_0^t \langle u, \partial_t \varphi \rangle_{D_+} \, d\tau$$

$$- \int_0^t \langle u \otimes u : \nabla \varphi \rangle_{D_+} \, d\tau + \langle p, \varphi \cdot n \rangle_{\partial D_+ \times [0,t]} = 0.$$

Since we have not restricted the values of φ on ∂D_+ we have a contribution from the boundary, namely

$$\langle p, \varphi_3 \rangle_{\partial D_+ \times [0,t]}.$$

We therefore require $p \in \mathcal{D}'(\partial D_+ \times [0,T])$, a distribution on the boundary, in our definition of a weak solution.

Definition 8.11 (Weak Solution on D_+)   A weak solution of the Euler equations on $D_+ \times [0,T]$ is a pair (u,p), where $u \in C_{\mathrm{w}}([0,T]; H_\sigma(D_+))$ and $p \in \mathcal{D}'(\partial D_+ \times [0,T])$ such that

$$\langle u(t), \varphi(t) \rangle_{D_+} - \langle u(0), \varphi(0) \rangle_{D_+} - \int_0^t \langle u(\tau), \partial_t \varphi(\tau) \rangle_{D_+} \, \mathrm{d}\tau$$

$$= \int_0^t \langle u(\tau) \otimes u(\tau) : \nabla \varphi(\tau) \rangle_{D_+} \, \mathrm{d}\tau - \langle p, \varphi \cdot n \rangle_{\partial D_+ \times [0,t]}, \quad (8.24)$$

for every $t \in [0,T]$ and for every $\varphi \in \mathcal{S}_\sigma^T(D_+)$.

Note that in the final term, $\varphi \cdot n = -\varphi_3$.

8.4.2 Half-plane reflection map

We introduce an extension u_E that takes a weak solution u defined in D_+ to one defined on the whole of D. Essentially we extend 'by reflection', with appropriate sign changes to ensure that u_R, the 'reflection' of u, is a weak solution on $D_- := \mathbb{T}^2 \times \mathbb{R}_-$. We can then show that $u_E := u + u_R$ is a weak solution on the whole of D (in the sense of Definition 8.2).

Given a vector-valued function $f \colon D_\pm \to \mathbb{R}^3$ we define $f_R \colon D_\mp \to \mathbb{R}^3$ by

$$f_R(x, y, z) := \begin{pmatrix} f_1(x, y, -z) \\ f_2(x, y, -z) \\ -f_3(x, y, -z) \end{pmatrix}$$

extending f and f_R by zero beyond their natural domain of definition, we set

$$f_E(x, y, z) := \begin{cases} f(x, y, z) + f_R(x, y, z) & z \neq 0 \\ \frac{1}{2}(f(x, y, z) + f_R(x, y, z)) & z = 0 \\ \quad = (f_1(x, y, 0), f_2(x, y, 0), 0). \end{cases}$$

Clearly $f_E = f + f_R$ almost everywhere.

Lemma 8.12 *If* $u \in H_\sigma(D_+)$ *then* $u_R \in H_\sigma(D_-)$ *and* $u_E \in H_\sigma(D)$.

Proof Since $u \in H_\sigma(D_+)$ there exists $u_n \in \mathcal{S}_{n,\sigma}(D_+)$ (see (8.21)) such that $u_n \to u$ in $L^2(D_+)$. Clearly $u_{n,R} \in \mathcal{S}_{n,\sigma}(D_-)$ and $u_{n,R} \to u_R$ in $L^2(D_-)$. Therefore $u_R \in H_\sigma(D_-)$. Further, $u_n + u_{n,R}$ trivially belongs to $S_\sigma(D)$ and is divergence free. Since $u_n + u_{n,R}$ converges to u_E in $L^2(D)$ we obtain the desired result. $\qquad \square$

Now we will show that, with an appropriate choice of the pressure, u_R is a weak solution of the Euler equations in the lower half space D_-. Note that we do not need to extend the pressure distribution p.

Theorem 8.13   *If (u, p) is a weak solution to the Euler equations on D_+ then (u_R, p) is a weak solution in D_-, i.e.*

$$\langle u_R(t), \varphi(t) \rangle_{D_-} - \langle u_R(0), \varphi(0) \rangle_{D_-} - \int_0^t \langle u_R(\tau), \partial_t \varphi(\tau) \rangle_{D_-} \, d\tau$$

$$= \int_0^t \langle u_R(\tau) \otimes u_R(\tau) : \nabla \varphi(\tau) \rangle_{D_-} \, d\tau - \langle p, \varphi \cdot n \rangle_{\partial D_- \times [0,t]}, \quad (8.25)$$

for every $t \in [0, T]$ and for every $\varphi \in \mathcal{S}_\sigma^T(D_-)$.

Note that now in the final term we have $\varphi \cdot n = \varphi_3$.

Proof   Notice first that any $\varphi \in \mathcal{S}_\sigma^T(D_-)$ can be written as ψ_R, where $\psi = \varphi_R \in \mathcal{S}_\sigma^T(D_+)$. Now, making the change of variables $(x_1, x_2, x_3) \mapsto (y_1, y_2, -y_3)$ in the linear term yields

$$\langle u_R, \psi_R \rangle_{D_-} = \langle u, \psi \rangle_{D_+}.$$

For the nonlinear term one can check case-by-case, with the same change of variables, that

$$\int_{D_-} [(u_R)_i (u_R)_j \partial_j (\psi_R)_i](x) \, dx = \int_{D_+} [u_i u_j \partial_j \psi_i](y) \, dy.$$

Finally for the pressure term we have

$$\langle p, \psi \cdot n \rangle_{\partial D_+} = \langle p, \psi_3 \rangle = -\langle p, \varphi_3 \rangle = \langle p, \varphi \cdot n \rangle_{\partial D_-},$$

since $\psi_3(x, y, 0) = -\varphi_3(x, y, 0)$. $\square$

By adding (8.24) and (8.25) it follows that u_E is a weak solution of the Euler equations on D.

Corollary 8.14   *The extension u_E is a weak solution of the Euler equations on D in the sense of Definition 8.2.*

Proof   For $\zeta \in \mathcal{S}_\sigma^T$ we can use $\zeta|_{D_+}$ as a test function in (8.24) and $\zeta|_{D_-}$ in (8.25) and add the two equations to obtain

$$\langle u_E(t), \zeta(t) \rangle_D - \langle u_E(0), \zeta(0) \rangle_D - \int_0^t \langle u_E(\tau), \partial_\tau \zeta(\tau) \rangle_D \, d\tau$$

$$= \int_0^t \langle u_E(\tau) \otimes u_E(\tau) : \nabla \zeta(\tau) \rangle_D \, d\tau,$$

where the pressure terms have cancelled due to the opposite signs of the normal in the two domains; but this is now the definition of a weak solution of the Euler equations in D.   □

Since u_E is a weak solution of the incompressible Euler equations on D, Corollary 8.9 guarantees that if $u_E \in L^3(0, T; L^3(D))$ and

$$\lim_{|y| \to 0} \frac{1}{|y|} \int_0^T \int_D |u_E(t, x + y) - u_E(t, x)|^3 \, dx \, dt = 0 \qquad (8.26)$$

then u_E conserves energy on $D \times [0, T]$. Due to the definition of u_E this implies that

$$\|u_E(t)\|_{L^2(D)}^2 - \|u_E(0)\|_{L^2(D)}^2 = 2\|u(t)\|_{L^2(D_+)}^2 - 2\|u(0)\|_{L^2(D_+)}^2 = 0,$$

i.e. we obtain energy conservation for u. We now find conditions on u alone (rather than $u_E = u + u_R$) that guarantee that (8.26) is satisfied.

8.5 Energy Conservation on D_+

Here we will prove our main result in Theorem 8.15: energy conservation on D_+ under certain assumptions on the weak solution u. The two conditions we need for u to conserve energy are similar to the conditions needed for Corollary 8.9 where we had no boundary. We will impose an extra continuity condition to deal with the presence of the boundary: we assume that there exists a $\delta > 0$ such that $u \in L^3(0, T; C^0(\mathbb{T}^2 \times [0, \delta]))$.

We note that, since $\partial D^+ = \mathbb{T}^2$ is compact, for each $t \in [0, T]$ there exists a non-decreasing function $w_t : [0, \infty) \to [0, \infty)$ with $w_t(0) = 0$ and continuous at 0, such that

$$|u(x + z, t) - u(x, t)| < w_t(|z|) \qquad (8.27)$$

whenever $x \in \partial D^+$ and $x + z \in D$ with $|z| \leq \delta$.

We can now provide conditions on u to ensure energy conservation.

Theorem 8.15   *Let* $u \in L^3(0, T; L^3(D_+))$ *be a weak solution of the Euler equations such that* $u \in L^3(0, T; C^0(\mathbb{T}^2 \times [0, \delta)))$ *for some* $\delta > 0$ *and*

$$\lim_{|y| \to 0} \frac{1}{|y|} \int_{t_1}^{t_2} \iint_{\mathbb{T}^2} \int_{|y|}^{\infty} |u(t, x + y) - u(t, x)|^3 \, dx_3 \, dx_1 \, dx_2 \, dt = 0; \quad (8.28)$$

then u *conserves energy on* $D_+ \times [t_1, t_2]$.

Proof We can split

$$\lim_{|y| \to 0} \frac{1}{|y|} \int_0^T \int_D |u_E(t, x + y) - u_E(t, x)|^3 \, \mathrm{d}x \, \mathrm{d}t = 0$$

into three integrals over the regions $A := \{x_3 > |y|\}$, $B := \{x_3 < -|y|\}$, and $C := \{|x_3| \le |y|\}$. We have

$$|u_E(t, x + y) - u_E(t, x)|^3$$
$$= [\mathbb{I}_A(x) + \mathbb{I}_B(x) + \mathbb{I}_C(x)] \, |u_E(t, x + y) - u_E(t, x)|^3.$$

For $\int_A$ we see that since $x_3 > 0$ and $x_3 + y_3 > 0$ then u_E is in fact u, thus after integrating and taking the limit it goes to zero by (8.28). For $\int_B$ a very similar argument (but with the change of variables $x_3 \mapsto -z_3$) gives the same outcome.

We are left with $\int_C$: we need to show that

$$\lim_{|y| \to 0} \frac{1}{|y|} \int_{t_1}^{t_2} \int_{\mathbb{T}^2} \int_{-|y|}^{|y|} |u_E(t, x + y) - u_E(t, x)|^3 \, \mathrm{d}x_3 \, \mathrm{d}x_2 \, \mathrm{d}x_1 \, \mathrm{d}t = 0.$$

We have assumed that $u \in L^3(0, T; C^0(\mathbb{T}^2 \times [0, \delta]))$ and so certainly

$$u_E \in L^3(0, T; L^\infty(\mathbb{T}^2 \times [-\delta, \delta])).$$

Then, since for all $|y| < \delta$ we have

$$\frac{1}{|y|} \int_{\mathbb{T}^2} \int_{-|y|}^{|y|} |u_E(t, x + y) - u_E(t, x)|^3 \, \mathrm{d}x_3 \, \mathrm{d}x_2 \, \mathrm{d}x_1 \le C \sup_{x \in \mathbb{T}^2 \times [0, \delta)} |u(t)|^3,$$

we can then move the limit inside the time integral using the Dominated Convergence Theorem, and it suffices to show that

$$\lim_{|y| \to 0} \frac{1}{|y|} \int_{\mathbb{T}^2} \int_{-|y|}^{|y|} |u_E(t, x + y) - u_E(t, x)|^3 \, \mathrm{d}x_3 \, \mathrm{d}x_2 \, \mathrm{d}x_1 = 0$$

for almost every $t \in (t_1, t_2)$.

As, $u(\cdot, t) \in C_{\partial D_+}$ and $u \cdot n = 0$ on the boundary, the boundary values are the same for u and u_R and so $u_E(\cdot, t) \in C_{\{z=0\}}$.

Now fix t and let $x' = (x_1, x_2, 0)$; then, using (8.27),

$$|u_E(t, x' + x_3 + y) - u_E(t, x' + x_3)|$$
$$\le |u_E(t, x' + x_3 + y) - u_E(t, x') + u_E(t, x') - u_E(t, x' + x_3)|$$
$$\le w_t(|y + x_3|) + w_t(|x_3|) \le 2w_t(2|y|)$$

and thus

$$\frac{1}{|y|} \iint_{\mathbb{T}^2} \int_{-|y|}^{|y|} |u_E(t, x+y) - u_E(t, x)|^3 \, \mathrm{d}x_3 \, \mathrm{d}x_2 \, \mathrm{d}x_1$$

$$\leq C \frac{1}{|y|} \iint_{\mathbb{T}^2} \int_{-|y|}^{|y|} |w_t(2|y|)|^3 \, \mathrm{d}x_3 \, \mathrm{d}x_2 \, \mathrm{d}x_1$$

$$\leq C \frac{1}{|y|} |\mathbb{T}^2| |y| |w_t(2|y|)|^3 \to 0$$

as $|y| \to 0$, which is what we required. □

8.6 Conclusion

Assuming the simple integral condition

$$\lim_{|y| \to 0} \frac{1}{|y|} \int_{t_1}^{t_2} \iint_{\mathbb{T}^2} \int_{|y|}^{\infty} |u(t, x+y) - u(t, x)|^3 \, \mathrm{d}x_3 \, \mathrm{d}x_1 \, \mathrm{d}x_2 \, \mathrm{d}t = 0,$$

which is similar to the weakest condition known on $\mathbb{R}^3$ or $\mathbb{T}^3$, and appropriate continuity in a neighbourhood of the boundary we have proved energy conservation of the incompressible Euler equations with a flat boundary of finite area. Further, our methods do not depend on the dimension so analogues hold in $\mathbb{T}^{d-1} \times \mathbb{R}_+$ for $d \geq 2$.

Based on the argument in Constantin et al. (1994) we were able to prove a similar result under the same conditions (Robinson, Rodrigo, & Skipper, 2018) using a definition of weak solution that entirely dispenses with the pressure.

8.7 Afterward: the result of Bardos & Titi on a general bounded domain

In more recent work, Bardos & Titi (2018) have proved energy conservation in a general C^2 bounded domain for solutions that satisfy the equations in the sense of distributions, provided that

$$u \in L^3(0, T; C^\alpha(\bar{\Omega})) \tag{8.29}$$

for some $\alpha > 1/3$. Their formulation of what it means to be a solution involves the pressure throughout the domain, and their argument requires certain regularity estimates on the pressure which our approach (both here and in Robinson et al., 2018) avoids by making use of the particular geometric properties of our simple domain.

On a bounded domain there are many choices of definition for a weak solution depending on the family of test functions used. The definition in line with that used in this paper would require the test functions to be incompressible but would not restrict their boundary values; we take any test function ψ in the space

$$\{\psi \in C^\infty(\Omega \times [0,T]) : \nabla \cdot \psi(\cdot,t) = 0 \text{ for every } t \in [0,T]\}. \tag{8.30}$$

As in Section 8.4.1, the resulting formulation would require a boundary pressure term $p \in \mathcal{D}'(\partial\Omega \times [0,T])$.

If we take a smaller class of test functions, requiring the boundary condition $\psi \cdot n = 0$ on $\partial\Omega$ (as in Robinson et al., 2018) then we can obtain a formulation that does not contain any explicit reference to the pressure and - since we take a smaller class of test functions - allows for more possible 'weak solutions'. More formally, let us define

$$C_{n,\sigma}^\infty(\Omega) := \{\psi \in C^\infty(\Omega) : \nabla \cdot \psi = 0 \text{ and } \psi \cdot n = 0 \text{ on } \partial\Omega\},$$

$$C_{n,\sigma}^\infty(\Omega \times [0,T]) := \{u \in C^\infty(\Omega \times [0,T]) : u(\cdot,t) \in C_{n,\sigma}^\infty(\Omega) \; \forall \, t \in [0,T]\},$$

and

$$H_\sigma(\Omega) := \text{the completion of } C_{n,\sigma}^\infty(\Omega) \text{ in the norm of } L^2(\Omega).$$

Taking the space-time inner product of the equations with a test function $\psi \in C_{n,\sigma}^\infty(\Omega \times [0,T])$, integrating by parts, and using the fact that $\psi \cdot n = 0$ on $\partial\Omega$ and is divergence free leads to the following definition.

Definition 8.16   A *weak solution* of the Euler equations on $\Omega \times [0,T]$ is a vector-valued function $u \in C_w([0,T]; H_\sigma(\Omega))$ such that

$$\langle u(t), \psi(t) \rangle - \langle u(0), \psi(0) \rangle - \int_0^t \langle u(\tau), \partial_t \psi(\tau) \rangle \, d\tau$$

$$= \int_0^t \langle u(\tau) \otimes u(\tau) : \nabla\psi(\tau) \rangle \, d\tau, \tag{8.31}$$

for every $t \in [0,T]$, for any $\psi \in C_{n,\sigma}^\infty(\Omega \times [0,T])$.

Bardos & Titi (2018) use a different class of test functions that are compactly supported in $\Omega \times (0,T)$, but are not incompressible. This leads them to consider solutions of the Euler equations 'in the sense of distributions', and requires the inclusion of a pressure term.

Definition 8.17 (Bardos & Titi, 2018)   Let $\Omega \subset \mathbb{R}^3$ be a bounded domain with a C^2 boundary. Then (u, p) is a *distributional solution* of the incompressible Euler equations in $\Omega \times (0, T)$ if $u \in L^\infty(0, T; H_\sigma(\Omega))$, $p \in \mathcal{D}'(\Omega \times [0, T])$ and

$$\int_0^T \langle u, \partial_t \varphi \rangle + \langle u \otimes u : \nabla \varphi \rangle + \langle p, \nabla \cdot \varphi \rangle \, \mathrm{d}t = 0 \qquad (8.32)$$

for every test function $\varphi \in \mathcal{D}(\Omega \times (0, T))$.

No regularity of p is supposed by Bardos & Titi (2018), so here we have assumed the minimum required to ensure that the final term on the left-hand side of (8.32) makes sense.

We now show that any weak solution in the sense of Definition 8.16 is also a distributional solution in the sense of Definition 8.17; so the result of Bardos & Titi (2018) provides energy conservation for 'weak solutions' under the assumption that $u \in L^3(0, T; C^\alpha)$, $\alpha > 1/3$.

Theorem 8.18   *Suppose that u is a weak solution in the sense of Definition 8.16 and let p satisfy*

$$\int_0^T \langle u \otimes u : \nabla(\nabla \sigma) \rangle \, \mathrm{d}\tau + \int_0^T \langle p, \Delta \sigma \rangle \, \mathrm{d}\tau = 0 \qquad (8.33)$$

for all $\sigma \in C^\infty(0, T; C^\infty(\Omega))$. Then the pair (u, p) is a distributional solution in the sense of Definition 8.17.

Proof   Take a weak solution u as in Definition 8.16, so that

$$\langle u(t), \psi(t) \rangle - \langle u(0), \psi(0) \rangle - \int_0^t \langle u(\tau), \partial_t \psi(\tau) \rangle \, \mathrm{d}\tau$$

$$= \int_0^t \langle u(\tau) \otimes u(\tau) : \nabla \psi(\tau) \rangle \, \mathrm{d}\tau,$$

for every $t \in [0, T]$, for any $\varphi \in C^\infty_{n,\sigma}(\Omega \times [0, T])$. Now we take p satisfying (8.33) and want to show that the pair (u, p) satisfies (8.32) for any choice of $\psi \in C^\infty([0, T]; \mathcal{D}(\Omega))$.

Given any $\psi \in \mathcal{D}(\Omega \times (0, T))$, we can perform a Helmholtz decomposition (see Theorem 2.16 in Robinson et al., 2016) so that

$$\psi = \varphi + \nabla \sigma \quad \text{where} \quad \varphi, \sigma \in \mathcal{D}((0, T); C^\infty(\Omega)), \qquad (8.34)$$

with $\nabla \cdot \varphi(\cdot, t) = 0$ and $\varphi(\cdot, t) \cdot n = 0$ and $\nabla \sigma(\cdot, t) \cdot n = 0$ on $\partial \Omega$.

When we use this decomposition in (8.32) we obtain

$$
-\int_0^T \langle u(\tau), \partial_t \varphi(\tau) \rangle \, \mathrm{d}\tau
$$

$$
= \int_0^T \langle u(\tau) \otimes u(\tau) : \nabla \varphi(\tau) \rangle \, \mathrm{d}\tau + \int_0^T \langle u(\tau) \otimes u(\tau) : \nabla(\nabla \sigma)(\tau) \rangle \, \mathrm{d}\tau
$$

$$
+ \int_0^T \langle p(\tau), \nabla \cdot \varphi(\tau) \rangle \, \mathrm{d}\tau + \int_0^T \langle p(\tau), \Delta \sigma(\tau) \rangle \, \mathrm{d}\tau. \quad (8.35)
$$

Since u satisfies (8.31) this simplifies to

$$
\int_0^T \langle u \otimes u : \nabla \nabla \sigma \rangle \, \mathrm{d}\tau + \int_0^T \langle p, \Delta \sigma \rangle \, \mathrm{d}\tau = 0, \quad (8.36)
$$

for all $\sigma \in C^\infty([0,T]; C^\infty(\Omega))$. Thus if p solves this equation then (8.32) holds. □

The above result assumes that one can find a function p satisfying (8.33). In fact the existence of such a p is a significant part of the argument in Bardos & Titi (2018): provided that $u \in L^3(0,T; C^\delta)$ for any $\delta > 0$ one can use the results from Chapters 5 and 6 in Krylov (1996) along with the observation that (8.33) is the weak formulation of the elliptic boundary value problem

$$
-\Delta p = \partial_i \partial_j (u_i u_j) \text{ in } \Omega, \qquad \nabla p \cdot n = -\frac{1}{2}(u_j \partial_j u_i) n_i \text{ on } \partial\Omega
$$

to show that there exists such a p with $p \in L^{3/2}(0,T; C^\delta)$.

Acknowledgements

JLR is currently supported by the European Research Council, grant no. 616797. JWDS was supported by EPSRC as part of the MASDOC DTC at the University of Warwick, Grant No. EP/HO23364/1.

References

Albiac, F. & Kalton, N.J. (2016) *Topics in Banach space theory.* Second edition. Springer-Verlag.

Bahouri, H., Chemin, J.-Y., & Danchin, R. (2011) *Fourier analysis and nonlinear partial differential equations.* Springer-Verlag.

Bardos, C. & Titi, E.S. (2018) Onsager's conjecture for the incompressible Euler equations in bounded domains. *Arch. Rat. Mech. Anal.* **228**, 197–207.

Cheskidov, A., Constantin, P., Friedlander, S., & Shvydkoy, R. (2008) Energy conservation and Onsager's conjecture for the Euler equations. *Nonlinearity* **21**, 1233–1252.

Constantin, P., E, W., & Titi, E.S. (1994) Onsager's conjecture on the energy conservation for solutions of Euler's equation. *Comm. Math. Phys.* **165**, 207–209.

Di Nezza, E., Palatucci, G., & Valdinoci, E. (2012) Hitchhiker's guide to the fractional Sobolev spaces. *Bull. Sci. Math.* **136**, 521-573.

Duchon, J. & Robert, R. (2000) Inertial energy dissipation for weak solutions of incompressible Euler and Navier-Stokes equations. *Nonlinearity* **13**, 249–255.

Evans, L.C. (1998) *Partial differential equations*. American Mathematical Society, Providence, RI.

Eyink, G. (1994) Energy dissipation without viscosity in ideal hydrodynamics I. Fourier analysis and local energy transfer. *Physica D: Nonlinear Phenomena* **78**, 222–240.

Eyink, G. & Sreenivasan, K. (2006) Onsager and the theory of hydrodynamic turbulence. *Reviews of modern physics* **78**, 87–135.

Jost, J. & Li-Jost, X. (1998) *Calculus of variations*. Cambridge University Press.

Krylov, N.V. (1996) *Lectures on elliptic and parabolic equations in Hölder spaces*. American Mathematical Society, Providence, RI.

Lemarié-Rieusset, P.G. (2002) *Recent developments in the Navier–Stokes problem*. CRC Press.

Onsager, L. (1949) Statistical hydrodynamics. *Il Nuovo Cimento (1943-1954)* **6**, 279–287.

Robinson, J.C., Rodrigo, J.L., & Sadowski, W. (2016) *The three-dimensional Navier–Stokes equations*. Cambridge University Press, Cambridge, UK.

Robinson, J.C., Rodrigo, J.L., & Skipper, J.W.D. (2018) Energy conservation for the 3D Euler equations on $\mathbb{T}^2 \times \mathbb{R}_+$ for weak solutions defined without reference to the pressure. *Asymptotic Analysis*, to appear.

Skipper, J.W.D. (2018) *Energy conservation for the Euler equations with boundaries*. PhD thesis, University of Warwick.

Shvydkoy, R. (2009) On the energy of inviscid singular flows. *J. Math. Anal. Appl.* **349**, 583–595.

Shvydkoy, R. (2010) Lectures on the Onsager conjecture. *Discrete Contin. Dyn. Syst. Ser. S* **3**, 473–496.

9

Regularity of Navier–Stokes flows with bounds for the velocity gradient along streamlines and an effective pressure

Chuong V. Tran

School of Mathematics and Statistics, University of St Andrews,
St Andrews, KY16 9SS. UK.
`cvt1@st-andrews.ac.uk`

Xinwei Yu
Department of Mathematical and Statistical Sciences, University of Alberta,
Edmonton, AB, T6G 2G1. Canada.
`xinweiyu@math.ualberta.ca`

Abstract

Regularity criteria for solutions of the three-dimensional Navier–Stokes equations are derived in this paper. Let

$$\Omega(t,q) := \left\{ x : |u(x,t)| > C(t,q) \, \|u\|_{L^{3q-6}(\mathbb{R}^3)} \right\} \cap \{ x : \widehat{u} \cdot \nabla |u| \neq 0 \},$$

$$\widetilde{\Omega}(t,q) := \left\{ x : |u(x,t)| \leq C(t,q) \, \|u\|_{L^{3q-6}(\mathbb{R}^3)} \right\} \cap \{ x : \widehat{u} \cdot \nabla |u| \neq 0 \},$$

where $q \geq 3$ and

$$C(t,q) := \left(\frac{\|u\|_{L^4(\mathbb{R}^3)}^2 \, \left\| |u|^{(q-2)/2} \, \nabla |u| \right\|_{L^2(\mathbb{R}^3)}}{cq \, \|u_0\|_{L^2(\mathbb{R}^3)} \, \|p + \mathcal{P}\|_{L^2(\widetilde{\Omega})} \, \left\| |u|^{(q-2)/2} \, \widehat{u} \cdot \nabla |u| \right\|_{L^2(\widetilde{\Omega})}} \right)^{2/(q-2)}.$$

Here $u_0 = u(x,0)$, $\mathcal{P}(x, |u|, t)$ is a pressure moderator of relatively broad form, $\widehat{u} \cdot \nabla |u|$ is the gradient of $|u|$ along streamlines, and $c = (2/\pi)^{2/3}/\sqrt{3}$ is the constant in the inequality $\|f\|_{L^6(\mathbb{R}^3)} \leq c \, \|\nabla f\|_{L^2(\mathbb{R}^3)}$. If

$$\left\| \widehat{u} \cdot \nabla |u| \right\|_{L^{3/2}(\Omega)} \leq \frac{2}{c^2 q^2} \frac{\|u\|_{L^{3q}(\mathbb{R}^3)}^2}{\|p + \mathcal{P}\|_{L^{3q/2}(\Omega)}},$$

$$\|p + \mathcal{P}\|_{L^{3/2}(\Omega)} \leq \frac{1}{c^2 q^2} \frac{\|u\|_{L^{3q}(\mathbb{R}^3)}^2}{\|p + \mathcal{P}\|_{L^{3q/2}(\Omega)}},$$

or

$$\|p + \mathcal{P}\|_{L^{(q+2)/2}(\Omega)} \leq \frac{1}{cq} \frac{\|u\|_{L^{3q}(\mathbb{R}^3)}^{q/2}}{\|u\|_{L^{q+2}(\Omega)}^{(q-2)/2}},$$

then $\|u\|_{L^q(\mathbb{R}^3)}$ decreases and no singularities can develop.

Published as part of *Partial Differential Equations in Fluid Mechanics*, edited by C.L. Fefferman, J.C. Robinson, & J.L. Rodrigo. © Cambridge University Press 2018.

9.1 Introduction

We study the Cauchy problem for the Navier–Stokes equations

$$\frac{\partial u}{\partial t} + (u \cdot \nabla)u + \nabla p = \Delta u, \qquad (9.1)$$

$$\nabla \cdot u = 0,$$

in $\mathbb{R}^3 \times (0, T)$ with $u(x, 0) = u_0(x)$ divergence free, sufficiently smooth and $\|u_0\|_{L^2} < \infty$. Given such $u_0(x)$, it is well known that a classical solution exists locally in time, at least up to some $t = T$. The question is whether the solution remains regular beyond T, particularly up to all $t \geq T$ (global regularity). Active research since the seminal study by Leray (1934) has not been able to provide any clue to a positive answer to this question. On the contrary, numerous results on partial regularity in the literature cast doubt on the capability of viscous effects in balancing their nonlinear counterparts, thereby lending some favour to a negative answer. Nonetheless, recent studies by Tran & Yu (2015, 2016, 2017) have examined the possibility of such a balance and suggested its plausibility on the basis of genuine depletion of nonlinearity in the pressure force.

Unlike vortex stretching, whose depletion of nonlinearity (if there is any) may not be readily recognisable, depletion of nonlinearity in the pressure force manifests itself plainly through several facets. First, the conservation of energy, though taken for granted and usually not considered a manifestation of nonlinear depletion of the pressure gradient, gives rise to a relatively high degree of regularity for the solution, namely $u \in L^2(0, T; H^1(\mathbb{R}^3))$. Second, generic anti-correlation between $|u|$ and $|\nabla|u||$ in the pressure term that drives the dynamics of $\|u\|_{L^q} := \|u\|_{L^q(\mathbb{R}^3)}$, for $q \geq 3$, can be expected to weaken the nonlinearity of the driving term to some extent. Furthermore, Tran & Yu (2016) have shown that the pressure in this term is determined up to a function $\mathcal{P}(x, |u|, t)$ of broad form and suggested that $\mathcal{P}(x, |u|, t)$ could be used to "moderate" the physical pressure p in regions of high velocity. This feature may provide classical energy methods with options in the search for optimal estimates of the driving term. For preliminary investigations concerning the depletion of nonlinearity in the pressure force and its moderation see Tran & Yu (2015, 2016, 2017).

This study extends the investigations in Tran & Yu (2015, 2016, 2017) in a straightforward manner, aiming at the ultimate goal of establishing a balance between viscous and nonlinear effects in Navier–Stokes flows.

Regularity criteria are derived by making use of the favourable features of the pressure force described in the preceding paragraph. The results are invariant with respect to the natural scaling of the Navier–Stokes equations and are expressible in terms of an "effective" (moderated) pressure $p + \mathcal{P}$ and the velocity gradient along streamlines in the regions of high velocity.

9.2 Preliminaries

The evolution of the local energy $|u|^2/2$ is governed by

$$\frac{\partial}{\partial t}\frac{|u|^2}{2} + u \cdot \nabla \frac{|u|^2}{2} + u \cdot \nabla p = \Delta \frac{|u|^2}{2} - |\nabla u|^2. \tag{9.2}$$

Multiplying (9.2) by $|u|^{q-2}$ and integrating we obtain the evolution equation for $\|u\|_{L^q}$

$$\frac{1}{q}\frac{d}{dt}\|u\|_{L^q}^q = (q-2)\int_{\mathbb{R}^3} p|u|^{q-2}\,\widehat{u}\cdot\nabla|u|\,dx$$
$$- (q-2)\left\||u|^{(q-2)/2}\nabla|u|\right\|_{L^2}^2 - \left\||u|^{(q-2)/2}\nabla u\right\|_{L^2}^2. \tag{9.3}$$

Here $\widehat{u}$ is the unit vector in the direction of u, thus $\widehat{u}\cdot\nabla|u|$ is the velocity gradient along streamlines. When $q = 2$, (9.3) reduces to

$$\frac{1}{2}\frac{d}{dt}\|u\|_{L^2}^2 = -\|\nabla u\|_{L^2}^2\,,$$

which readily implies the a priori estimate

$$\int_0^T \|\nabla u\|_{L^2}^2\,dt \leq \frac{\|u_0\|_{L^2}^2}{2}. \tag{9.4}$$

Another a priori estimate available to us is the Calderón–Zygmund inequality

$$\|p\|_{L^r} \leq c_0(r)\|u\|_{L^{2r}}^2\,, \tag{9.5}$$

where $c_0(r)$ is a constant and $r \in (1, \infty)$. We will frequently make use of Sobolev's inequality

$$\|f\|_{L^6} \leq c\|\nabla f\|_{L^2}^2\,, \tag{9.6}$$

where $c = (2/\pi)^{2/3}/\sqrt{3}$ (Talenti, 1976), together with Hölder's and Young's inequalities.

Given the classical result of Ladyzhenskaya (1967), Prodi (1959), Serrin (1962) and Escauriaza, Seregin, & Sverák (2003) that $\|u\|_{L^q} < \infty$, for

$q \geq 3$, implies regularity, this study takes a direct approach to the problem by monitoring the evolution of $\|u\|_{L^q}$ governed by (9.3). We consider conditions under which the driving term in (9.3) can be in balance with the corresponding dissipation terms, thereby allowing for no growth of $\|u\|_{L^q}$ and hence regularity. For the reader's convenience, we reproduce below a previous result by Tran & Yu (2016, 2017) that constitutes an integral part of the present study.

Lemma 9.1   *Let*

$$\mathcal{P}(x, |u|, t) := \sum_{i=1}^{n} f_i(x, t) g_i(|u|, t), \qquad (9.7)$$

where $u \cdot \nabla f_i(x, t) = 0$ and $g_i(\xi, t) \in C^1$, then

$$\int_{\mathbb{R}^3} \mathcal{P} \, |u|^{q-2} \, \widehat{u} \cdot \nabla |u| \, dx = 0. \qquad (9.8)$$

Proof   Let

$$h_i(|u|, t) = \frac{1}{|u|^{q-2}} \int_0^{|u|} \xi^{q-3} g_i(\xi, t) \, d\xi.$$

Then by virtue of $\nabla \cdot u = 0$ and the hypothesis $u \cdot \nabla f_i = 0$, we have

$$\nabla \cdot \left(f_i h_i |u|^{q-2} \, u \right) = f_i u \cdot \nabla \int_0^{|u|} \xi^{q-3} g_i(\xi, t) \, d\xi$$

$$= f_i g_i |u|^{q-2} \, \widehat{u} \cdot \nabla |u|. \qquad (9.9)$$

Summing over i and integrating the resulting equation over $\mathbb{R}^3$ proves the lemma. $\square$

Adding the vanishing term in Lemma 9.1 to (9.3) yields

$$\frac{1}{q} \frac{d}{dt} \|u\|_{L^q}^q = (q - 2) \int_{\mathbb{R}^3} (p + \mathcal{P}) |u|^{q-2} \, \widehat{u} \cdot \nabla |u| \, dx$$

$$- (q - 2) \left\| |u|^{(q-2)/2} \nabla |u| \right\|_{L^2}^2 - \left\| |u|^{(q-2)/2} \nabla u \right\|_{L^2}^2. \qquad (9.10)$$

The function $\mathcal{P}$ has been called a pressure moderator, via which the effective pressure $p + \mathcal{P}$ can be varied within the admissible class for optimal estimates of the driving term in (9.10). The condition $u \cdot \nabla f_i(x, t) = 0$ requires f_i to be independent of x along each streamline. This is the only constraint on $\mathcal{P}$ as g_i is quite arbitrary, albeit having the same vanishing conditions at infinity as u and p.

As in Tran & Yu (2016, 2017), we determine a maximal subset of $\mathbb{R}^3$, say $\widetilde{\Omega}$, whose contribution to the driving term in (9.10) can a priori be neutralised by the terms due to viscosity, leaving the contribution from $\mathbb{R}^3 \setminus \widetilde{\Omega}$ solely responsible for possible growth of $\|u\|_{L^q}$. More specifically, we wish to determine $\widetilde{\Omega}$ such that

$$\int_{\widetilde{\Omega}} (p + \mathcal{P}) |u|^{q-2} \,\widehat{u} \cdot \nabla |u| \,dx \leq \frac{1}{2} \left\| |u|^{(q-2)/2} \, \nabla |u| \right\|_{L^2}^2 . \tag{9.11}$$

To this end, consider

$$\widetilde{\Omega}(t, q) := \{ x : |u(x, t)| \leq C(t, q) \, \|u\|_{L^{3q-6}} \} \cap \{ x : \widehat{u} \cdot \nabla |u| \neq 0 \}, \tag{9.12}$$

where $q \geq 3$. Clearly, $\widetilde{\Omega}$ becomes larger for greater C. Equivalently, $\mathbb{R}^3 \setminus \widetilde{\Omega}$ becomes smaller for greater C. Therefore, the problem reduces to maximising C subject to the constraint (9.11).

Since $|u(x, t)| \leq C \, \|u\|_{L^{3q-6}}$ for $x \in \widetilde{\Omega}$ we have

$$\int_{\widetilde{\Omega}} (p + \mathcal{P}) |u|^{q-2} \,\widehat{u} \cdot \nabla |u| \,dx$$

$$\leq C^{(q-2)/2} \|u\|_{L^{3q-6}}^{(q-2)/2} \|p + \mathcal{P}\|_{L^2(\widetilde{\Omega})} \left\| |u|^{(q-2)/2} \,\widehat{u} \cdot \nabla |u| \right\|_{L^2(\widetilde{\Omega})}$$

$$\leq C^{(q-2)/2} c_1 \|u\|_{L^4}^2 \|u\|_{L^{3q-6}}^{(q-2)/2} \left\| |u|^{(q-2)/2} \,\widehat{u} \cdot \nabla |u| \right\|_{L^2(\widetilde{\Omega})}$$

$$\leq C^{(q-2)/2} c_1 \|u\|_{L^2} \|u\|_{L^{3q}}^{q/2} \left\| |u|^{(q-2)/2} \,\widehat{u} \cdot \nabla |u| \right\|_{L^2(\widetilde{\Omega})}, \tag{9.13}$$

where $c_1(t)$ is given by $c_1 := \|p + \mathcal{P}\|_{L^2(\widetilde{\Omega})} / \|u\|_{L^4}^4$ and the interpolation estimates

$$\|u\|_{L^4} \leq \|u\|_{L^2}^{(3q-4)/(6q-4)} \|u\|_{L^{3q}}^{3q/(6q-4)}$$

and

$$\|u\|_{L^{3q-6}} \leq \|u\|_{L^2}^{4/[(3q-2)(q-2)]} \|u\|_{L^{3q}}^{q(3q-8)/[(3q-2)(q-2)]}$$

have been used. Now by applying Sobolev's inequality to $|u|^{q/2}$ we obtain

$$\|u\|_{L^{3q}}^{q/2} \leq \frac{cq}{2} \left\| |u|^{(q-2)/2} \, \nabla |u| \right\|_{L^2} . \tag{9.14}$$

Substituting (9.14) and $\|u\|_{L^2} \leq \|u_0\|_{L^2}$ into (9.13) yields

$$\int_{\widetilde{\Omega}} (p + \mathcal{P}) |u|^{q-2} \,\widehat{u} \cdot \nabla |u| \,dx$$

$$\leq \frac{C^{(q-2)/2} c_1 cq \|u_0\|_{L^2}}{2} \frac{\left\| |u|^{(q-2)/2} \,\widehat{u} \cdot \nabla |u| \right\|_{L^2(\widetilde{\Omega})}}{\left\| |u|^{(q-2)/2} \, \nabla |u| \right\|_{L^2}} \left\| |u|^{(q-2)/2} \, \nabla |u| \right\|_{L^2}^2$$

$$:= F \left\| |u|^{(q-2)/2} \, \nabla |u| \right\|_{L^2}^2 . \tag{9.15}$$

It can be seen that both c_1 and $\left\| |u|^{(q-2)/2} \,\widehat{u} \cdot \nabla |u| \right\|_{L^2(\widetilde{\Omega})}$ are increasing functions of C. Hence the prefactor F of $\left\||u|^{(q-2)/2} \nabla|u|\right\|_{L^2}^2$ in (9.15) is an increasing function of C. Furthermore, $F(C)$ is continuous. Indeed, given C_0, if $C > C_0$ then

$$\widetilde{\Omega}(C)\backslash\widetilde{\Omega}(C_0) = \{x : C_0 \|u\|_{L^{3q-6}} < |u| \le C \|u\|_{L^{3q-6}}\}\cap\{x : \widehat{u} \cdot \nabla|u| \ne 0\}.$$

It follows that

$$\lim_{C\searrow C_0} \widetilde{\Omega}(C) \setminus \widetilde{\Omega}(C_0) = \emptyset.$$

Hence by the Dominated Convergence Theorem we have

$$\lim_{C\searrow C_0} |\widetilde{\Omega}(C) \setminus \widetilde{\Omega}(C_0)| = 0. \tag{9.16}$$

On the other hand, if $C < C_0$ then

$$\widetilde{\Omega}(C_0)\backslash\widetilde{\Omega}(C) = \{x : C \|u\|_{L^{3q-6}} < |u| \le C_0 \|u\|_{L^{3q-6}}\}\cap\{x : \widehat{u} \cdot \nabla|u| \ne 0\}.$$

So

$$\lim_{C\nearrow C_0} \widetilde{\Omega}(C_0) \setminus \widetilde{\Omega}(C) = \{x : |u| = C_0 \|u\|_{L^{3q-6}}\} \cap \{x : \widehat{u} \cdot \nabla|u| \ne 0\}.$$

It can be seen (cf. Exercise 18 on page 308, Chapter 5 of Evans, 2010) that $\nabla|u| = 0$ a.e. on the level set $\{x : |u| = C_0 \|u\|_{L^{3q-6}}\}$. Hence by the Dominated Convergence Theorem we have

$$\lim_{C\nearrow C_0} |\widetilde{\Omega}(C_0) \setminus \widetilde{\Omega}(C)| = 0. \tag{9.17}$$

Now (9.16) and (9.17) readily imply that

$$\lim_{C\to C_0} \int_{\widetilde{\Omega}(C)} f(x)\,\mathrm{d}x = \int_{\widetilde{\Omega}(C_0)} f(x)\,\mathrm{d}x,$$

for integrable $f(x)$. Therefore, $F(C)$ is continuous.

Setting $F(C) = 1/2$ uniquely determines C and satisfies (9.11). This result gives us the following lemma.

Lemma 9.2 *Let $\mathcal{P}$ be a pressure moderator as in lemma 9.1. Let $\widetilde{\Omega}(t,q)$ be implicitly defined by (9.12), where $C(t,q)$ is given by*

$$C(t,q) = \left(\frac{\|u\|_{L^4}^2 \left\||u|^{(q-2)/2} \nabla|u|\right\|_{L^2}}{cq \|u_0\|_{L^2} \|p + \mathcal{P}\|_{L^2(\widetilde{\Omega})} \left\||u|^{(q-2)/2} \widehat{u} \cdot \nabla|u|\right\|_{L^2(\widetilde{\Omega})}} \right)^{2/(q-2)}, \tag{9.18}$$

then (9.11) holds.

Let $\Omega(t, q)$ be defined by

$$\Omega(t, q) := \{x : |u(x, t)| > C(t, q)\, \|u\|_{L^{3q-6}}\} \cap \{x : \widehat{u} \cdot \nabla |u| \neq 0\}, \quad (9.19)$$

for $q \geq 3$. Upon invoking lemma 9.2, (9.10) reduces to

$$\frac{1}{q}\frac{d}{dt}\|u\|_{L^q}^q \leq (q-2)\int_\Omega (p + \mathcal{P})|u|^{q-2}\,\widehat{u} \cdot \nabla |u|\,dx$$
$$- \frac{q-2}{2}\left\||u|^{(q-2)/2}\nabla|u|\right\|_{L^2}^2 - \left\||u|^{(q-2)/2}\nabla u\right\|_{L^2}^2. \quad (9.20)$$

Remark In the absence of a pressure moderator, we have $c_1 \leq c_0$.

Remark The size of the "pre-singular" set $\Omega(t, q)$ is determined by several factors. The roles of q and $\|u\|_{L^{3q-6}}$ are rather obvious. Given all else finite, $C(t, q) \to 1$ in the limit $q \to \infty$. As a result, Ω becomes empty in that limit. In the event of singularity development, i.e. $\|u\|_{L^{3q-6}} \to \infty$ for $q \geq 3$, Ω collapses to a set of zero measure (see Caffarelli, Kohn, & Nirenberg (1982) for more detail on this fact). The factor

$$R(t, q) := \frac{\left\||u|^{(q-2)/2}\,\nabla|u|\right\|_{L^2}}{\left\||u|^{(q-2)/2}\,\widehat{u} \cdot \nabla|u|\right\|_{L^2(\widetilde{\Omega})}} \quad (9.21)$$

would blow up as can be seen in what follows, thereby playing a significant role in such a collapse.

Remark Tran & Yu (2016, 2017) have used a similar definition for the set Ω with R and $\|p + \mathcal{P}\|_{L^2(\widetilde{\Omega})}$ replaced by unity and $\|p + \mathcal{P}\|_{L^2}$, respectively. Apparently, the present Ω can be much smaller in size when $R \gg 1$.

9.3 Results

We are now in a position to state and prove the main results of this study.

Proposition 9.3 *Let u and p solve the Navier–Stokes equations (9.1), let $\mathcal{P}$ be given as in Lemma 9.2, and let Ω be defined by (9.19). Furthermore, suppose that $c_1 ¿ 0$ and*

$$c_2 := \frac{\|p + \mathcal{P}\|_{L^q(\Omega)}}{\|u\|_{L^{2q}}^2} < \infty.$$

If $R < \infty$, where R is defined in (9.21), then $\|u\|_{L^q} < \infty$.

Proof From (9.14) and (9.20) we have

$$\frac{1}{q(q-2)} \frac{\mathrm{d}}{\mathrm{d}t} \|u\|_{L^q}^q \leq \int_{\Omega} (p+\mathcal{P})|u|^{q-2}\widehat{u}\cdot\nabla|u|\,\mathrm{d}x - \frac{2}{c^2 q^2}\|u\|_{L^{3q}}^q$$

$$\leq \|p+\mathcal{P}\|_{L^q(\Omega)} \left\||u|^{(q-2)/2}\right\|_{L^{2q/(q-2)}(\Omega)} \left\||u|^{(q-2)/2}\widehat{u}\cdot\nabla|u|\right\|_{L^2(\Omega)}$$

$$- \frac{2}{c^2 q^2}\|u\|_{L^{3q}}^q$$

$$\leq c_2 R \|u\|_{L^{2q}}^2 \|u\|_{L^q}^{(q-2)/2} \left\||u|^{(q-2)/2}\widehat{u}\cdot\nabla|u|\right\|_{L^2(\widetilde{\Omega})} - \frac{2}{c^2 q^2}\|u\|_{L^{3q}}^q$$

$$\leq c_2 C^{(q-2)/2} R \|u\|_{L^{2q}}^2 \|u\|_{L^q}^{(q-2)/2} \|u\|_{L^{3q-6}}^{(q-2)/2} \|\widehat{u}\cdot\nabla|u|\|_{L^2(\widetilde{\Omega})}$$

$$- \frac{2}{c^2 q^2}\|u\|_{L^{3q}}^q$$

$$\leq \frac{c_2 R^2}{cqc_1 \|u_0\|_{L^2}} \|u\|_{L^q}^{q/2} \|u\|_{L^{3q}}^{q/2} \|\widehat{u}\cdot\nabla|u|\|_{L^2(\widetilde{\Omega})} - \frac{2}{c^2 q^2}\|u\|_{L^{3q}}^q$$

$$\leq \frac{c_2^2 R^4}{8c_1^2 \|u_0\|_{L^2}^2} \|u\|_{L^q}^q \|\widehat{u}\cdot\nabla|u|\|_{L^2(\widetilde{\Omega})}^2 , \qquad (9.22)$$

where Hölder's inequality, the interpolation estimates

$$\|u\|_{L^{2q}} \leq \|u\|_{L^q}^{1/4} \|u\|_{L^{3q}}^{3/4}$$

and

$$\|u\|_{L^{3q-6}} \leq \|u\|_{L^q}^{1/(q-2)} \|u\|_{L^{3q}}^{(q-3)/(q-2)}$$

and Young's inequality have been used. It follows that

$$\frac{\mathrm{d}}{\mathrm{d}t} \|u\|_{L^q} \leq \frac{c_2^2 R^4 (q-2)}{8c_1^2 \|u_0\|_{L^2}^2} \|u\|_{L^q} \|\nabla u\|_{L^2}^2 ,$$

which immediately yields

$$\|u\|_{L^q} \leq \exp\left\{ \frac{q-2}{8\|u_0\|_{L^2}^2} \int_0^t \frac{c_2^2 R^4}{c_1^2} \|\nabla u\|_{L^2}^2 \,\mathrm{d}\tau \right\} \|u_0\|_{L^q} .$$

This, together with (9.4) and the hypotheses of the proposition, readily implies $\|u\|_{L^q} < \infty$, and the proof is completed. □

Remark In the absence of $\mathcal{P}$, we have $c_2 \leq c_0$.

Remark From the very definition of Ω, we have

$$\|u\|_{L^3(\Omega)} > C \|u\|_{L^{3q-6}} |\Omega|^{1/3} = \frac{R^{2/(q-2)} \|u\|_{L^{3q-6}} |\Omega|^{1/3}}{(cqc_1 \|u_0\|_{L^2})^{2/(q-2)}} . \qquad (9.23)$$

On the other hand, by Hölder's inequality we have

$$\|u\|_{L^3(\Omega)} \le \|u\|_{L^{3q-6}(\Omega)} |\Omega|^{(q-3)/(3q-6)}. \qquad (9.24)$$

Comparing (9.23) and (9.24) immediately yields

$$R < cqc_1 \|u_0\|_{L^2} |\Omega|^{-1/6}.$$

Interestingly, singularities would require a relatively mild divergence of R, less rapid than $|\Omega|^{-1/6}$ as $|\Omega| \to 0$. If R grows no less rapidly than $|\Omega|^{-1/6}$, then viscous effects become stronger than their nonlinear counterparts ($\Omega = \emptyset$) and no singularities can develop.

It is possible to further reduce Ω in the driving term of (9.20). For example, the contribution to this term from the subset of Ω, where

$$(p + \mathcal{P})\widehat{u} \cdot \nabla|u| \le \frac{1}{4} |\nabla|u||^2 + \frac{1}{q-2} |\nabla u|^2, \quad \text{for } x \in \Omega, \qquad (9.25)$$

can be cancelled out by $(q-2) \left\||u|^{(q-2)/2}\nabla|u|\right\|_{L^2}^2 / 4 + \left\||u|^{(q-2)/2}\nabla u\right\|_{L^2}^2$. As another example, consider the following estimate of the driving term in (9.20):

$$\int_\Omega (p + \mathcal{P})|u|^{q-2}\widehat{u} \cdot \nabla|u| \, dx$$

$$\le \|p + \mathcal{P}\|_{L^\infty(\Omega)} \left\||u|^{(q-2)/2}\right\|_{L^2(\Omega)} \left\||u|^{(q-2)/2}\widehat{u} \cdot \nabla|u|\right\|_{L^2}$$

$$\le \|p + \mathcal{P}\|_{L^\infty(\Omega)} \|u\|_{L^{3q}}^{(q-2)/2} |\Omega|^{(q+1)/(3q)} \left\||u|^{(q-2)/2}\widehat{u} \cdot \nabla|u|\right\|_{L^2}$$

$$\le \|p + \mathcal{P}\|_{L^\infty(\Omega)} \|u\|_{L^{3q}}^{-1} |\Omega|^{(q+1)/(3q)} \frac{cq}{2} \left\||u|^{(q-2)/2}\widehat{u} \cdot \nabla|u|\right\|_{L^2}^2,$$

where (9.14) and Hölder's inequality have been used. It follows that

$$\int_\Omega (p + \mathcal{P})|u|^{q-2}\widehat{u} \cdot \nabla|u| \, dx \le \frac{1}{4} \left\||u|^{(q-2)/2}\nabla|u|\right\|_{L^2}^2$$

if

$$|p + \mathcal{P}| \le \frac{\|u\|_{L^{3q}}}{2cq|\Omega|^{(q+1)/(3q)}}, \quad \text{for } x \in \Omega. \qquad (9.26)$$

So the contribution to the driving term in (9.20) from the subset of Ω in which (9.26) holds can be cancelled out by $(q-2) \left\||u|^{(q-2)/2}\nabla|u|\right\|_{L^2}^2 / 4$, i.e. half of the first dissipation term in (9.20).

When $|u|$ scales as $|\Omega|^{-1/3}$, the right-hand side of (9.26) scales as $|\Omega|^{-2/3}$. The set satisfying (9.26) can be substantial. However, it is not clear how the reduction of Ω on the basis of the above analysis can improve the subsequent results. Hence, we will not make use of such reduction in the present study.

Theorem 9.4 *Let u and p solve the Navier–Stokes equations* (9.1) *and $\mathcal{P}$ and Ω be given as in Lemma 9.3. If*

$$\|\widehat{u} \cdot \nabla|u|\|_{L^{3/2}(\Omega)} \leq \frac{2}{c^2 q^2} \frac{\|u\|_{L^{3q}}^2}{\|p + \mathcal{P}\|_{L^{3q/2}(\Omega)}}, \qquad (9.27)$$

then $\|u\|_{L^q}$ decreases and regularity follows.

Proof From the first equation of (9.22) we have

$$\frac{1}{q(q-2)} \frac{\mathrm{d}}{\mathrm{d}t} \|u\|_{L^q}^q \leq \int_\Omega (p + \mathcal{P})|u|^{q-2}\, \widehat{u} \cdot \nabla|u|\, \mathrm{d}x - \frac{2}{c^2 q^2} \|u\|_{L^{3q}}^q \qquad (9.28)$$

$$\leq \|p + \mathcal{P}\|_{L^{3q/2}(\Omega)} \|u\|_{L^{3q}(\Omega)}^{q-2} \|\widehat{u} \cdot \nabla|u|\|_{L^{3/2}(\Omega)} - \frac{2}{c^2 q^2} \|u\|_{L^{3q}}^q$$

$$\leq \frac{\|p + \mathcal{P}\|_{L^{3q/2}(\Omega)}}{\|u\|_{L^{3q}}^2} \|u\|_{L^{3q}}^q \|\widehat{u} \cdot \nabla|u|\|_{L^{3/2}(\Omega)} - \frac{2}{c^2 q^2} \|u\|_{L^{3q}}^q \,,$$

where Hölder's inequality has been used. Invoking criterion (9.27) completes the proof. □

Remark Since $\|u\|_{L^{3q}}^2 / \|p\|_{L^{3q/2}(\Omega)} \geq 1/c_0$, in the absence of $\mathcal{P}$ we have the criterion

$$\|\widehat{u} \cdot \nabla|u|\|_{L^{3/2}(\Omega)} \leq \frac{2}{c^2 q^2 c_0}$$

as an immediate corollary of theorem 9.4.

Remark Although $|u|$ and $\widehat{u} \cdot \nabla|u|$ are completely decoupled in the above estimates, the anti-correlation between them is not lost in the final result. The reason is that the norm $\|\widehat{u} \cdot \nabla|u|\|_{L^{3/2}(\Omega)}$ is over the high velocity region Ω, effectively tying peak $|u|$ and vanishing $\widehat{u} \cdot \nabla|u|$ together.

Remark When $\mathcal{P} = 0$ or $\mathcal{P} \propto |u|^2$, the criterion (9.27) remains unchanged under the scalings $p(x,t) \mapsto \lambda^2 p(\lambda x, \lambda^2 t)$, $u(x,t) \mapsto \lambda u(\lambda x, \lambda^2 t)$ that render (9.1) invariant.

Remark A highly desirable feature of (9.27), not captured by previous regularity criteria in terms of the velocity gradient by Cao & Titi (2011) and Zhou (2002), is that $\widehat{u} \cdot \nabla|u|$ vanishes over a nontrivial set in the vicinity of Ω. Indeed, let ℓ be a streamline passing through the set $\{x : |u| > C \|u\|_{L^{3q-6}}\}$. On $\ell \cap \{x : |u| > C \|u\|_{L^{3q-6}}\}$, $|u|$ achieves a maximum at some point, where its directional derivative along ℓ vanishes, i.e. $\widehat{u} \cdot \nabla |u| = 0$. It follows that $\widehat{u} \cdot \nabla |u| = 0$ over a surface in $\{x : |u| > C \left\| u \right\|_{L^{3q-6}} \}$. Another interesting feature of (9.27) is the utility of the pressure moderator $\mathcal{P}$, which could be employed to minimize the effective pressure $p + \mathcal{P}$ within Ω. This issue is discussed further later.

For $q = 4$, it is possible (as shown below) to replace $\left\| u \right\|_{L^{12}}^{2}$ in (9.27) by its equivalent counterpart $\left\| \Delta u \right\|_{L^2}$. Here "equivalent" is understood in the sense that $\left\| \Delta u \right\|_{L^2} / \left\| u \right\|_{L^{12}}^2$ is scale invariant. The trick involves a combination of the evolution equations for the two equivalent dynamical quantities $\left\| u \right\|_{L^4}^4$ and $\left\| \nabla u \right\|_{L^2}^2$, allowing one to cancel the driving term of the latter by one of the dissipation terms of the former. This trick has been used previously by Tran & Yu (2016).

Theorem 9.5   *Let u and p solve the Navier–Stokes equations* (9.1) *and $\mathcal{P}$ and Ω be given as in Lemma 9.3. If*

$$\left\| \widehat{u} \cdot \nabla |u| \right\|_{L^{3/2}(\Omega)} \le \frac{1}{2c} \frac{\left\| \Delta u \right\|_{L^2}}{\left\| p + \mathcal{P} \right\|_{L^6(\Omega)}}, \qquad (9.29)$$

then at least one of $\left\| u \right\|_{L^4}$ and $\left\| \nabla u \right\|_{L^2}$ decreases and regularity follows.

Proof   By retaining both dissipation terms and setting $q = 4$, (9.28) becomes

$$\frac{1}{8} \frac{d}{dt} \left\| u \right\|_{L^4}^4 \le \int_\Omega (p + \mathcal{P})|u|^2 \, \widehat{u} \cdot \nabla |u| \, dx - \frac{1}{8c^2} \left\| u \right\|_{L^{12}}^4 - \frac{1}{2} \left\| |u| \nabla u \right\|_{L^2}^2$$

$$\le \left\| p + \mathcal{P} \right\|_{L^6(\Omega)} \left\| u \right\|_{L^{12}(\Omega)}^2 \left\| \widehat{u} \cdot \nabla |u| \right\|_{L^{3/2}(\Omega)} - \frac{1}{8c^2} \left\| u \right\|_{L^{12}}^4$$

$$\qquad - \frac{1}{2} \left\| |u| \nabla u \right\|_{L^2}^2$$

$$\le 2c^2 \left\| p + \mathcal{P} \right\|_{L^6(\Omega)}^2 \left\| \widehat{u} \cdot \nabla |u| \right\|_{L^{3/2}(\Omega)}^2 - \frac{1}{2} \left\| |u| \nabla u \right\|_{L^2}^2 . \quad (9.30)$$

On the other hand, the evolution of $\left\| \nabla u \right\|_{L^2}$ is governed by

$$\frac{1}{2} \frac{d}{dt} \left\| \nabla u \right\|_{L^2}^2 = \int_{\mathbb{R}^3} \Delta u \cdot (u \cdot \nabla) u \, dx - \left\| \Delta u \right\|_{L^2}^2$$

$$\le \left\| \Delta u \right\|_{L^2} \left\| |u| \nabla u \right\|_{L^2} - \left\| \Delta u \right\|_{L^2}^2$$

$$\le \left\| \Delta u \right\|_{L^2} \left\| |u| \nabla u \right\|_{L^2} - \left\| \Delta u \right\|_{L^2}^2$$

$$\le \frac{1}{2} \left\| |u| \nabla u \right\|_{L^2} - \frac{1}{2} \left\| \Delta u \right\|_{L^2}^2 , \qquad (9.31)$$

where Hölder's and Young's inequalities have been used. Summing (9.30) and (9.31) yields

$$\frac{\mathrm{d}}{\mathrm{d}t}\left(\frac{1}{8}\|u\|_{L^4}^4 + \frac{1}{2}\|\nabla u\|_{L^2}^2\right) \le 2c^2 \|p + \mathcal{P}\|_{L^6(\Omega)}^2 \|\widehat{u} \cdot \nabla|u|\|_{L^{3/2}(\Omega)}^2$$

$$-\frac{1}{2}\|\Delta u\|_{L^2}^2. \qquad (9.32)$$

By the hypothesis of the theorem the right-hand side of (9.32) is non-positive, and the proof is complete. □

Remark By Hölder's and Sobolev's inequalities we have

$$\|\widehat{u} \cdot \nabla|u|\|_{L^{3/2}(\Omega)} \le |\Omega|^{1/2} \|\widehat{u} \cdot \nabla|u|\|_{L^6(\Omega)} \le c|\Omega|^{1/2} \|\Delta u\|_{L^2}.$$

Hence we have the following criterion

$$|\Omega|^{1/2} \le \frac{1}{2c^2 \|p + \mathcal{P}\|_{L^6(\Omega)}}$$

as a corollary of Theorem 9.5.

Theorem 9.6 *Let u and p solve the Navier–Stokes equations (9.1) and $\mathcal{P}$ and Ω be given as in Lemma 9.3. If*

$$\|p + \mathcal{P}\|_{L^{(q+2)/2}(\Omega)} \le \frac{1}{cq} \frac{\|u\|_{L^{3q}}^{q/2}}{\|u\|_{L^{q+2}(\Omega)}^{(q-2)/2}}, \qquad (9.33)$$

then $\|u\|_{L^q}$ decreases and regularity follows.

Proof As in the proof of Theorem 9.4, we have

$$\frac{1}{q(q-2)}\frac{\mathrm{d}}{\mathrm{d}t}\|u\|_{L^q}^q$$

$$\le \int_\Omega (p+\mathcal{P})|u|^{q-2}\,\widehat{u}\cdot\nabla|u|\,\mathrm{d}x - \frac{1}{cq}\|u\|_{L^{3q}}^{q/2}\left\||u|^{(q-2)/2}\,\nabla|u|\right\|_{L^2}$$

$$\le \|p+\mathcal{P}\|_{L^{(q+2)/2}(\Omega)}\left\||u|^{(q-2)/2}\right\|_{L^{2(q+2)/(q-2)}(\Omega)}\left\||u|^{(q-2)/2}\,\widehat{u}\cdot\nabla|u|\right\|_{L^2(\Omega)}$$

$$-\frac{1}{cq}\|u\|_{L^{3q}}^{q/2}\left\||u|^{(q-2)/2}\,\nabla|u|\right\|_{L^2}$$

$$\le \|p+\mathcal{P}\|_{L^{(q+2)/2}(\Omega)}\|u\|_{L^{q+2}(\Omega)}^{(q-2)/2}\left\||u|^{(q-2)/2}\,\widehat{u}\cdot\nabla|u|\right\|_{L^2(\Omega)}$$

$$-\frac{1}{cq}\|u\|_{L^{3q}}^{q/2}\left\||u|^{(q-2)/2}\,\nabla|u|\right\|_{L^2} \qquad (9.34)$$

where Hölder's inequality has been used. Invoking criterion (9.33) completes the proof. □

Tran & Yu (2017) have derived a criterion similar to (9.27), where $\widehat{u} \cdot \nabla|u|$ is replaced by $p + \mathcal{P}$ and for a larger version of Ω. That result goes through without change for the present form of Ω as can be seen in what follows.

Theorem 9.7 *Let u and p solve the Navier–Stokes equations (9.1) and $\mathcal{P}$ and Ω be given as in Lemma 9.3. If*

$$\|p + \mathcal{P}\|_{L^{3/2}(\Omega)} \leq \frac{1}{c^2 q^2} \frac{\|u\|_{L^{3q}}^2}{\|p + \mathcal{P}\|_{L^{3q/2}(\Omega)}}, \tag{9.35}$$

then $\|u\|_{L^q}$ decreases and regularity follows.

Proof As in the proof of Theorem 9.6, we have

$$\frac{1}{q(q-2)} \frac{\mathrm{d}}{\mathrm{d}t} \|u\|_{L^q}^q$$

$$\leq \int_\Omega (p + \mathcal{P})|u|^{q-2} \, \widehat{u} \cdot \nabla|u| \, \mathrm{d}x - \frac{1}{cq} \|u\|_{L^{3q}}^{q/2} \left\| |u|^{(q-2)/2} \nabla|u| \right\|_{L^2}$$

$$\leq \|p + \mathcal{P}\|_{L^{3/2}(\Omega)}^{1/2} \left\| (p + \mathcal{P})|u|^{q-2} \right\|_{L^3(\Omega)}^{1/2} \left\| |u|^{(q-2)/2} \nabla|u| \right\|_{L^2(\Omega)}$$

$$\quad - \frac{1}{cq} \|u\|_{L^{3q}}^{q/2} \left\| |u|^{(q-2)/2} \nabla|u| \right\|_{L^2}$$

$$\lesssim \|p + \mathcal{P}\|_{L^{3/2}(\Omega)}^{1/2} \|p + \mathcal{P}\|_{L^{3q/2}(\Omega)}^{1/2} \|u\|_{L^{3q}(\Omega)}^{(q-2)/2} \left\| |u|^{(q-2)/2} \nabla|u| \right\|_{L^2(\Omega)}$$

$$\quad - \frac{1}{cq} \|u\|_{L^{3q}}^{q/2} \left\| |u|^{(q-2)/2} \nabla|u| \right\|_{L^2}$$

$$\leq c_3^{1/2} \|p + \mathcal{P}\|_{L^{3/2}(\Omega)}^{1/2} \|u\|_{L^{3q}}^{q/2} \left\| |u|^{(q-2)/2} \nabla|u| \right\|_{L^2(\Omega)}$$

$$\quad - \frac{1}{cq} \|u\|_{L^{3q}}^{q/2} \left\| |u|^{(q-2)/2} \nabla|u| \right\|_{L^2}, \tag{9.36}$$

where (9.14) and Hölder's inequality have been used and

$$c_3 := \|p + \mathcal{P}\|_{L^{3q/2}(\Omega)} \, / \, \|u\|_{L^{3q}}^2 \, .$$

Invoking criterion (9.35) completes the proof. □

Remark As for c_1 and c_2, we have $c_3 \leq c_0$ in the absence of $\mathcal{P}$.

Remark Similar to (9.27), criteria (9.29), (9.33) and (9.35) remain unchanged under the invariant scaling of the Navier–Stokes equations when $\mathcal{P} = 0$ or $\mathcal{P} \propto |u|^2$.

Remark Criteria (9.33) and (9.35) appear to have opposite regimes of optimality. Intuitively, (9.33) is better in the limit of large q and moderate $\|u\|_{L^{3q}}$ while (9.35) is better in the limit of large $\|u\|_{L^{3q}}$ and moderate q.

We conclude this study by considering two simple forms of the pressure moderator $\mathcal{P}$. First, let

$$\mathcal{P} := -\frac{\int_\Omega p \, dx}{\int_\Omega |u|^2 \, dx} \, |u|^2. \tag{9.37}$$

Evidently, we have

$$\int_\Omega (p + \mathcal{P}) \, dx = 0.$$

This means that $p + \mathcal{P}$ vanishes on a surface in Ω, making it plausible that

$$\|p + \mathcal{P}\|_{L^r(\Omega)} \ll \|p\|_{L^r(\Omega)}.$$

This in turn implies

$$\frac{\|u\|_{L^{2r}}^2}{\|p + \mathcal{P}\|_{L^r(\Omega)}} \gg 1,$$

which is desirable for each of the criteria (9.27), (9.29), (9.33), and (9.35). Since the Calderón–Zygmund inequality (9.5) does not apply to the border case $r = 1$, we have no a priori control over the ratio $\int_\Omega p \, dx / \int_\Omega |u|^2 \, dx$, which may affect c_1 and Ω. Hence one should be mindful of this issue when making use of the moderator (9.37). Note, however, that the ratio R in the definition of Ω would blow up in the singularity picture. This effectively broadens the admissible class of $\mathcal{P}$. Indeed, a moderator with $R/c_1 > 0$ means $C > 0$, thereby rendering a well-defined Ω.

Second, consider a simple case in which $|u|$ peaks at a single point, say $x_0(t)$. Let $\mathcal{P} := c'(t)|u|^2$ be such that $p + \mathcal{P} = 0$ at x_0, i.e.

$$p(x_0, t) + c'(t)|u(x_0, t)|^2 = 0. \tag{9.38}$$

For moderate $|u(x_0, t)|$, $|c'|$ can be expected to have a moderate value as well. In the limit of large q, the set Ω collapses upon x_0, where both $p + \mathcal{P}$ and $\hat{u} \cdot \nabla |u|$ vanish. The left-hand and right-hand sides of criterion (9.27) scale as $|\hat{u} \cdot \nabla |u|| |\Omega|^{2/3}$ and $1/(q^2 |p + \mathcal{P}| |\Omega|^{2/(3q)})$, respectively. Evidently, $\|u\|_{L^q}$ decays if $|p + \mathcal{P}| |\hat{u} \cdot \nabla |u|| |\Omega|^{2(1+1/q)/3}$ approaches zero more rapidly than $1/q^2$. On the other hand, the left-hand and right-hand sides of criterion (9.33) scale as $|p + \mathcal{P}| |\Omega|^{2/(q+2)}$ and $1/(q|\Omega|^{(q-2)/[2(q+2)]})$, respectively. As a result, $\|u\|_{L^q}$ decays if $|p + \mathcal{P}| |\Omega|^{1/2}$ approaches zero more rapidly than $1/q$. Meanwhile, criterion (9.25) means that $\|u\|_{L^q}$ decays if $|p + \mathcal{P}||\widehat{u} \cdot \nabla|u||$ approaches zero more rapidly than $1/q$. In the limit of large $|u(x_0, t)|$, the existence of a finite c' in (9.38) would become questionable since the Calderón–Zygmund inequality (9.5) does not apply to the limiting case $r = \infty$. Nonetheless, as discussed in the preceding paragraph, $|c'|$ in (9.38) could be as large as necessary, provided that $R/c_1 > 0$. The existence of such c' would ensure $p + \mathcal{P} = 0$ at x_0. However, unlike in the limit $q \to \infty$, $p + \mathcal{P}$ in Ω would not necessarily tend to zero in the present case. On the contrary, the possibility that $|p + \mathcal{P}|$ (as well as $|\widehat{u} \cdot \nabla|u||$) in Ω grows along with $|u(x_0, t)|$ is to be considered. After all, this undesirable scenario is at the heart of the regularity problem. The argument for regularity, on the basis of criterion (9.35) for example, is that $|p + \mathcal{P}||\Omega|^{(1+1/q)/3}$ grows less rapidly than $\|u\|_{L^{3q}}$. As another example, criterion (9.25) means that regularity is secured if in Ω the quantity $|p + \mathcal{P}||\widehat{u} \cdot \nabla|u||$ grows less rapidly than $|\nabla u|^2$.

9.4 Conclusion

We have investigated conditions under which a balance between viscous and nonlinear effects in Navier–Stokes flows is possible. It is well known that standard energy methods have not been able to establish an unconditional balance, which appears to require hyperviscosity represented by $-(-\Delta)^\alpha$, for $\alpha \geq 5/4$, rather than the usual viscosity. However, viscous effects may turn out to be adequate, provided that the nonlinearity of the pressure force is depleted to a certain extent. The present study has explored this possibility in some depth, focusing on a manifestation of nonlinear depletion through generic anti-correlation between $|u|$ and $|\widehat{u} \cdot \nabla|u||$ in the driving term in the evolution equation for $\|u\|_{L^q}$. Several regularity criteria have been derived and discussed. An important finding is that the net pressure force responsible for growing $\|u\|_{L^q}$, for $q \geq 3$, is confined within the set $\{x : |u| > C \|u\|_{L^{3q-6}}\}$. In the singularity picture, this set becomes vanishingly small because of not only $\|u\|_{L^{3q-6}} \to \infty$ but also $C \to \infty$. The latter feature is promising and deserves further investigations.

It is interesting that the physical pressure in the driving term in the governing equation for $\|u\|_{L^q}$ is determined up to a fictitious pressure $\mathcal{P}(x, |u|, t)$ of relatively broad form, which we have called a pressure moderator. This allows us to vary $\mathcal{P}$ within its admissible class for optimal estimates of the said term. While the full potential and effectiveness of $\mathcal{P}$ remain to be explored, we have considered in this study two simple but promising forms of $\mathcal{P}$. Remarkably, the triple-correlation (driving) term $\int_\Omega (p + \mathcal{P})|u|^{q-2}\, \widehat{u} \cdot \nabla |u|\, dx$ with a suitable form of $\mathcal{P}$ has an anti-correlation between $|u|^{q-2}$ and each of $|\widehat{u}\cdot\nabla|u||$ and $|p+\mathcal{P}|$. Conceivably, this is a highly favourable feature for a balance between nonlinear and viscous effects, hence regularity.

References

Caffarelli, L., Kohn, R., & Nirenberg, L. (1982) Partial regularity of suitable weak solutions of the Navier–Stokes equations. *Commun. Pure Appl. Math.* **35**, 771–831.

Cao, C., & Titi, E.S. (2011) Global regularity criterion for the 3D Navier–Stokes equations involving one entry of the velocity gradient tensor. *Arch. Rational Mech. Anal.* **202**, 919–932.

Escauriaza, L., Seregin, G.A., & Sverák, V. (2003) $L_{3,\infty}$-solutions of Navier–Stokes equations and backward uniqueness. *Uspekhi Mat. Nauk.* **58**, 3–44.

Evans, L.C. (2010) *Partial Differential Equations*, 2nd ed. American. Math. Soc., Providence.

Ladyzhenskaya, O.A. (1967) On uniqueness and smoothness of generalized solutions to the Navier–Stokes equations. *Zapiski Nauchn. Seminar POMI* **5**, 169–185.

Leray, J. (1934) On the motion of a viscous liquid filling space. *Acta. Math.* **63**, 193–248.

Prodi, G. (1959) Un teorema di unicità per le equazioni di Navier–Stokes. *Ann. Math. Pura Appl.* **4**, 173–182.

Serrin, J. (1962) On the interior regularity of weak solutions of the Navier–Stokes equations. *Arch. Rational Mech. Anal.* **9**, 187–195.

Talenti, G. (1976) Best constants in Sobolev inequality. *Ann. Mat. Pura Appl.* **110**, 353–362.

Tran, C.V. & and Yu, X. (2015) Depletion of nonlinearity in the pressure force driving Navier–Stokes flows. *Nonlinearity* **28**, 1295–1306.

Tran, C.V. & Yu, X. (2016) Pressure moderation and effective pressure in Navier–Stokes flows. *Nonlinearity* **29**, 2990–3005.

Tran, C.V. & Yu, X. (2017) Regularity of Navier–Stokes flows with bounds for the pressure. *Appl. Math. Lett.* **67**, 21–27.

Zhou, Y. (2002) A new regularity criterion for the Navier–Stokes equations in terms of the gradient of one velocity component. *Methods Appl. Anal.* **9**, 563–578.

10

A direct approach to Gevrey regularity on the half-space

Igor Kukavica

Department of Mathematics
University of Southern California
Los Angeles, CA 90089
kukavica@usc.edu

Vlad Vicol
Department of Mathematics
Princeton University
Princeton, NJ 08544
vvicol@math.princeton.edu

Abstract

We consider the inhomogeneous heat and Stokes equations on the half space and prove an instantaneous space-time analytic regularization result, uniformly up to the boundary of the half space.

10.1 Introduction

The main goal of this paper is to study Gevrey regularity for the Dirichlet problem for the inhomogeneous heat equation

$$\partial_t u - \Delta u = f, \qquad \text{in } \Omega, \tag{10.1}$$

$$u = 0, \qquad \text{on } \partial\Omega, \tag{10.2}$$

with the initial condition

$$u(x,0) = u_0(x), \qquad \text{in } \Omega, \tag{10.3}$$

posed in the half space

$$\Omega = \left\{ x = (x_1, \dots, x_d) \in \mathbb{R}^d : x_d > 0 \right\}. \tag{10.4}$$

We will focus on obtaining regularization estimates up to the boundary of the domain. Throughout the paper we denote by $\bar{\partial}$ the vector of tangential derivatives $\bar{\partial} = (\partial_1, \dots, \partial_{d-1})$.

Published as part of *Partial Differential Equations in Fluid Mechanics*, edited by C.L. Fefferman, J.C. Robinson, & J.L. Rodrigo. © Cambridge University Press 2018.

In this paper we prove that, from an initial datum with finite Sobolev regularity, the solution to (10.1)–(10.4) instantly becomes real analytic, jointly in space and time, with an analyticity radius that is uniform up to the boundary of $\partial\Omega$. The result holds under the assumption that the force is real analytic in space-time. In order to state the main result of the paper, we first introduce some notation.

For $r \geq 1$, define the index sets

$$B = \big\{(i,j,k) : i,j,k \in \mathbb{N}_0, i+j+k \geq r\big\},$$
$$B^c = \mathbb{N}_0^3 \backslash B. \tag{10.5}$$

Fix $T > 0$, let $0 < \widetilde{\varepsilon}, \bar{\varepsilon}, \varepsilon \leq 1$, and define the sum

$$\varphi(u) = \sum_B \frac{(i+j+k)^r \varepsilon^i \widetilde{\varepsilon}^j \bar{\varepsilon}^k}{(i+j+k)!} \|t^{i+j+k-r} \partial_t^i \partial_d^j \bar{\partial}^k u\|_{L^2_{x,t}([0,T]\times\Omega)}$$
$$+ \sum_{B^c} \|\partial_t^i \partial_d^j \bar{\partial}^k u\|_{L^2_{x,t}([0,T]\times\Omega)} \tag{10.6}$$
$$= \bar{\varphi}(u) + \varphi_0(u).$$

Here $\bar{\partial}^k$ for $k \in \mathbb{N}_0$ used above, is defined as follows

$$\|\partial_t^i \partial_d^j \bar{\partial}^k u\|_{L^2_{x,t}} = \sum_{\alpha \in \mathbb{N}_0^{d-1}, |\alpha|=k} \|\partial_t^i \partial_d^j \partial^\alpha u\|_{L^2_{x,t}}.$$

We note that $\varphi_0(u)$ is the $H^{r-1}([0,T] \times \Omega)$ norm of u.

In particular, it is well known (see e.g. Lions & Magenes, 1972b and Evans, 1998) that for smooth and compatible initial datum which vanishes on $\partial\Omega$ (for instance $u_0 \in H_0^1(\Omega) \cap H^{2(r-1)}(\Omega)$ is sufficient), and force $f \in H_{t,x}^{2(r-2)+}((0,T) \times \Omega)$ we have that

$$\varphi_0(u) \lesssim \|u_0\|_{H^{2(r-1)}(\Omega)} + \|f\|_{H^{2(r-2)+}((0,T)\times\Omega)}. \tag{10.7}$$

The following is our main result.

Theorem 10.1 *Let $T > 0$ and $r \geq 1$. Then there exist $\varepsilon, \widetilde{\varepsilon}, \bar{\varepsilon} \in (0,1]$, which only depend on T, r, and the dimension d, such that for any initial data $u_0 \in H_0^1(\Omega) \cap H^{2(r-1)}(\Omega)$ that satisfies the compatibility conditions, and f sufficiently smooth, the solution u of (10.1)–(10.3) satisfies the* *estimate*

$$\varphi(u) \lesssim \varphi_0(u)$$

$$+ \sum_{i+j+k \geq (r-2)_+} \frac{(i+j+k+2)^r \varepsilon^i \tilde{\varepsilon}^{j+2} \bar{\varepsilon}^k}{(i+j+k+2)!} \|t^{i+j+k+2-r} \partial_t^i \partial_d^j \bar{\partial}^k f\|_{L^2_{x,t}((0,T) \times \Omega)}$$

$$+ \sum_{i+k \geq (r-2)_+} \frac{(i+k+2)^r \varepsilon^i \bar{\varepsilon}^{k+2}}{(i+k+2)!} \|t^{i+k+2-r} \partial_t^i \bar{\partial}^k f\|_{L^2_{x,t}((0,T) \times \Omega)}$$

$$+ \sum_{i \geq r-1} \frac{(i+1)^r \varepsilon^{i+1}}{(i+1)!} \|t^{i+1-r} \partial_t^i f\|_{L^2_{x,t}((0,T) \times \Omega)}.$$

(10.8)

Here we assume on f that the norms in (10.7) above and (10.8) below are finite.

Remark It is clear from the proof of Theorem 10.1 that the Gevrey-s regularization, jointly in space and time, also follows. For this purpose, simply replace $(i+j+k)!$ with $((i+j+k)!)^s$, where $s > 1$, throughout the paper.

Combining (10.7) and (10.8), we conclude that for an initial datum u_0 of finite Sobolev regularity, and real analytic force f, the solution u of (10.1)–(10.3) becomes real-analytic, jointly in space and time, for any $t \in (0, T]$, and with an analyticity radius that is uniform up to $\partial\Omega$. This latter point concerning the fact that the analyticity radius does not vanish as one approaches $\partial\Omega$ is the main result of the paper.

For a direct method for proving analyticity in the interior, see Bradshaw, Grujić, & Kukavica (2015) and Grujić & Kukavica (1999).

The method used to prove Theorem 10.1 extends easily to the case of the inhomogeneous Stokes system on the half space

$$\partial_t u - \Delta u + \nabla p = f, \quad \text{in } \Omega$$
$$\nabla \cdot u = 0, \quad \text{in } \Omega$$

(10.9)

with the Dirichlet boundary condition

$$u = 0, \quad \text{on } \partial\Omega.$$

(10.10)

For $T > 0$, smooth and compatible initial datum u_0, and sufficiently smooth force f, it is known (see e.g. Solonnikov, 1977, Chapter III) that $\varphi_0(u)$ is a priori bounded in terms of u_0 and f, with an estimate similar to (10.7) above. With the notation of Theorem 10.1 we have the following result

Theorem 10.2 *Let $T > 0$ and $r \geq 1$. Then there exist $\varepsilon, \widetilde{\varepsilon}, \bar{\varepsilon} \in (0, 1]$, depending only on T, r, and d, such that for any sufficiently smooth u_0 satisfying appropriate compatibility conditions, and sufficiently smooth f, the solution u of the Cauchy problem for (10.9)–(10.10) satisfies the estimate (10.8).*

The motivation for treating the inhomogeneous linear heat (10.1) and Stokes (10.9) equations comes from the study of semi linear problems (i.e. $f = f(t, x, u, \nabla u)$) for the heat and Stokes equations on domains with boundary, with Dirichlet boundary conditions. These nonlinear problems will be addressed in our forthcoming work, see Camliyurt, Kukavica, & Vicol (2017).

In the absence of boundaries, analytic and Gevrey-class regularization results are well known, even in the context of nonlinear problems. We refer the reader to Foias & Temam (1989) for results concerning the Navier–Stokes equations and to Kukavica & Vicol (2011), Levermore & Oliver (1997), and Prüss, Saal, & Simonett (2007) for results relating to other models.

On domains with boundary, analyticity up to the boundary has been obtained in the fundamental works of Kinderlehrer & Nirenberg (1978) and Komatsu (1979); see also Giga (1983) and Komatsu (1980).

These classical works achieve real-analyticity based on an induction scheme on the number of derivatives. By comparison, our proof is based directly on classical energy inequalities for the heat and Laplace equations on Ω. Moreover, Theorem 10.1 obtains the analyticity in the time variable concomitantly.

We believe that this transparent proof is going to be useful in establishing real-analytic and Gevrey regularization results for PDEs with different types of boundary conditions for which the methods of Foias & Temam (1989) work in the absence of boundaries. In fact, our motivation for the present paper comes from the simplicity of the approach in Foias & Temam (1989) in the case when the boundaries are not present.

For other works relating to the use of the Gevrey regularity method of Foias and Temam, see Biswas (2005), Biswas & Swanson (2007), Ferrari & Titi (1998), Oliver & Titi (2001), while Grujić & Kukavica (1998) and Lemarié-Rieusset (2000) explore an alternative approach to analyticity in domains without the boundary. We also refer the interested reader to Giga (1983), Kato & Masuda (1986), Kukavica & Vicol (2011), Lee & Lin Guo (2001), Prüss et al. (2007) for some other results on analyticity, and to Friz, Kukavica, & Robinson (2001) for some applications.

272   *I. Kukavica & V. Vicol*

The main idea in the proof of (10.8) is to split the sum defining $\bar{\varphi}(u)$ into several sub-sums, and on each one perform a *derivative reduction estimate*, as described in Section 10.3. These estimates follow from the classical maximal regularity theory for Laplace's equation in Ω (cf. Section 10.2), and an energy estimate for solutions of (10.1)–(10.3).

The paper is organised as follows. We begin Section 10.2 with some preliminary results. Section 10.3 contains the results on derivative reduction estimates, which are the key ingredient in the proof of Theorem 10.1, contained in Section 10.4. Finally, we discuss the details of the necessary modifications to prove Theorem 10.2 in Section 10.5.

10.2 Preliminaries

We use the following notational agreement: if the domain in the Sobolev space is not indicated, it is either Ω or $\Omega \times (0,T)$, depending on the context, with $T > 0$ a fixed parameter.

We recall (cf. Lions & Magenes, 1972a) a simple statement on interpolation, which asserts that

$$\|\nabla u\|_{L^2(\Omega)} \lesssim \|u\|_{L^2(\Omega)}^{1/2}\|u\|_{\dot{H}^2(\Omega)}^{1/2} + \|u\|_{L^2(\Omega)} \tag{10.11}$$

holds for $u \in H^2(\Omega)$. The proof uses a Sobolev extension operator and the interpolation inequality in $\mathbb{R}^d$.

In addition to (10.11), we shall also use the H^2 regularity for Laplace's equation

$$\Delta u = g, \quad \text{in } \Omega. \tag{10.12}$$

For the problem (10.12), we have

$$\|u\|_{\dot{H}^2(\Omega)} \lesssim \|g\|_{L^2(\Omega)} + \|\bar{\partial}u\|_{\dot{H}^1(\Omega)} + \|u\|_{L^2(\Omega)}. \tag{10.13}$$

If, in addition, $u\big|_{\partial\Omega} = 0$, then we have

$$\|u\|_{\dot{H}^2(\Omega)} \lesssim \|g\|_{L^2(\Omega)}. \tag{10.14}$$

The estimate (10.14) follows from the H^2 inequality for the odd extension of u to $\mathbb{R}^d$, while the bound (10.13) follows from the H^2 regularity for the problem (10.12) (cf. Lions & Magenes, 1972a), the trace theorem, and the interpolation inequality (10.11) as follows:

$$\|u\|_{H^2} \lesssim \|g\|_{L^2} + \|u\|_{H^{3/2}(\partial\Omega)} \lesssim \|g\|_{L^2} + \|\bar{\partial}u\|_{H^{1/2}(\partial\Omega)}$$
$$\lesssim \|g\|_{L^2} + \|\bar{\partial}u\|_{H^1} \lesssim \|g\|_{L^2} + \|\bar{\partial}u\|_{\dot{H}^1} + \|u\|_{H^1}$$
$$\lesssim \|g\|_{L^2} + \|\bar{\partial}u\|_{\dot{H}^1} + \|u\|_{L^2} + \|u\|_{L^2}^{1/2}\|u\|_{\dot{H}^2}^{1/2}.$$

Finally, estimate (10.13) follows from the ε-Young inequality.

10.3 Derivative reduction

In this section we give the proofs of the normal, tangential, and time derivative reduction estimates which are the main ingredients in the proof of Theorem 10.1.

Let u be a smooth solution of (10.1)–(10.3). Throughout this section we will require that the nonnegative integers i, j, k obey the constraint $i + j + k \geq r$.

10.3.1 Normal derivative reduction

We start by estimating the L^2 norm of $t^{i+j+k-r}\partial_t^i\partial_d^j\bar{\partial}^k u$. For $j \geq 2$ we will show that

$$\|t^{i+j+k-r}\partial_t^i\partial_d^j\bar{\partial}^k u\|_{L_{x,t}^2} \lesssim \|t^{i+j+k-r}\partial_t^{i+1}\partial_d^{j-2}\bar{\partial}^k u\|_{L_{x,t}^2}$$
$$+ \|t^{i+j+k-r}\partial_t^i\partial_d^{j-1}\bar{\partial}^{k+1} u\|_{L_{x,t}^2}$$
$$+ \|t^{i+j+k-r}\partial_t^i\partial_d^{j-2}\bar{\partial}^{k+2} u\|_{L_{x,t}^2}$$
$$+ \|t^{i+j+k-r}\partial_t^i\partial_d^{j-2}\bar{\partial}^k u\|_{L_{x,t}^2}$$
$$+ \|t^{i+j+k-r}\partial_t^i\partial_d^{j-2}\bar{\partial}^k f\|_{L_{x,t}^2}. \quad (10.15)$$

This inequality allows us to reduce the number of vertical derivatives (∂_d) in the Gevrey (analytic) norm.

On the other hand, for $j = 1$ and $k \geq 1$ we will prove that

$$\|t^{i+1+k-r}\partial_t^i\partial_d\bar{\partial}^k u\|_{L_{x,t}^2} \lesssim \|t^{i+1+k-r}\partial_t^{i+1}\bar{\partial}^{k-1} u\|_{L_{x,t}^2}$$
$$+ \|t^{i+1+k-r}\partial_t^i\bar{\partial}^{k-1} f\|_{L_{x,t}^2}, \quad (10.16)$$

while for $j = 1$ and $k = 0$, we will show

$$\|t^{i+1-r}\partial_t^i\nabla u\|_{L_{x,t}^2} \lesssim \|t^{i+1-r}\partial_t^i u\|_{L_{x,t}^2}^{1/2}\|t^{i+1-r}\partial_t^{i+1} u\|_{L_{x,t}^2}^{1/2}$$
$$+ \|t^{i+1-r}\partial_t^i u\|_{L_{x,t}^2} + \|t^{i+1-r}\partial_t^i f\|_{L_{x,t}^2} \quad (10.17)$$

whenever $i \geq r$.

The remainder of this subsection contains the proofs of estimates (10.15)–(10.17).

Proof of (10.15) We first use (10.1) to compute

$$\Delta(t^{i+j+k-r}\partial_t^i\partial_d^{j-2}\bar{\partial}^k u) = t^{i+j+k-r}\partial_t^i\partial_d^{j-2}\bar{\partial}^k \Delta u$$
$$= t^{i+j+k-r}\partial_t^{i+1}\partial_d^{j-2}\bar{\partial}^k u - t^{i+j+k-r}\partial_t^i\partial_d^{j-2}\bar{\partial}^k f.$$
(10.18)

Using the H^2-regularity estimate (10.13), we get

$$\|t^{i+j+k-r}\partial_t^i\partial_d^j\bar{\partial}^k u\|_{L^2} \leq \|t^{i+j+k-r}\partial_t^i\partial_d^{j-2}\bar{\partial}^k u\|_{\dot{H}^2}$$
$$\lesssim \|t^{i+j+k-r}\partial_t^{i+1}\partial_d^{j-2}\bar{\partial}^k u\|_{L^2}$$
$$+ \|t^{i+j+k-r}\partial_t^i\partial_d^{j-2}\bar{\partial}^k f\|_{L^2}$$
$$+ \|t^{i+j+k-r}\partial_t^i\partial_d^{j-2}\bar{\partial}^{k+1} u\|_{\dot{H}^1}$$
$$+ \|t^{i+j+k-r}\partial_t^i\partial_d^{j-2}\bar{\partial}^k u\|_{L^2}$$

and (10.15) follows. □

Proof of (10.16) Let $k \geq 1$. First, set $j = 2$ in (10.18) eliminating the ∂_d derivatives. Replacing k by $k - 1$ we obtain

$$\Delta(t^{i+1+k-r}\partial_t^i\bar{\partial}^{k-1} u) = t^{i+1+k-r}\partial_t^i\bar{\partial}^{k-1}\Delta u$$
$$= t^{i+1+k-r}\partial_t^{i+1}\bar{\partial}^{k-1} u - t^{i+1+k-r}\partial_t^i\bar{\partial}^{k-1} f.$$

Since $\partial_t^i\bar{\partial}^{k-1} u|_{\partial\Omega} = 0$, we may apply (10.14) to this equation, leading to

$$\|t^{i+1+k-r}\partial_t^i\partial_d\bar{\partial}^k u\|_{L^2} \leq \|t^{i+1+k-r}\partial_t^i\bar{\partial}^{k-1} u\|_{\dot{H}^2}$$
$$\lesssim \|t^{i+1+k-r}\partial_t^{i+1}\bar{\partial}^{k-1} u\|_{L^2}$$
$$+ \|t^{i+1+k-r}\partial_t^i\bar{\partial}^{k-1} f\|_{L^2},$$

and we obtain (10.16). □

Proof of (10.17) First, we have

$$\Delta(t^{i+1-r}\partial_t^i u) = t^{i+1-r}\partial_t^i\Delta u = t^{i+1-r}\partial_t^{i+1} u - t^{i+1-r}\partial_t^i f.$$

Using that the H^1 norm may be interpolated as in (10.11), and applying the H^2 regularity estimate (10.14), which may be used since $\partial_t^i u|_{\partial\Omega} = 0$, we get

$$\|t^{i+1-r}\partial_t^i u\|_{\dot{H}^1} \lesssim \|t^{i+1-r}\partial_t^i u\|_{\dot{H}^2}^{1/2} \|t^{i+1-r}\partial_t^i u\|_{L^2}^{1/2} + \|t^{i+1-r}\partial_t^i u\|_{L^2}$$

$$\lesssim \left(\|t^{i+1-r}\partial_t^{i+1} u\|_{L^2} + \|t^{i+1-r}\partial_t^i f\|_{L^2}\right)^{1/2} \|t^{i+1-r}\partial_t^i u\|_{L^2}^{1/2}$$
$$+ \|t^{i+1-r}\partial_t^i u\|_{L^2}$$

$$\lesssim \|t^{i+1-r}\partial_t^{i+1} u\|_{L^2}^{1/2} \|t^{i+1-r}\partial_t^i u\|_{L^2}^{1/2} + \|t^{i+1-r}\partial_t^i f\|_{L^2}$$
$$+ \|t^{i+1-r}\partial_t^i u\|_{L^2}.$$

Inequality (10.17) follows. $\qquad\square$

10.3.2 Tangential derivative reduction

In order to reduce the number of tangential derivatives, we will show that for $k \geq 2$ we have

$$\|t^{i+k-r}\partial_t^i \bar{\partial}^k u\|_{L^2_{x,t}} \lesssim \|t^{i+k-r}\partial_t^{i+1}\bar{\partial}^{k-2} u\|_{L^2_{x,t}}$$
$$+ \|t^{i+k-r}\partial_t^i \bar{\partial}^{k-2} f\|_{L^2_{x,t}}, \qquad (10.19)$$

while for $k = 1$, we have

$$\|t^{i+1-r}\partial_t^i \bar{\partial} u\|_{L^2_{x,t}} \lesssim \|t^{i+1-r}\partial_t^i u\|_{L^2_{x,t}}^{1/2} \|t^{i+1-r}\partial_t^{i+1} u\|_{L^2_{x,t}}^{1/2}$$
$$+ \|t^{i+1-r}\partial_t^i u\|_{L^2_{x,t}} + \|t^{i+1-r}\partial_t^i f\|_{L^2_{x,t}} \qquad (10.20)$$

for all $i \geq r$.

Proof of (10.19) We first set $j = 2$ in (10.18), eliminating all the ∂_d derivatives, and therefore obtaining

$$\Delta(t^{i+2+k-r}\partial_t^i \bar{\partial}^k u) = t^{i+2+k-r}\partial_t^i \bar{\partial}^k \Delta u$$
$$= t^{i+2+k-r}\partial_t^{i+1}\bar{\partial}^k u - t^{i+2+k-r}\partial_t^i \bar{\partial}^k f.$$

Since $t^{i+2+k-r}\partial_t^i \bar{\partial}^k u$ vanishes on $\partial\Omega$, using (10.14) we have

$$\|t^{i+2+k-r}\partial_t^i \bar{\partial}^{k+2} u\|_{L^2} \leq \|t^{i+2+k-r}\partial_t^i \bar{\partial}^k u\|_{\dot{H}^2}$$
$$\lesssim \|t^{i+2+k-r}\partial_t^{i+1}\bar{\partial}^k u\|_{L^2}$$
$$+ \|t^{i+2+k-r}\partial_t^i \bar{\partial}^k f\|_{L^2}$$

for $k \geq 0$.

The bound (10.19) follows upon replacing $k + 2$ with k. $\qquad\square$

Proof of (10.20) This inequality is obtained by replacing ∇ with $\bar{\partial}$ in inequality (10.17). $\qquad\square$

10.3.3 The time-derivative reduction

Similarly to the previous section we want to obtain derivative reduction estimates, but this time involving only time. Here we claim that for all $i \geq r + 1$ we have

$$\|t^{i-r}\partial_t^i u\|_{L^2_{x,t}} \lesssim (i-r)\|t^{i-1-r}\partial_t^{i-1}u\|_{L^2_{x,t}} + \|t^{i-r}\partial_t^{i-1}f\|_{L^2_{x,t}}. \quad (10.21)$$

Proof of (10.21)   For the system (10.1), we have the energy inequality

$$\|\partial_t u\|_{L^2_{x,t}} + \|\nabla u\|_{L^2_x}\Big|_{t=T} \lesssim \|\nabla u\|_{L^2_x}\Big|_{t=0} + \|f\|_{L^2_{x,t}}, \quad (10.22)$$

which is obtained by testing (10.1) with $\partial_t u$, which obeys the homogeneous Dirichlet boundary condition (10.3). We apply estimate (10.22) to the equation

$$\partial_t(t^{i-r}\partial_t^{i-1}u) - \Delta(t^{i-r}\partial_t^{i-1}u) = (i-r)t^{i-1-r}\partial_t^{i-1}u + t^{i-r}\partial_t^{i-1}f.$$

If $i \geq r+1$, then $t^{i-r}\partial_t^{i-1}u\big|_{t=0} = 0$, so that the initial value term in (10.22) vanishes, and we obtain

$$\|\partial_t(t^{i-r}\partial_t^{i-1}u)\|_{L^2} \lesssim \|(i-r)t^{i-1-r}\partial_t^{i-1}u\|_{L^2} + \|t^{i-r}\partial_t^{i-1}f\|_{L^2}.$$

The inequality (10.21) now follows from the product rule. $\qquad\square$

10.4 Proof of Theorem 10.1

With the notation introduced in (10.6), our goal reduces to establishing an inequality of the form

$$\bar{\varphi}(u) \leq C\varphi_0(u) + \frac{1}{2}\varphi(u) + C\|f\|, \quad (10.23)$$

where $\|f\|$ denotes a suitable (semi)-norm of f.

To see this, observe that adding $\varphi_0(u)$ to both sides of the above estimate, and absorbing the $\varphi(u)/2$ term on the left side yields

$$\varphi(u) \lesssim \varphi_0(u) + \|f\|,$$

which concludes the proof, in view of (10.7) and the assumed regularity of the force f.

We split the sum $\bar{\varphi}(u)$ according to the values of i, j, and k, so that the inequalities in Section 10.3 may be applied. Recall the definition of

B in (10.5). We split the first sum in (10.6) as

$$\bar{\varphi}(u) = \sum_{\ell=1}^{6} S_\ell,$$

where

$$S_\ell = \sum_{B_\ell} \frac{(i+j+k)^r \varepsilon^i \tilde{\varepsilon}^j \bar{\varepsilon}^k}{(i+j+k)!} \|t^{i+j+k-r} \partial_t^i \partial_d^j \bar{\partial}^k u\|_{L^2_{x,t}}, \qquad \ell = 1,\ldots,6,$$

with the sets of indices for the corresponding summations given by

$$B_1 = \{(i,j,k) \in B : j \geq 2\},$$
$$B_2 = \{(i,j,k) \in B : j = 1, k \geq 1\},$$
$$B_3 = \{(i,j,k) \in B : j = 1, k = 0\},$$
$$B_4 = \{(i,j,k) \in B : j = 0, k \geq 2\},$$
$$B_5 = \{(i,j,k) \in B : j = 0, k = 1\},$$
$$B_6 = \{(i,j,k) \in B : j = 0, k = 0\}.$$

As we will now show, the above sums are bounded according to (10.15), (10.16), (10.17), (10.19), (10.20), and (10.21), respectively.

10.4.1 The S_1 term

We start with S_1. Notice that B_1 corresponds to a set of indices with $j \geq 2$, therefore allowing us to use (10.15) as follows.

$$
\begin{aligned}
S_1 \quad &\lesssim \sum_{B_1} \frac{(i+j+k)^r \varepsilon^i \tilde{\varepsilon}^j \bar{\varepsilon}^k}{(i+j+k)!} \|t^{i+j+k-r} \partial_t^{i+1} \partial_d^{j-2} \bar{\partial}^k u\|_{L^2_{x,t}} \\
&+ \sum_{B_1} \frac{(i+j+k)^r \varepsilon^i \tilde{\varepsilon}^j \bar{\varepsilon}^k}{(i+j+k)!} \|t^{i+j+k-r} \partial_t^i \partial_d^{j-1} \bar{\partial}^{k+1} u\|_{L^2_{x,t}} \\
&+ \sum_{B_1} \frac{(i+j+k)^r \varepsilon^i \tilde{\varepsilon}^j \bar{\varepsilon}^k}{(i+j+k)!} \|t^{i+j+k-r} \partial_t^i \partial_d^{j-2} \bar{\partial}^{k+2} u\|_{L^2_{x,t}} \\
&+ \sum_{B_1} \frac{(i+j+k)^r \varepsilon^i \tilde{\varepsilon}^j \bar{\varepsilon}^k}{(i+j+k)!} \|t^{i+j+k-r} \partial_t^i \partial_d^{j-2} \bar{\partial}^k u\|_{L^2_{x,t}} \\
&+ \sum_{B_1} \frac{(i+j+k)^r \varepsilon^i \tilde{\varepsilon}^j \bar{\varepsilon}^k}{(i+j+k)!} \|t^{i+j+k-r} \partial_t^i \partial_d^{j-2} \bar{\partial}^k f\|_{L^2_{x,t}}.
\end{aligned}
$$

By relabeling, we obtain

$$S_1 \lesssim \sum_{(i-1,j+2,k)\in B_1} \frac{(i+j+k+1)^r \varepsilon^{i-1}\widetilde{\varepsilon}^{j+2}\bar{\varepsilon}^k}{(i+j+k+1)!} \|t^{i+j+k+1-r}\partial_t^i\partial_d^j\bar{\partial}^k u\|_{L_{x,t}^2}$$

$$+ \sum_{(i,j+1,k-1)\in B_1} \frac{(i+j+k)^r \varepsilon^i\widetilde{\varepsilon}^{j+1}\bar{\varepsilon}^{k-1}}{(i+j+k)!} \|t^{i+j+k-r}\partial_t^i\partial_d^j\bar{\partial}^k u\|_{L_{x,t}^2}$$

$$+ \sum_{(i,j+2,k-2)\in B_1} \frac{(i+j+k)^r \varepsilon^i\widetilde{\varepsilon}^{j+2}\bar{\varepsilon}^{k-2}}{(i+j+k)!} \|t^{i+j+k-r}\partial_t^i\partial_d^j\bar{\partial}^k u\|_{L_{x,t}^2}$$

$$+ \sum_{(i,j+2,k)\in B_1} \frac{(i+j+k+2)^r \varepsilon^i\widetilde{\varepsilon}^{j+2}\bar{\varepsilon}^k}{(i+j+k+2)!} \|t^{i+j+k+2-r}\partial_t^i\partial_d^j\bar{\partial}^k u\|_{L_{x,t}^2}$$

$$+ \sum_{(i,j+2,k)\in B_1} \frac{(i+j+k+2)^r \varepsilon^i\widetilde{\varepsilon}^{j+2}\bar{\varepsilon}^k}{(i+j+k+2)!} \|t^{i+j+k+2-r}\partial_t^i\partial_d^j\bar{\partial}^k f\|_{L_{x,t}^2}.$$

Therefore, since $t \leq T$, we get

$$S_1 \lesssim \frac{T\widetilde{\varepsilon}^2}{\varepsilon} \sum_{(i-1,j+2,k)\in B_1} \frac{(i+j+k+1)^r \varepsilon^i\widetilde{\varepsilon}^j\bar{\varepsilon}^k}{(i+j+k+1)!} \|t^{i+j+k-r}\partial_t^i\partial_d^j\bar{\partial}^k u\|_{L_{x,t}^2}$$

$$+ \frac{\widetilde{\varepsilon}}{\varepsilon} \sum_{(i,j+1,k-1)\in B_1} \frac{(i+j+k)^r \varepsilon^i\widetilde{\varepsilon}^j\bar{\varepsilon}^k}{(i+j+k)!} \|t^{i+j+k-r}\partial_t^i\partial_d^j\bar{\partial}^k u\|_{L_{x,t}^2}$$

$$+ \frac{\widetilde{\varepsilon}^2}{\varepsilon^2} \sum_{(i,j+2,k-2)\in B_1} \frac{(i+j+k)^r \varepsilon^i\widetilde{\varepsilon}^j\bar{\varepsilon}^k}{(i+j+k)!} \|t^{i+j+k-r}\partial_t^i\partial_d^j\bar{\partial}^k u\|_{L_{x,t}^2}$$

$$+ T^2\widetilde{\varepsilon}^2 \sum_{(i,j+2,k)\in B_1} \frac{(i+j+k+2)^r \varepsilon^i\widetilde{\varepsilon}^j\bar{\varepsilon}^k}{(i+j+k+2)!} \|t^{i+j+k-r}\partial_t^i\partial_d^j\bar{\partial}^k u\|_{L_{x,t}^2}$$

$$+ \widetilde{\varepsilon}^2 \sum_{(i,j+2,k)\in B_1} \frac{(i+j+k+2)^r \varepsilon^i\widetilde{\varepsilon}^j\bar{\varepsilon}^k}{(i+j+k+2)!} \|t^{i+j+k+2-r}\partial_t^i\partial_d^j\bar{\partial}^k f\|_{L_{x,t}^2},$$

and thus

$$S_1 \lesssim \left(\frac{T\widetilde{\varepsilon}^2}{\varepsilon} + \frac{\widetilde{\varepsilon}}{\varepsilon} + \frac{\widetilde{\varepsilon}^2}{\varepsilon^2} + T^2\widetilde{\varepsilon}^2 \right) \varphi(u)$$

$$+ \widetilde{\varepsilon}^2 \sum_{(i,j+2,k)\in B_1} \frac{(i+j+k+2)^r \varepsilon^i\widetilde{\varepsilon}^j\bar{\varepsilon}^k}{(i+j+k+2)!} \|t^{i+j+k+2-r}\partial_t^i\partial_d^j\bar{\partial}^k f\|_{L_{x,t}^2}.$$

$$(10.24)$$

10.4.2 The S_2 term

In order to treat S_2 we use (10.16). Note that $j = 1$ and $k \geq 1$. We have

$$S_2 \lesssim \sum_{B_2} \frac{(i+1+k)^r \varepsilon^i \widetilde{\varepsilon} \bar{\varepsilon}^k}{(i+1+k)!} \|t^{i+1+k-r} \partial_t^{i+1} \bar{\partial}^{k-1} u\|_{L^2_{x,t}}$$

$$+ \sum_{B_2} \frac{(i+1+k)^r \varepsilon^i \widetilde{\varepsilon} \bar{\varepsilon}^k}{(i+1+k)!} \|t^{i+1+k-r} \partial_t^i \bar{\partial}^{k-1} f\|_{L^2_{x,t}}$$

and then, by relabeling,

$$S_2 \lesssim \sum_{(i-1,j,k+1)\in B_2} \frac{(i+1+k)^r \varepsilon^{i-1} \widetilde{\varepsilon} \bar{\varepsilon}^{k+1}}{(i+1+k)!} \|t^{i+1+k-r} \partial_t^i \bar{\partial}^k u\|_{L^2_{x,t}}$$

$$+ \sum_{(i,j,k+1)\in B_2} \frac{(i+2+k)^r \varepsilon^i \widetilde{\varepsilon} \bar{\varepsilon}^{k+1}}{(i+2+k)!} \|t^{i+2+k-r} \partial_t^i \bar{\partial}^k f\|_{L^2_{x,t}}.$$

Therefore, we obtain

$$S_2 \lesssim \frac{T \bar{\varepsilon} \widetilde{\varepsilon}}{\varepsilon} \sum_{(i-1,j,k+1)\in B_2} \frac{(i+1+k)^r \varepsilon^i \bar{\varepsilon}^k}{(i+1+k)!} \|t^{i+k-r} \partial_t^i \bar{\partial}^k u\|_{L^2_{x,t}}$$

$$+ \bar{\varepsilon} \widetilde{\varepsilon} \sum_{(i,j,k+1)\in B_2} \frac{(i+k+2)^r \varepsilon^i \bar{\varepsilon}^k}{(i+k+2)!} \|t^{i+k+2-r} \partial_t^i \bar{\partial}^k f\|_{L^2_{x,t}}$$

from where

$$S_2 \lesssim \frac{T \bar{\varepsilon} \widetilde{\varepsilon}}{\varepsilon} \varphi(u) + \bar{\varepsilon} \widetilde{\varepsilon} \sum_{(i,j,k+1)\in B_2} \frac{(i+k+2)^r \varepsilon^i \bar{\varepsilon}^k}{(i+k+2)!} \|t^{i+k+2-r} \partial_t^i \bar{\partial}^k f\|_{L^2_{x,t}}.$$

$$(10.25)$$

10.4.3 The S_3 term

For S_3, we use (10.17) and write (note that $j = 1$ and $k = 0$)

$$S_3 \lesssim \sum_{(i,j,k)\in B_3} \frac{(i+j+k)^r \varepsilon^i \widetilde{\varepsilon}^j \bar{\varepsilon}^k}{(i+j+k)!} \|t^{i+1-r} \partial_t^i u\|_{L^2_{x,t}}^{1/2} \|t^{i+1-r} \partial_t^{i+1} u\|_{L^2_{x,t}}^{1/2}$$

$$+ \sum_{(i,j,k)\in B_3} \frac{(i+j+k)^r \varepsilon^i \widetilde{\varepsilon}^j \bar{\varepsilon}^k}{(i+j+k)!} \|t^{i+1-r} \partial_t^i u\|_{L^2_{x,t}}$$

$$+ \sum_{(i,j,k)\in B_3} \frac{(i+j+k)^r \varepsilon^i \widetilde{\varepsilon}^j \bar{\varepsilon}^k}{(i+j+k)!} \|t^{i+1-r} \partial_t^i f\|_{L^2_{x,t}}$$

which implies by the Cauchy–Schwarz inequality that

$$
S_3 \lesssim \frac{T^{1/2}\widetilde{\varepsilon}}{\varepsilon^{1/2}} \left(\sum_{(i-1,j,k)\in B_3} \frac{i^r \varepsilon^i}{i!} \|t^{i-r}\partial_t^i u\|_{L^2_{x,t}} \right)^{1/2}
$$

$$
\times \left(\sum_{(i,j,k)\in B_3} \frac{(i+1)^r \varepsilon^i}{(i+1)!} \|t^{i-r}\partial_t^i u\|_{L^2_{x,t}} \right)^{1/2}
$$

$$
+ T\widetilde{\varepsilon} \sum_{(i,j,k)\in B_3} \frac{(i+1)^r \varepsilon^i}{(i+1)!} \|t^{i-r}\partial_t^i u\|_{L^2_{x,t}}
$$

$$
+ \widetilde{\varepsilon} \sum_{(i,j,k)\in B_3} \frac{(i+1)^r \varepsilon^i}{(i+1)!} \|t^{i+1-r}\partial_t^i f\|_{L^2_{x,t}}.
$$

Thus we get

$$
S_3 \lesssim \left(\frac{T^{1/2}\widetilde{\varepsilon}}{\varepsilon^{1/2}} + T\widetilde{\varepsilon} \right) \varphi(u)
$$

$$
+ \widetilde{\varepsilon} \sum_{(i,j,k)\in B_3} \frac{(i+1)^r \varepsilon^i}{(i+1)!} \|t^{i+1-r}\partial_t^i f\|_{L^2_{x,t}}. \tag{10.26}
$$

10.4.4 The S_4 term

We use (10.19) to estimate S_4, noting that B_4 corresponds precisely to the case in which $j = 0$ and $k \geq 2$. We have

$$
S_4 \lesssim \sum_{(i,j,k)\in B_4} \frac{(i+j+k)^r \varepsilon^i \widetilde{\varepsilon}^j \bar{\varepsilon}^k}{(i+j+k)!} \|t^{i+k-r}\partial_t^{i+1}\bar{\partial}^{k-2} u\|_{L^2_{x,t}}
$$

$$
+ \sum_{(i,j,k)\in B_4} \frac{(i+j+k)^r \varepsilon^i \widetilde{\varepsilon}^j \bar{\varepsilon}^k}{(i+j+k)!} \|t^{i+k-r}\partial_t^i \bar{\partial}^{k-2} f\|_{L^2_{x,t}}.
$$

By relabeling, we have

$$
S_4 \lesssim \sum_{(i-1,j,k+2)\in B_4} \frac{(i+j+k+1)^r \varepsilon^{i-1}\bar{\varepsilon}^{k+2}}{(i+j+k+1)!} \|t^{i+k+1-r}\partial_t^i \bar{\partial}^k u\|_{L^2_{x,t}}
$$

$$
+ \sum_{(i,j,k+2)\in B_4} \frac{(i+j+k+2)^r \varepsilon^i \bar{\varepsilon}^{k+2}}{(i+j+k+2)!} \|t^{i+k+2-r}\partial_t^i \bar{\partial}^k f\|_{L^2_{x,t}}
$$

and thus

$$S_4 \lesssim \frac{\bar{\varepsilon}^2 T}{\varepsilon} \sum_{(i-1,j,k+2)\in B_4} \frac{(i+j+k+1)^r \varepsilon^i \bar{\varepsilon}^k}{(i+j+k+1)!} \|t^{i+k-r} \partial_t^i \bar{\partial}^k u\|_{L^2_{x,t}}$$

$$+ \bar{\varepsilon}^2 \sum_{(i,j,k+2)\in B_4} \frac{(i+j+k+2)^r \varepsilon^i \bar{\varepsilon}^k}{(i+j+k+2)!} \|t^{i+k+2-r} \partial_t^i \bar{\partial}^k f\|_{L^2_{x,t}}.$$

We obtain

$$S_4 \lesssim \frac{\bar{\varepsilon}^2 T}{\varepsilon} \varphi(u) + \bar{\varepsilon}^2 \sum_{(i,j,k+2)\in B_4} \frac{(i+j+k+2)^r \varepsilon^i \bar{\varepsilon}^k}{(i+j+k+2)!} \|t^{i+k+2-r} \partial_t^i \bar{\partial}^k f\|_{L^2_{x,t}}.$$

$$(10.27)$$

10.4.5 The S_5 term

For S_5 (note that $j = 0$ and $k = 1$), we use (10.20) and write

$$S_5 \lesssim \sum_{(i,j,k)\in B_5} \frac{(i+j+k)^r \varepsilon^i \bar{\bar{\varepsilon}}^j \bar{\varepsilon}^k}{(i+j+k)!} \|t^{i+1-r} \partial_t^i u\|_{L^2_{x,t}}^{1/2} \|t^{i+1-r} \partial_t^{i+1} u\|_{L^2_{x,t}}^{1/2}$$

$$+ \sum_{(i,j,k)\in B_5} \frac{(i+j+k)^r \varepsilon^i \bar{\bar{\varepsilon}}^j \bar{\varepsilon}^k}{(i+j+k)!} \|t^{i+1-r} \partial_t^i u\|_{L^2_{x,t}}$$

$$+ \sum_{(i,j,k)\in B_5} \frac{(i+j+k)^r \varepsilon^i \bar{\bar{\varepsilon}}^j \bar{\varepsilon}^k}{(i+j+k)!} \|t^{i+1-r} \partial_t^i f\|_{L^2_{x,t}}.$$

We again relabel and use the Cauchy-Schwarz inequality to deduce

$$S_5 \lesssim \frac{T^{1/2} \bar{\varepsilon}}{\varepsilon^{1/2}} \left(\sum_{(i-1,j,k)\in B_5} \frac{i^r \varepsilon^i}{i!} \|t^{i-r} \partial_t^i u\|_{L^2_{x,t}} \right)^{1/2}$$

$$\times \left(\sum_{(i,j,k)\in B_5} \frac{(i+1)^r \varepsilon^i}{(i+1)!} \|t^{i-r} \partial_t^i u\|_{L^2_{x,t}} \right)^{1/2}$$

$$+ T\bar{\varepsilon} \sum_{(i,j,k)\in B_5} \frac{(i+1)^r \varepsilon^i}{(i+1)!} \|t^{i-r} \partial_t^i u\|_{L^2_{x,t}}$$

$$+ \bar{\varepsilon} \sum_{(i,j,k)\in B_5} \frac{(i+1)^r \varepsilon^i}{(i+1)!} \|t^{i+1-r} \partial_t^i f\|_{L^2_{x,t}}.$$

We conclude

$$S_5 \lesssim \left(\frac{T^{1/2}\bar{\varepsilon}}{\varepsilon^{1/2}} + T\bar{\varepsilon} \right) \varphi(u)$$

$$+ \bar{\varepsilon} \sum_{(i,j,k)\in B_5} \frac{(i+1)^r \varepsilon^i}{(i+1)!} \|t^{i+1-r}\partial_t^i f\|_{L^2_{x,t}}. \tag{10.28}$$

10.4.6 The S_6 term

Lastly, for S_6 (note that $j = k = 0$), we use (10.21) and write

$$S_6 \lesssim \sum_{(i,j,k)\in B_6} \frac{(i-r)(i+j+k)^r \varepsilon^i \bar{\varepsilon}^j \bar{\varepsilon}^k}{(i+j+k)!} \|t^{i-1-r}\partial_t^{i-1}u\|_{L^2_{x,t}}$$

$$+ \sum_{(i,j,k)\in B_6} \frac{(i+j+k)^r \varepsilon^i \bar{\varepsilon}^j \bar{\varepsilon}^k}{(i+j+k)!} \|t^{i-r}\partial_t^{i-1} f\|_{L^2_{x,t}}$$

which we can rewrite as

$$S_6 \lesssim \sum_{(i+1,j,k)\in B_6} \frac{i^r(i-r+1)\varepsilon^{i+1}}{(i+1)!} \|t^{i-r}\partial_t^i u\|_{L^2_{x,t}}$$

$$+ \sum_{(i+1,j,k)\in B_6} \frac{(i+1)^r \varepsilon^{i+1}}{(i+1)!} \|t^{i+1-r}\partial_t^i f\|_{L^2_{x,t}}$$

and therefore

$$S_6 \lesssim \varepsilon\varphi(u)$$

$$+ \varepsilon \sum_{(i+1,j,k)\in B_6} \frac{(i+1)^r \varepsilon^i}{(i+1)!} \|t^{i+1-r}\partial_t^i f\|_{L^2_{x,t}}. \tag{10.29}$$

10.4.7 Conclusion of the proof

Combining the bounds (10.24), (10.25), (10.26), (10.27), (10.28), and (10.29) that we have obtained for $\bar{\varphi}$ we conclude

$$\bar{\varphi}(u) \lesssim \left(\frac{T(\tilde{\varepsilon}^2 + \bar{\varepsilon}^2)}{\varepsilon} + \frac{\tilde{\varepsilon}}{\varepsilon} + \frac{\tilde{\varepsilon}^2}{\varepsilon^2} + T^2\tilde{\varepsilon}^2 \right.$$

$$\left. + \frac{T\bar{\varepsilon}\tilde{\varepsilon}}{\varepsilon} + \frac{T^{1/2}(\tilde{\varepsilon} + \bar{\varepsilon})}{\varepsilon^{1/2}} + T(\tilde{\varepsilon} + \bar{\varepsilon}) + \varepsilon \right) \varphi(u) + \varphi_0(u)$$

$$+ \sum_{(i,j+2,k) \in B_1} \frac{(i+j+k+2)^r \varepsilon^i \tilde{\varepsilon}^{j+2} \bar{\varepsilon}^k}{(i+j+k+2)!} \| t^{i+j+k+2-r} \partial_t^i \partial_d^j \bar{\partial}^k f \|_{L^2_{x,t}}$$

$$+ \sum_{(i,j,k+1) \in B_2} \frac{(i+k+2)^r \varepsilon^i \tilde{\varepsilon} \bar{\varepsilon}^{k+1}}{(i+k+2)!} \| t^{i+2+k-r} \partial_t^i \bar{\partial}^k f \|_{L^2_{x,t}}$$

$$+ \sum_{(i,j,k) \in B_3} \frac{(i+1)^r \varepsilon^i \tilde{\varepsilon}}{(i+1)!} \| t^{i+1-r} \partial_t^i f \|_{L^2_{x,t}}$$

$$+ \sum_{(i,j,k+2) \in B_4} \frac{(i+k+2)^r \varepsilon^i \bar{\varepsilon}^{k+2}}{(i+k+2)!} \| t^{i+k+2-r} \partial_t^i \bar{\partial}^k f \|_{L^2_{x,t}}$$

$$+ \sum_{(i,j,k) \in B_5} \frac{(i+1)^r \varepsilon^i \tilde{\varepsilon}}{(i+1)!} \| t^{i+1-r} \partial_t^i f \|_{L^2_{x,t}}$$

$$+ \sum_{(i+1,j,k) \in B_6} \frac{(i+1)^r \varepsilon^{i+1}}{(i+1)!} \| t^{i+1-r} \partial_t^i f \|_{L^2_{x,t}} . \tag{10.30}$$

Denote by C the implicit dimensional constant in the $\lesssim$ symbol in (10.30). Let $T > 0$ be arbitrary. First, choose a sufficiently small $\varepsilon = \varepsilon(C) > 0$ such that

$$\varepsilon \le \frac{1}{6C}. \tag{10.31}$$

Next, we choose a sufficiently small $\bar{\varepsilon} = \bar{\varepsilon}(\varepsilon, T, C) = \bar{\varepsilon}(T, C) > 0$ such that

$$\frac{T\bar{\varepsilon}^2}{\varepsilon} + \frac{T^{1/2}\bar{\varepsilon}}{\varepsilon^{1/2}} + T\bar{\varepsilon} \le \frac{1}{6C}. \tag{10.32}$$

Finally, choose $\tilde{\varepsilon} = \tilde{\varepsilon}(\bar{\varepsilon}, \varepsilon, T, C) = \tilde{\varepsilon}(T, C) > 0$ such that

$$\frac{T\tilde{\varepsilon}^2}{\varepsilon} + \frac{\tilde{\varepsilon}}{\varepsilon} + \frac{\tilde{\varepsilon}^2}{\varepsilon^2} + T^2\tilde{\varepsilon}^2 + \frac{T\bar{\varepsilon}\tilde{\varepsilon}}{\varepsilon} + \frac{T^{1/2}\tilde{\varepsilon}}{\varepsilon^{1/2}} + T\tilde{\varepsilon} \le \frac{1}{6C}. \tag{10.33}$$

For convenience, in addition to the constraints (10.31)–(10.33) we may also ensure that

$$\tilde{\varepsilon} \le \bar{\varepsilon} \le \varepsilon \le 1. \tag{10.34}$$

Reorganizing the terms in (10.30), the choices (10.31)–(10.34) imply that

$$\bar{\varphi}(u) \le \frac{1}{2}\varphi(u) + C\varphi_0(u)$$

$$+ C \sum_{(i,j+2,k)\in B_1} \frac{(i+j+k+2)^r \varepsilon^i \widetilde{\varepsilon}^{j+2} \bar{\varepsilon}^k}{(i+j+k+2)!} \|t^{i+j+k+2-r} \partial_t^i \partial_d^j \bar{\partial}^k f\|_{L_{x,t}^2}$$

$$+ C \left(\widetilde{\varepsilon}\bar{\varepsilon} \sum_{(i,j,k+1)\in B_2} + \bar{\varepsilon}^2 \sum_{(i,j,k+2)\in B_4} \right)$$

$$\times \left(\frac{(i+k+2)^r \varepsilon^i \bar{\varepsilon}^k}{(i+k+2)!} \|t^{i+2+k-r} \partial_t^i \bar{\partial}^k f\|_{L_{x,t}^2} \right)$$

$$+ C \left(\widetilde{\varepsilon} \sum_{(i,j,k)\in B_3} + \bar{\varepsilon} \sum_{(i,j,k)\in B_5} + \varepsilon \sum_{(i+1,j,k)\in B_6} \right)$$

$$\times \left(\frac{(i+1)^r \varepsilon^i}{(i+1)!} \|t^{i+1-r} \partial_t^i f\|_{L_{x,t}^2} \right)$$

$$+ C \sum_{(i,j+2,k)\in B_1} \frac{(i+j+k+2)^r \varepsilon^i \widetilde{\varepsilon}^{j+2} \bar{\varepsilon}^k}{(i+j+k+2)!} \|t^{i+j+k+2-r} \partial_t^i \partial_d^j \bar{\partial}^k f\|_{L_{x,t}^2}$$

$$(10.35)$$

which in turn yields

$$\bar{\varphi}(u) \le \frac{1}{2}\varphi(u) + C\varphi_0(u)$$

$$+ 2C \sum_{i+j+k\ge(r-2)_+} \frac{(i+j+k+2)^r \varepsilon^i \widetilde{\varepsilon}^{j+2} \bar{\varepsilon}^k}{(i+j+k+2)!} \|t^{i+j+k+2-r} \partial_t^i \partial_d^j \bar{\partial}^k f\|_{L_{x,t}^2}$$

$$+ 2C \sum_{i+k\ge(r-2)_+} \frac{(i+k+2)^r \varepsilon^i \bar{\varepsilon}^{k+2}}{(i+k+2)!} \|t^{i+k+2-r} \partial_t^i \bar{\partial}^k f\|_{L_{x,t}^2}$$

$$+ 3C \sum_{i\ge(r-1)_+} \frac{(i+1)^r \varepsilon^{i+1}}{(i+1)!} \|t^{i+1-r} \partial_t^i f\|_{L_{x,t}^2} . \qquad (10.36)$$

The estimate (10.36) is precisely the desired bound (10.23), thereby completing the proof of the a priori estimates. These a priori estimates may be made rigorous by working with truncated sums in $\bar{\varphi}(u)$, such that $i + 2j + 2k \le N$, which is allowed by parabolic regularity, and passing to the limit as N tends to ∞. We omit the details here.

10.5 Derivative reduction for the Stokes problem and the proof of Theorem 10.2

The proof of Theorem 10.2 is identical to the proof of Theorem 10.1, except that the derivative reduction estimates in Section 10.3 need to be adapted to the case of the Stokes system (10.9). To avoid redundancy, for the Stokes system we only present the arguments for these reduction estimates.

The reductions are mainly based on two H^2 inequalities. Namely, for the stationary Stokes system

$$-\Delta u + \nabla p = g$$
$$\nabla \cdot u = 0, \tag{10.37}$$

we have

$$\|u\|_{\dot{H}^2(\Omega)} + \|p\|_{\dot{H}^1} \lesssim \|g\|_{L^2(\Omega)} + \|\bar{\partial} u\|_{\dot{H}^1(\Omega)} + \|u\|_{L^2(\Omega)}. \tag{10.38}$$

If also (10.10) holds, i.e., if $u\big|_{\partial\Omega} = 0$, then the above estimate becomes

$$\|u\|_{\dot{H}^2(\Omega)} \lesssim \|g\|_{L^2(\Omega)}. \tag{10.39}$$

10.5.1 Normal derivative reduction for the Stokes operator

In all the inequalities below, we require $i + j + k \geq r$. We first claim

$$\|t^{i+j+k-r}\partial_t^i\partial_d^j\bar{\partial}^k u\|_{L^2_{x,t}} + \|t^{i+j+k-r}\partial_t^i\partial_d^{j-1}\bar{\partial}^k p\|_{L^2}$$
$$\lesssim \|t^{i+j+k-r}\partial_t^{i+1}\partial_d^{j-2}\bar{\partial}^k u\|_{L^2_{x,t}} + \|t^{i+j+k-r}\partial_t^i\partial_d^{j-1}\bar{\partial}^{k+1} u\|_{L^2_{x,t}}$$
$$+ \|t^{i+j+k-r}\partial_t^i\partial_d^{j-2}\bar{\partial}^{k+2} u\|_{L^2_{x,t}} + \|t^{i+j+k-r}\partial_t^i\partial_d^{j-2}\bar{\partial}^k u\|_{L^2_{x,t}}$$
$$+ \|t^{i+j+k-r}\partial_t^i\partial_d^{j-2}\bar{\partial}^k f\|_{L^2_{x,t}}, \qquad j \geq 2 \tag{10.40}$$

which allows us to reduce the number of vertical derivatives (∂_d) in the Gevrey (analytic) norm. On the other hand, for $j = 1$, we have

$$\|t^{i+1+k-r}\partial_t^i\partial_d\bar{\partial}^k u\|_{L^2_{x,t}} + \|t^{i+1+k-r}\partial_t^i\bar{\partial}^k p\|_{L^2_{x,t}}$$
$$\lesssim \|t^{i+1+k-r}\partial_t^{i+1}\bar{\partial}^{k-1} u\|_{L^2_{x,t}} + \|t^{i+1+k-r}\partial_t^i\bar{\partial}^{k-1} f\|_{L^2_{x,t}}, \qquad k \geq 1. \tag{10.41}$$

For $j = 1$ and $k = 0$, we claim

$$\|t^{i+1-r}\partial_t^i \nabla u\|_{L^2_{x,t}} \lesssim \|t^{i+1-r}\partial_t^i u\|_{L^2_{x,t}}^{1/2} \|t^{i+1-r}\partial_t^{i+1} u\|_{L^2_{x,t}}^{1/2}$$
$$+ \|t^{i+1-r}\partial_t^i u\|_{L^2} + \|t^{i+1-r}\partial_t^i f\|_{L^2_{x,t}}, \qquad i \geq r. \tag{10.42}$$

10.5.2 Tangential derivative reduction for the Stokes operator

In order to reduce the number of tangential derivatives, we claim

$$\|t^{i+k-r}\partial_t^i \bar{\partial}^k u\|_{L^2_{x,t}} + \|t^{i+k-r}\partial_t^i \bar{\partial}^{k-1} p\|_{L^2_{x,t}}$$
$$\lesssim \|t^{i+k-r}\partial_t^{i+1}\bar{\partial}^{k-2} u\|_{L^2_{x,t}} + \|t^{i+k-r}\partial_t^i \bar{\partial}^{k-2} f\|_{L^2_{x,t}}, \qquad k \geq 2 \tag{10.43}$$

while for $k = 1$, we use a special case (replace ∇ with $\bar{\partial}$) of the inequality (10.42)

$$\|t^{i+1-r}\partial_t^i \bar{\partial} u\|_{L^2_{x,t}} \lesssim \|t^{i+1-r}\partial_t^i u\|_{L^2_{x,t}}^{1/2} \|t^{i+1-r}\partial_t^{i+1} u\|_{L^2_{x,t}}^{1/2}$$
$$+ \|t^{i+1-r}\partial_t^i u\|_{L^2} + \|t^{i+1-r}\partial_t^i f\|_{L^2_{x,t}}, \qquad i \geq r. \tag{10.44}$$

10.5.3 Time derivative reduction for the Stokes operator

Here we have, for $i - 1 \geq r$

$$\|t^{i-r}\partial_t^i u\|_{L^2_{x,t}} \lesssim (i - r)\|t^{i-1-r}\partial_t^{i-1} u\|_{L^2_{x,t}} + \|t^{i-r}\partial_t^{i-1} f\|_{L^2_{x,t}}. \tag{10.45}$$

The proofs of the inequalities (10.40)–(10.45) are the same as for those in Section 10.3, except that instead of appealing to (10.13)–(10.14), we use (10.38)–(10.39). Moreover, the energy inequality (10.22) is the same for the Stokes system, since $\nabla \cdot u = 0$. Further details are omitted.

Acknowledgments

IK was supported in part by the NSF grant DMS-1615239, while VV was supported in part by the NSF grant DMS-1652134 and an Alfred P. Sloan Fellowship in mathematics.

References

Biswas, A. (2005) Local existence and Gevrey regularity of 3-D Navier-Stokes equations with l_p initial data. *J. Differential Equations* **215** no. 2, 429–447.

Biswas, A. & Swanson, D. (2007) Gevrey regularity of solutions to the 3D Navier–Stokes equations with weighted l_p initial data. *Indiana Univ. Math. J.* **56** no. 3, 1157–1188.

Bradshaw, Z., Grujić, Z., & Kukavica, I. (2015) Local analyticity radii of solutions to the 3D Navier-Stokes equations with locally analytic forcing. *J. Differential Equations* **259** no. 8, 3955–3975.

Camliyurt, G., Kukavica, I. & Vicol, V.C. (2018) Analytic and Gevrey space-time regularity for solutions of the Navier–Stokes equations (in preparation).

Evans, L.C. (1998) *Partial differential equations* Graduate Studies in Mathematics **19** American Mathematical Society, Providence, RI.

Ferrari, A.B. & Titi, E.S. (1998) Gevrey regularity for nonlinear analytic parabolic equations. *Comm. Partial Differential Equations* **23** no. 1-2, 1–16.

Foias, C. & Temam, R. (1989) Gevrey class regularity for the solutions of the Navier-Stokes equations. *J. Funct. Anal.* **87** no. 2, 359–369.

Friz, P.K., Kukavica, I., & Robinson, J.C. (2001) Nodal parametrisation of analytic attractors. *Discrete Contin. Dynam. Systems* **7** no. 3, 643–657.

Giga, Y. (1983) Time and spatial analyticity of solutions of the Navier-Stokes equations. *Comm. Partial Differential Equations* **8**, 929–948.

Grujić, Z. & Kukavica, I. (1998) Space analyticity for the Navier-Stokes and related equations with initial data in L^p. *J. Funct. Anal.* **152** no. 2, 447–466.

Grujić, Z. & Kukavica, I. (1999) Space analyticity for the nonlinear heat equation in a bounded domain. *J. Differential Equations* **154** no. 1, 42–54.

Kato, T. & Masuda, K. (1986) Nonlinear evolution equations and analyticity. I. *Ann. Inst. H. Poincaré Anal. Non Linéaire* **3**, 455–467.

Kinderlehrer, D. & Nirenberg, L. (1978) Analyticity at the boundary of solutions of nonlinear second-order parabolic equations. *Comm. Pure Appl. Math.* **31** no. 3, 283–338.

Komatsu, G. (1979) Analyticity up to the boundary of solutions of nonlinear parabolic equations. *Comm. Pure Appl. Math.* **32** no. 5, 669–720.

Komatsu, G. (1980) Global analyticity up to the boundary of solutions of the Navier-Stokes equation. *Comm. Pure Appl. Math.* **33**, no. 4, 545–566.

Kukavica, I. & Vicol, V.C. (2011) The domain of analyticity of solutions to the three-dimensional Euler equations in a half space. *Discrete Contin. Dyn. Syst.* **29**, no. 1, 285–303.

Lemarié-Rieusset, P.G. (2000) Une remarque sur l'analyticité des solutions milds des équations de Navier-Stokes dans $\mathbf{R}^3$. *C. R. Acad. Sci. Paris Sér. I Math.* **330** no. 3, 183–186.

Lee, P.-L. & Lin Guo, Y.-J. (2001) Gevrey class regularity for parabolic equations. *Differential Integral Equations* **14**, 989–1004.

Levermore, C.D. & Oliver, M. (1997) Analyticity of solutions for a generalized Euler equation. *J. Differential Equations* **133**, no. 2, 321–339.

Lions, J.-L. & Magenes, E. (1972a) *Non-homogeneous boundary value problems and applications. Vol. I.* Springer-Verlag, New York. Translated from the French by P. Kenneth, Die Grundlehren der mathematischen Wissenschaften, Band 181.

Lions, J.-L. & Magenes, E. (1972b) *Non-homogeneous boundary value problems and applications. Vol. II.* Springer-Verlag, New York. Translated from the French by P. Kenneth, Die Grundlehren der mathematischen Wissenschaften, Band 182.

Oliver, M. & Titi, E.S. (2001) On the domain of analyticity of solutions of second order analytic nonlinear differential equations. *J. Differential Equations* **174**, no. 1, 55–74.

Prüss, J., Saal, J., & Simonett, G. (2007) Existence of analytic solutions for the classical Stefan problem. *Math. Ann.* **338**, no. 3, 703–755.

Sammartino, M. & Caflisch, R.E. (1998) Zero viscosity limit for analytic solutions, of the Navier-Stokes equation on a half-space. I. Existence for Euler and Prandtl equations. *Comm. Math. Phys.* **192**, no. 2, 433–461.

Solonnikov, V. A. (1977) Estimates for solutions of nonstationary Navier–Stokes equations. *Journal of Mathematical Sciences* **8**, no. 4, 467–529.

11
Weak-Strong Uniqueness in Fluid Dynamics

Emil Wiedemann

Institute of Applied Mathematics, Leibniz University Hannover,
Welfengarten 1, 30167 Hannover. Germany.
`wiedemann@ifam.uni-hannover.de`

Abstract

We give a survey of recent results on weak-strong uniqueness for compressible and incompressible Euler and Navier–Stokes equations, and
also make some new observations. The importance of the weak-strong
uniqueness principle stems, on the one hand, from the instances of non-
uniqueness for the Euler equations exhibited in the past years; and on
the other hand from the question of convergence of singular limits, for
which weak-strong uniqueness represents an elegant tool.

11.1 Introduction

The Euler and Navier–Stokes equations can be viewed as the fundamental partial differential equations of the continuum mechanics of fluids.
Known for centuries, their rigorous mathematical treatment has however
remained vastly incomplete. For a system of time-dependent differential
equations supposed to describe classical mechanics, one would expect
existence and uniqueness of solutions given suitable initial (and possibly
boundary) conditions. In addition, in any real situation the data is necessarily subject to measurement errors, and so one should require small
initial perturbations to cause only small perturbations at later times.
These three properties – existence, uniqueness, and stability of solutions
– are known as the **well-posedness** of a partial differential equation. Of
course, to make sense of the notion of well-posedness, one has to make
various specifications of the problem, like the function spaces one works
in.

In the realm of fluid dynamics, one often talks about **laminar** ver-

Published as part of *Partial Differential Equations in Fluid Mechanics*, edited by
C.L. Fefferman, J.C. Robinson, & J.L. Rodrigo. © Cambridge University Press 2018.

sus **turbulent** flows, the former being characterised by a smooth, predictable dynamics of the fluid, while the latter features chaotic and irregular behaviour. We will not become more precise about this distinction, but the reader should imagine a laminar flow when we talk about a "smooth" or "strong" solution (in a class like C^1 or better), and a turbulent one when we mention "weak" solutions (which could be as bad as a generic L^2 function).

For both the Euler and the Navier–Stokes equations (compressible or incompressible), in the physically most relevant case of three space dimensions, well-posedness is known only locally in time and in function spaces of high regularity (specifically, $C^{1,\alpha}$ or Sobolev spaces that embed into $C^{1,\alpha}$). We refer the reader to the well-known textbooks in the field, e.g. Constantin & Foias (1988); Temam (2001); Majda & Bertozzi (2002); Marchioro & Pulvirenti (1994); Lions (1996, 1998); Feireisl (2004); Galdi (2000); Robinson, Rodrigo, & Sadowski (2016). Thus, a laminar flow remains laminar at least for a short time. Whether or not blowup can actually occur after a finite time is in fact one of the Clay Foundation's Millennium Problems.

On the other hand, it is known that well-posedness can fail for the three-dimensional Euler equations in spaces of lower regularity, like the Hölder spaces C^α (Bardos & Titi, 2010) or critical Sobolev spaces (Bourgain & Li, 2015). This means that a flow initially belonging to such a class may drop out of this class instantaneously.

More shockingly, at lower regularity one may in fact observe **non-uniqueness**: two distinct solutions of the same equation can arise from the same initial data. This was shown for the first time by Scheffer (1993) (see also Shnirelman, 1997), who constructed a nontrivial weak solution of the 2D incompressible Euler equations (in L^2) with compact support in time. More recently, De Lellis and Székelyhidi recovered and extended these examples in a series of groundbreaking papers (among them De Lellis & Székelyhidi, Jr, 2009, 2013). Their adaptation of Gromov's technique of **convex integration** led to various important new results, like non-uniqueness of weak solutions satisfying additional energy (in)equalities (De Lellis & Székelyhidi, Jr, 2010; Székelyhidi & Wiedemann, 2012; Daneri & Székelyhidi, 2017), global existence and non-uniqueness for arbitrary initial data in L^2 (Wiedemann, 2011), non-uniqueness of entropy solutions for the isentropic compressible Euler equations (De Lellis & Székelyhidi, Jr, 2010; Chiodaroli, 2014; Chiodaroli, De Lellis, & Kreml, 2015), and, most recently, the proof of Onsager's Conjecture on energy-dissipating Hölder continuous solutions

(Isett, 2017; Buckmaster et al., 2017). It remains open, however, whether weak solutions of the Euler equations exist for all finite-energy data.

For the Navier–Stokes equations, at least the global existence of weak solutions is known both for the incompressible case (Leray, 1934) and for the compressible case (Lions, 1998; Feireisl, Novotný, & Petzeltová, 2001). The uniqueness of weak solutions of the incompressible Navier–Stokes equations in the sense of distributions has recently been disproved by Buckmaster & Vicol (2017), but the uniqueness of Leray–Hopf solutions, which live in the appropriate energy space, remains an open problem. An interesting indication of non-uniqueness of Leray–Hopf solutions has been demonstrated recently (Guillod & Šverák, 2017). For the compressible Navier–Stokes system, the uniqueness of weak solutions is completely open.

In the light of these results, is there no hope to obtain unique weak solutions? There are two conceivable remedies: either one imposes additional admissibility criteria to single out a unique solution, or one contents oneself with a conditional form of uniqueness. The first strategy is still at a very speculative stage and does not currently promise much success. In addition, it may be objected that the concept of a deterministic solution is rather meaningless in the turbulent regime, where any uniquely determined solution may be so unstable that it would be practically useless.

The second strategy is **weak-strong uniqueness** and was established by Leray (1934), Prodi (1959), and Serrin (1963) for the incompressible Navier–Stokes equations, and by Dafermos (2010) for conservation laws. The general principle can be stated as follows:

If there exists a strong solution, then any weak solution with the same initial data coincides with it.

Note carefully that this is not merely a uniqueness result of the standard form "If two solutions of the same type share the same initial data, then they are identical", as encountered in classical well-posedness theories. Rather, the strong solution is unique even when compared to all *weak* solutions. We may thus slightly reformulate the weak-strong uniqueness principle:

Strong solutions are unique in the potentially much larger class of weak solutions.

The meaning of "strong" and "weak" will be made precise in each particular situation. In any case, it is crucial that the notion of weak solution include some sort of **energy inequality**: if this ingredient is removed, then already the Scheffer–Shnirelman construction yields a counterexample to weak-strong uniqueness. Another restriction is that for *inviscid* models we need to work on domains without boundaries, see Section 11.5 below.

For the moment, however, note that weak-strong uniqueness is only helpful in the laminar regime, as it presupposes the existence of a strong ("laminar") solution. Very roughly, one could say that the property of weak-strong uniqueness indicates that the laminar and turbulent regimes are in a sense disjoint: *either* the flow is laminar and represented by a unique strong solution, *or* it is turbulent and represented by possibly very many weak solutions.

There is another reason for the recently refreshed interest in weak-strong uniqueness: the principle can come in very handy for the convergence of singular limits, such as vanishing viscosity, hydrodynamic limits, or numerical approximation schemes. The generic question, which is very classical, can be stated as follows:

Suppose the limit system has a strong solution. Does the singular limit converge to this solution?

Suppose, for instance, one considers the incompressible Euler equations with smooth initial data, so that by classical well-posedness results there exists a smooth solution at least for a certain time period. It is arguably rather difficult to show directly that the solutions of the Navier–Stokes equations converge to this solution as the viscosity tends to zero[1]. But it may be much easier to show that the Navier–Stokes solutions converge to some weak kind of solution to the Euler equations. Then weak-strong uniqueness kicks in: indeed, the weak solution will automatically be the smooth one, and thus convergence of the viscosity limit to the smooth solution of Euler is shown.

Unfortunately, it is presently not even possible to show convergence of a zero viscosity sequence to a weak solution of the Euler equations in the sense of distributions. Instead, one has to relax even further the notion of solution. This was done by P.-L. Lions (1996), who essentially built the weak-strong uniqueness property into his definition of **dissipative solutions**.[2] It is easy to show the convergence of a vanishing viscosity

---

[1] For the incompressible Navier–Stokes-to-Euler limit without boundaries, there are several classical results on this question. However the question may be much harder for other systems of fluid mechanics.

[2] The terminology has become somewhat ambiguous, as "dissipative solution" has recently often been used to refer to a solution in the sense of distributions that

sequence to such a dissipative solution and thus to the strong solution, if the latter exists. For applications to hydrodynamic limits, see Saint-Raymond (2009) and references therein.

We prefer here to work with so-called **measure-valued solutions**, as introduced by DiPerna & Majda (1987) for the incompressible Euler equations. This notion of solution also enjoys the weak-strong uniqueness property (as shown by Brenier, De Lellis, & Székelyhidi, Jr., 2011) and can therefore be considered "equivalent" to the dissipative solutions of Lions. Weak-strong uniqueness for measure-valued solutions has recently been proved for various other equations of fluid dynamics (Gwiazda, Świerczewska-Gwiazda, & Wiedemann, 2017; Feireisl et al., 2016; Březina & Mácha, 2017; Březina & Feireisl, 2017) and applied to the viscosity limit of nonlocal Euler equations (Březina & Mácha, 2017) and a finite element/finite volume scheme for the compressible Navier–Stokes equations (Feireisl & Lukáčová-Medvid'ová, 2017). The latter result in particular shows that the weak-strong uniqueness principle can be of concrete practical use in numerical analysis (and certainly much more is to be explored in this direction).

The aim of this paper is to give a survey of the main ideas in the topic, rather than a complete overview of all available results. There are however two original contributions: in Section 11.3, we give a slight simplification and generalisation of the result of Brenier et al. (2011), adapting their arguments to the framework of Feireisl et al. (2016); and in Section 11.6 we present an alternative proof of weak-strong uniqueness for weak solutions of the incompressible Euler equations, inspired by the notion of statistical solutions in the sense of Fjordholm, Lanthaler, & Mishra (2017).

The article is organised as follows: in Section 11.2 we introduce the relative energy method, which is at the heart of all our arguments, for the simplest case of the incompressible Euler equations, and also give the calculation for the isentropic Euler system. Section 11.3 presents the measure-valued framework in the formalism of Feireisl et al. (2016) and applies this setup to the incompressible Euler equations. In Section 11.4 we discuss the incompressible and compressible Navier–Stokes equations, noting that the techniques of Sections 11.3 and 11.4 can be combined to yield weak-strong uniqueness for measure-valued solutions of the compressible Navier–Stokes equations as in Feireisl et al. (2016). Our treatment of the viscous term in the compressible case is based on

satisfies some form of the energy inequality, or to a solution that satisfies such an inequality in a strict way.

Feireisl, Jin, & Novotný (2012). For Euler models, physical boundaries
pose an additional difficulty, as weak-strong uniqueness may then fail.
We discuss a possible remedy in Section 11.5. Section 11.6 then gives an
alternative proof of weak-strong uniqueness for the incompressible Euler
equations.

11.2 The Relative Energy Method

In this section we demonstrate the relative energy method, which forms
the basis of all proofs of weak-strong uniqueness, in the simplest case of
the incompressible Euler equations, and then in the slightly more tedious
case of the isentropic compressible Euler equations, on the torus. Recall
that the incompressible Euler equations are

$$\partial_t u + \operatorname{div}(u \otimes u) + \nabla p = 0, \qquad \operatorname{div} u = 0, \tag{11.1}$$

and a vector field $u \in L^2(\mathbb{T}^d \times [0,T])$ is called a **weak solution** of this
system with initial datum u^0 if, for almost every $\tau \in (0,T)$ and every
divergence-free vector field $\varphi \in C^1(\mathbb{T}^d \times [0,T]; \mathbb{R}^d)$, we have

$$\int_0^\tau \int_{\mathbb{T}^d} \partial_t \varphi(x,t) \cdot u(x,t) + \nabla \varphi(x,t) : (u \otimes u)(x,t) \, \mathrm{d}x \, \mathrm{d}t$$

$$= \int_{\mathbb{T}^d} u(x,\tau) \cdot \varphi(x,\tau) - u^0(x) \cdot \varphi(x,0) \, \mathrm{d}x,$$

and if u is weakly divergence free in the sense that

$$\int_{\mathbb{T}^d} u(x,\tau) \cdot \nabla \psi(x) \, \mathrm{d}x = 0$$

for almost every $\tau \in (0,T)$ and every $\psi \in C^1(\mathbb{T}^d)$. Note that the pressure
disappeared from the equations in the weak formulation, as it is merely
a Lagrange multiplier. Once a weak solution u has been found, one can
find a suitable pressure by solving the Poisson equation

$$-\Delta p = \operatorname{div} \operatorname{div}(u \otimes u).$$

We will call a weak solution u **globally admissible**[3] provided that
$u \in L^\infty((0,T); L^2(\mathbb{T}^d))$ and

$$\frac{1}{2} \int_{\mathbb{T}^d} |u(x,\tau)|^2 \, \mathrm{d}x \leq \frac{1}{2} \int_{\mathbb{T}^3} |v^0(x)|^2 \, \mathrm{d}x \quad \text{for almost every } \tau \in (0,T).$$

---

[3] Global admissibility is a weaker condition than the local variant of the energy
inequality, which resembles the entropy condition in the theory of conservation
laws.

In fact one can show (see the Appendix in De Lellis & Székelyhidi, Jr, 2010) that a globally admissible weak solution can be altered on a set of times of measure zero so that it becomes continuous from τ into $L^2(\mathbb{T}^d)$ equipped with the weak topology. In particular $u(\cdot, \tau)$ is well-defined for *every* time $\tau \in [0, T]$, and the initial datum is the weak limit of $u(\cdot, \tau)$ as $\tau \searrow 0$.

We can now state and prove our prototypical weak-strong uniqueness result.

Theorem 11.1 *Let* $u \in L^\infty((0, T); L^2(\mathbb{T}^d))$ *be a weak solution and let* $U \in C^1(\mathbb{T}^d \times [0, T])$ *be a strong solution of* (11.1) *that share the same initial datum* u^0. *Assume in addition that*

$$\frac{1}{2} \int_{\mathbb{T}^d} |u(x, \tau)|^2 \, dx \leq \frac{1}{2} \int_{\mathbb{T}^d} |u^0(x)|^2 \, dx$$

for almost every $\tau \in (0, T)$. *Then* $u(x, \tau) = U(x, \tau)$ *for almost every* $(x, \tau) \in \mathbb{T}^d \times (0, T)$.

Proof We define for almost every $\tau \in (0, T)$ the **relative energy**

$$E_{\text{rel}}(\tau) := \frac{1}{2} \int_{\mathbb{T}^d} |u(x, \tau) - U(x, \tau)|^2 \, dx;$$

and compute

$$E_{\text{rel}}(\tau) = \frac{1}{2} \int_{\mathbb{T}^3} |U(x, \tau)|^2 \, dx + \frac{1}{2} \int_{\mathbb{T}^d} |u(x, \tau)|^2 \, dx - \int_{\mathbb{T}^d} u(x, \tau) \cdot U(x, \tau) \, dx$$

$$\leq \frac{1}{2} \int_{\mathbb{T}^3} |u^0(x)|^2 dx + \frac{1}{2} \int_{\mathbb{T}^d} |u^0(x)|^2 \, dx - \int_{\mathbb{T}^d} u(x, \tau) \cdot U(x, \tau) \, dx$$

$$= \int_{\mathbb{T}^3} |u^0(x)|^2 \, dx - \int_{\mathbb{T}^3} u^0(x) \cdot u^0(x) \, dx$$

$$\quad - \int_0^\tau \int_{\mathbb{T}^d} [\partial_t U(x, t) \cdot u(x, t) + \nabla U(x, t) : (u \otimes u)(x, t)] \, dx \, dt$$

$$= - \int_0^\tau \int_{\mathbb{T}^d} [\partial_t U(x, t) \cdot u(x, t) + \nabla U(x, t) : (u \otimes u)(x, t)] \, dx \, dt$$

$$= \int_0^\tau \int_{\mathbb{T}^d} [\text{div}(U \otimes U)(x, t) \cdot u(x, t) - \nabla U(x, t) : (u \otimes u)(x, t)] \, dx \, dt$$

$$= \int_0^\tau \int_{\mathbb{T}^d} [(U - u)(x, t) \cdot \nabla_{\text{sym}} U(x, t)(u - U)(x, t)] \, dx \, dt$$

$$\leq \int_0^\tau \|\nabla_{\text{sym}} U(t)\|_{L^\infty(\mathbb{T}^d)} \int_{\mathbb{T}^d} |u - U|^2(x, t) \, dx \, dt$$

$$= 2 \int_0^\tau \|\nabla_{\text{sym}} U(t)\|_{L^\infty(\mathbb{T}^d)} E_{\text{rel}}(t) \, dt,$$

and the claim follows from Grönwall's inequality. In this calculation, we used the weak energy inequality for u, the energy equality for U, the definition of u being a weak solution, the fact that U is a strong solution, and the identities

$$\operatorname{div}(U \otimes U) = \nabla UU,$$

$$\nabla U : (u \otimes u) = u \cdot \nabla Uu,$$

and

$$\int_{\mathbb{T}^d} (U - u) \cdot \nabla UU \, \mathrm{d}x = 0,$$

the first and third of these equalities making use of the divergence-free property of U and u. Note also that $(x, Ax) = (x, A_{\mathrm{sym}}x)$ for any $x \in \mathbb{R}^d$ and $A \in \mathbb{R}^{d \times d}$, so that indeed it suffices to consider the symmetric gradient.   □

The Grönwall estimate suggests that the property $\nabla_{sym}U \in L^1_t L^\infty_x$ suffices for the strong solution, and we do not need to assume $U \in C^1_{t,x}$. This is probably true (and is in fact claimed in Brenier et al., 2011) but requires some further approximation arguments that the author considers non-trivial. We will not pursue this problem here.

Let us demonstrate the usefulness of the relative energy method for a somewhat more complicated fluid model, the **isentropic compressible Euler equations** with adiabatic exponent $\gamma > 1$:

$$\partial_t(\rho u) + \operatorname{div}(\rho u \otimes u) + \nabla \rho^\gamma = 0,$$
$$\partial_t \rho + \operatorname{div}(\rho u) = 0. \tag{11.2}$$

Again we can make sense of these equations in the sense of distributions: a pair (ρ, u) is a **weak solution** of the isentropic system with initial datum (ρ^0, u^0) if $\rho \in L^\gamma(\mathbb{T}^d \times (0, T))$, $\rho|u|^2 \in L^1(\mathbb{T}^d \times (0, T))$, and

$$\int_0^\tau \int_{\mathbb{T}^d} \partial_t \psi \rho + \nabla \psi \cdot \rho u \, \mathrm{d}x \, \mathrm{d}t + \int_{\mathbb{T}^d} \psi(x, 0)\rho^0 - \psi(x, \tau)\rho(x, \tau) \, \mathrm{d}x = 0,$$

$$\int_0^\tau \int_{\mathbb{T}^d} \partial_t \varphi \cdot \rho u + \nabla \varphi : (\rho u \otimes u) + \operatorname{div} \varphi \rho^\gamma \, \mathrm{d}x \, \mathrm{d}t$$

$$+ \int_{\mathbb{T}^d} \varphi(x, 0) \cdot \rho^0 u^0 - \varphi(x, \tau) \cdot \rho u(x, \tau) \, \mathrm{d}x = 0 \tag{11.3}$$

for almost every $\tau \in (0, T)$, every $\psi \in C^1(\mathbb{T}^d \times [0, T])$, and every $\varphi \in C^1(\mathbb{T}^d \times [0, T]; \mathbb{R}^d)$. By analogy with the incompressible situation, we say the solution is **globally admissible** if for almost every $\tau \in (0, T)$

$$E(\tau) \leq E^0, \tag{11.4}$$

where the energy E is defined as

$$E(\tau) = \frac{1}{2} \int_{\mathbb{T}^d} \rho(x, \tau) |u(x, \tau)|^2 \, \mathrm{d}x + \frac{1}{\gamma - 1} \rho(x, \tau)^\gamma \, \mathrm{d}x,$$

and

$$E^0 = \frac{1}{2} \int_{\mathbb{T}^d} \rho^0(x) |u^0(x)|^2 \, \mathrm{d}x + \frac{1}{\gamma - 1} \rho^0(x)^\gamma \, \mathrm{d}x.$$

Note that the global admissibility condition is weaker than the local energy equality (the so-called **entropy condition**) usually invoked in the study of compressible Euler systems.

As an analogue of the preceding theorem, we have the following result.

Theorem 11.2 *Let $R, U \in C^1(\mathbb{T}^d \times [0, T])$ be a solution of* (11.2) *with initial data ρ^0, u^0 such that $\rho^0 \geq c > 0$ and $R \geq c > 0$. If (ρ, u) is an admissible solution with the same initial data, then $\rho \equiv R$ and $u \equiv U$ almost everywhere on $\mathbb{T}^d \times (0, T)$.*

Proof Let us first define for a.e. $\tau \in [0, T]$ the **relative energy** between (R, U) and (ρ, u) as

$$E_{\mathrm{rel}}(\tau) = \int_{\mathbb{T}^d} \frac{1}{2} \rho |u - U|^2 + \frac{1}{\gamma - 1} \rho^\gamma - \frac{\gamma}{\gamma - 1} R^{\gamma - 1} \rho + R^\gamma \, \mathrm{d}x.$$

Note that, as $\gamma > 1$, the map $|\cdot|^\gamma$ is strictly convex, which implies that the relative energy is always non-negative with equality if and only if $\rho \equiv R$ and $u \equiv U$. Thus $E_{\mathrm{rel}}(\tau) = 0$ for a.e. τ implies Theorem 11.2.

We set $\varphi = U$ in the momentum equation (the second equation of (11.3)) in order to obtain

$$\int_{\mathbb{T}^d} \rho u \cdot U(\tau) \, \mathrm{d}x = \int_{\mathbb{T}^d} \rho^0 |u^0|^2 \, \mathrm{d}x + \int_0^\tau \int_{\mathbb{T}^d} \rho u \cdot \partial_t U + (\rho u \otimes u) : \nabla U \, \mathrm{d}x \, \mathrm{d}t$$

$$+ \int_0^\tau \int_{\mathbb{T}^d} \rho^\gamma \operatorname{div} U \, \mathrm{d}x \, \mathrm{d}t. \tag{11.5}$$

Similarly, putting $\psi = \frac{1}{2} |U|^2$ and then $\psi = \gamma R^{\gamma - 1}$ in the first equation of (11.3) gives

$$\frac{1}{2} \int_{\mathbb{T}^d} |U(\tau)|^2 \rho(\tau, x) \, \mathrm{d}x = \int_0^\tau \int_{\mathbb{T}^d} U \cdot \partial_t U \rho + \nabla U U \cdot \rho u \, \mathrm{d}x \, \mathrm{d}t$$

$$+ \int_{\mathbb{T}^d} \frac{1}{2} |u^0|^2 \rho^0 \, \mathrm{d}x \tag{11.6}$$

and

$$\int_{\mathbb{T}^d} \gamma R^{\gamma-1}(\tau)\rho(\tau)\,\mathrm{d}x = \int_0^\tau \int_{\mathbb{T}^d} \gamma(\gamma-1)R^{\gamma-2}\partial_t R\rho$$

$$+ \gamma(\gamma-1)R^{\gamma-2}\nabla R \cdot \rho u\,\mathrm{d}x\,\mathrm{d}t + \int_{\mathbb{T}^d} \gamma(\rho^0)^\gamma\,\mathrm{d}x, \quad (11.7)$$

respectively. Next, write the relative energy as

$$E_{\mathrm{rel}}(\tau) = \int_{\mathbb{T}^d} \frac{1}{2}\rho|u|^2 + \frac{1}{\gamma-1}\rho^\gamma\,\mathrm{d}x + \int_{\mathbb{T}^d} R^\gamma\,\mathrm{d}x + \frac{1}{2}\int_{\mathbb{T}^d} |U|^2\rho\,\mathrm{d}x$$

$$- \int_{\mathbb{T}^d} U\cdot \rho u\,\mathrm{d}x - \int_{\mathbb{T}^d} \frac{\gamma}{\gamma-1}R^{\gamma-1}\rho\,\mathrm{d}x$$

$$= E(\tau) + \int_{\mathbb{T}^d} R^\gamma\,\mathrm{d}x + \frac{1}{2}\int_{\mathbb{T}^d} |U|^2\rho\,\mathrm{d}x - \int_{\mathbb{T}^d} U\cdot \rho u\,\mathrm{d}x$$

$$- \int_{\mathbb{T}^d} \frac{\gamma}{\gamma-1}R^{\gamma-1}\rho\,\mathrm{d}x$$

(all integrands evaluated at time τ). Using the balances (11.5), (11.6), and (11.7) for the last three integrals, we obtain

$$E_{\mathrm{rel}}(\tau) = E(\tau) + \int_{\mathbb{T}^d} R^\gamma\,\mathrm{d}x$$

$$+ \int_0^\tau \int_{\mathbb{T}^d} U\cdot \partial_t U\rho + \nabla U U \cdot \rho u\,\mathrm{d}x\,\mathrm{d}t + \int_{\mathbb{T}^d} \frac{1}{2}|u^0|^2\rho^0\,\mathrm{d}x$$

$$- \int_{\mathbb{T}^d} \rho^0|u^0|^2\,\mathrm{d}x - \int_0^\tau \int_{\mathbb{T}^d} \rho u\cdot \partial_t U + (\rho u\otimes u):\nabla U\,\mathrm{d}x\,\mathrm{d}t$$

$$- \int_0^\tau \int_{\mathbb{T}^d} \rho^\gamma \operatorname{div} U\,\mathrm{d}x\,\mathrm{d}t$$

$$- \int_0^\tau \int_{\mathbb{T}^d} (\gamma R^{\gamma-2}\partial_t R\rho + \gamma R^{\gamma-2}\nabla R\cdot \rho u)\,\mathrm{d}x\,\mathrm{d}t - \int_{\mathbb{T}^d} \frac{\gamma}{\gamma-1}(\rho^0)^\gamma\,\mathrm{d}x,$$

and using (11.4) we see, for a.e. τ,

$$E_{\mathrm{rel}}(\tau) \le - \int_{\mathbb{T}^d} (\rho^0)^\gamma\,\mathrm{d}x + \int_0^\tau \int_{\mathbb{T}^d} R^\gamma\,\mathrm{d}x$$

$$+ \int_0^\tau \int_{\mathbb{T}^d} U\cdot \partial_t U\rho + \nabla U U \cdot \rho u\,\mathrm{d}x\,\mathrm{d}t$$

$$- \int_0^\tau \int_{\mathbb{T}^d} \rho u\cdot \partial_t U + (\rho u\otimes u):\nabla U\,\mathrm{d}x\,\mathrm{d}t \qquad (11.8)$$

$$- \int_0^\tau \int_{\mathbb{T}^d} \rho^\gamma \operatorname{div} U\,\mathrm{d}x\,\mathrm{d}t$$

$$- \int_0^\tau \int_{\mathbb{T}^d} (\gamma R^{\gamma-2}\partial_t R\rho + \gamma R^{\gamma-2}\nabla R\cdot \rho u)\,\mathrm{d}x\,\mathrm{d}t.$$

Let us collect some terms and write

$$\int_{\mathbb{T}^d} R^\gamma \, dx - \int_{\mathbb{T}^d} (\rho^0)^\gamma dx - \int_0^\tau \int_{\mathbb{T}^d} \gamma R^{\gamma-2} \partial_t R \rho \, dx \, dt$$

$$= \int_0^\tau \int_{\mathbb{T}^d} \frac{d}{dt} R^\gamma - \gamma R^{\gamma-2} \partial_t R \rho \, dx \, dt$$

$$= \int_0^\tau \int_{\mathbb{T}^d} \gamma R^{\gamma-1} \partial_t R - \gamma R^{\gamma-2} \partial_t R \rho \, dx \, dt \qquad (11.9)$$

$$= \int_0^\tau \int_{\mathbb{T}^d} \gamma R^{\gamma-2} \partial_t R (R - \rho) \, dx \, dt$$

and

$$\int_0^\tau \int_{\mathbb{T}^d} U \cdot \partial_t U \rho + \nabla U U \cdot \rho u - \rho u \cdot \partial_t U - (\rho u \otimes u) : \nabla U \, dx \, dt$$

$$= \int_0^\tau \int_{\mathbb{T}^d} \partial_t U \cdot \rho(U - u) + \nabla U : (\rho u \otimes (U - u)) \, dx \, dt. \qquad (11.10)$$

Insert equalities (11.9) and (11.10) into (11.8) to arrive at

$$E_{\text{rel}}(\tau) \le \int_0^\tau \int_{\mathbb{T}^d} \gamma R^{\gamma-2} \partial_t R (R - \rho) \, dx \, dt$$

$$+ \int_0^\tau \int_{\mathbb{T}^d} \partial_t U \cdot \rho(U - u) + \nabla U : (\rho u \otimes (U - u)) \, dx \, dt$$

$$- \int_0^\tau \int_{\mathbb{T}^d} \rho^\gamma \operatorname{div} U \, dx \, dt - \int_0^\tau \int_{\mathbb{T}^d} \gamma R^{\gamma-2} \nabla R \cdot \rho u \, dx \, dt. \qquad (11.11)$$

For the last two integrals, we have, using the divergence theorem,

$$- \int_0^\tau \int_{\mathbb{T}^d} \rho^\gamma \operatorname{div} U \, dx \, dt - \int_0^\tau \int_{\mathbb{T}^d} \gamma R^{\gamma-2} \nabla R \cdot \rho u \, dx \, dt$$

$$= \int_0^\tau \int_{\mathbb{T}^d} -\rho^\gamma \operatorname{div} U + \gamma R^{\gamma-2} \nabla R \cdot (RU - \rho u)$$

$$- \gamma R^{\gamma-2} \nabla R \cdot RU \, dx \, dt$$

$$= \int_0^\tau \int_{\mathbb{T}^d} (R^\gamma - \rho^\gamma) \operatorname{div} U + \gamma R^{\gamma-2} \nabla R \cdot (RU - \rho u) \, dx \, dt.$$

Plugging this back into (11.11) and observing that, by the mass equation for (R, U),

$$\gamma R^{\gamma-2} \partial_t R (R - \rho) + \gamma R^{\gamma-2} \operatorname{div} U R (R - \rho) + \gamma R^{\gamma-2} \nabla R \cdot RU$$

$$= \gamma R^{\gamma-2} U \cdot \nabla R \rho,$$

we obtain

$$E_{\mathrm{rel}}(\tau) \leq \int_0^\tau \int_{\mathbb{T}^d} \gamma R^{\gamma-2} \cdot \nabla R \rho (U - u) \, \mathrm{d}x \, \mathrm{d}t$$

$$+ \int_0^\tau \int_{\mathbb{T}^d} \partial_t U \cdot \rho (U - u) + \nabla U : (\rho u \otimes (U - u)) \, \mathrm{d}x \, \mathrm{d}t$$

$$- \int_0^\tau \int_{\mathbb{T}^d} \gamma R^{\gamma-1} \operatorname{div} U (R - \rho) \, \mathrm{d}x \, \mathrm{d}t$$

$$+ \int_0^\tau \int_{\mathbb{T}^d} (R^\gamma - \rho^\gamma) \operatorname{div} U \, \mathrm{d}x \, \mathrm{d}t. \tag{11.12}$$

The expression in the third line is rewritten as

$$\partial_t U \cdot \rho (U - u) + \nabla U : (\rho u \otimes (U - u))$$
$$= \partial_t U \cdot \rho (U - u) + \nabla U : (\rho U \otimes (U - u)) + \nabla U : (\rho (u - U) \otimes (U - u)), \tag{11.13}$$

and the integral of the last expression as well as the last line in (11.12) can both be estimated by

$$C \|U\|_{C^1} \int_0^\tau E_{\mathrm{rel}}(t) \, \mathrm{d}t. \tag{11.14}$$

For the other terms in (11.13) we get, invoking the momentum equation for (R, U),

$$\partial_t U \cdot \rho (U - u) + \nabla U : U \otimes \rho (U - u)$$
$$= \frac{1}{R} (\partial_t (RU) + \operatorname{div}(RU \otimes U)) \cdot \rho (U - u) \tag{11.15}$$
$$= -\gamma R^{\gamma-2} \nabla R \cdot \rho (U - u).$$

Putting together (11.12), (11.14), and (11.15), we obtain

$$E_{\mathrm{rel}}(\tau) \leq C \|U\|_{C^1} \int_0^\tau E_{\mathrm{rel}}(t) \, \mathrm{d}t.$$

From Grönwall's inequality it then follows that $E_{\mathrm{rel}}(\tau) = 0$ for a.e. t. $\quad\square$

Let us summarise the strategy we used in both examples:

1. Define a **relative energy** between a weak and a strong solution that is non-negative and that vanishes identically if and only if the two solutions coincide.
2. Using the weak formulation of the equations with suitable expressions of the strong solution as test functions, rewrite the relative energy at time τ in terms of the initial data and a time integral from 0 to τ.

3. Cancel the initial terms by virtue of the energy equality for the strong solution and the admissibility criterion for the weak solution.

4. Using the pointwise form of the equations and the C^1 bounds for the strong solution, "absorb" the terms in the time integral into the relative energy.

5. Apply Grönwall's inequality to conclude that the relative energy vanishes identically.

11.3 Dissipative Measure-Valued Solutions

The problem with the results from the previous section is that they state weak-strong uniqueness only within a class for which existence is not known. Indeed, the existence of admissible weak solutions of the Euler equations for any initial data is still open (both in the compressible and in the incompressible case). For the application to singular limits as indicated in the introduction, it is therefore desirable to extend the notion of weak solution. We treat here only the incompressible system.

A conceivable approach to obtaining solutions to the Euler equations is to consider the **Navier–Stokes equations**,

$$\partial_t u + \operatorname{div}(u \otimes u) + \nabla p = \nu \Delta u,$$

$$\operatorname{div} u = 0,$$

for which the existence of weak solutions is classically known, and to let the viscosity $\nu \searrow 0$. The naïve hope is that the corresponding solutions will converge (at least weakly in the sense of distributions) to a weak solution of the Euler equations. However, this convergence might be obstructed by the appearance of **oscillations** and **concentrations**. For instance, the sequence $u_n(x) = \sin(nx)$ converges weakly to zero, but the sequence of the squares u_n^2 does *not* converge to the square of the limit (which is zero). In other words, weak convergence and nonlinearities do not, in general, commute. Let us mention that it is not known whether oscillations or concentrations can actually form in the viscosity sequence: the question is considered extremely difficult.

One way to handle such issues is to consider **measure-valued solutions**. We present here the formalism of Feireisl et al. (2016), which in turn is inspired by Demoulini, Stuart, & Tzavaras (2012). Our framework is slightly simpler and more general than the ones commonly used (DiPerna & Majda, 1987; Alibert & Bouchitté, 1997) and arguably represents the most general concept of solution still having the weak-strong uniqueness property (cf. the discussion in Feireisl et al., 2016). Another advantage is that this framework allows us to deal with viscosity terms.

Let u_ν denote a weak (Leray) solution[4] for the Navier–Stokes equations with viscosity ν. If we identify the vector field u_ν with the family

$$\{\delta_{u_\nu}(x,t)\}_{(x,t)\in\mathbb{T}^d\times(0,T)}$$

of (Dirac) probability measures, we can pass to the weak-$*$ limit in the space of parametrised probability measures along a subsequence. This limit is then compatible with nonlinearities. The facilitated convergence argument is thus "traded" for the probabilistic relaxation: indeed, the limit measure may no longer be Dirac, so that the exact value of the velocity field is ignored and only its probability distribution is described.

To make this discussion more rigorous, we define a *Young measure* to be a family

$$\{\nu_{x,t}\}_{(x,t)\in\mathbb{T}^d\times(0,T)}$$

of probability measures on $\mathbb{R}^d$ that is **weakly-$*$ measurable**, i.e. the map

$$(x,t)\mapsto\int_{\mathbb{R}^d}f(z)\,\mathrm{d}\nu_{x,t}(z)$$

is Borel measurable for any $f\in C_b(\mathbb{R}^d)$. Then the basic weak convergence statement (which follows basically from the Banach–Alaoglu Theorem) is as follows.

Theorem 11.3 (Fundamental Theorem of Young Measures) *Let $\{u_n\}$ be an L^2-bounded sequence of maps $\mathbb{T}^d\times(0,T)\to\mathbb{R}^d$. Then there is a subsequence (still denoted $\{u_n\}$) that generates a Young measure ν, in the sense that*

$$f\circ u_n \rightharpoonup \langle\nu,f\rangle := \int_{\mathbb{R}^d}f(z)\,\mathrm{d}\nu(z)$$

for any $f\in C(\mathbb{R}^d)$ for which $\{f\circ u_n\}_{n\in\mathbb{N}}$ is equiintegrable. Moreover,

$$\int_0^T\int_{\mathbb{T}^d}\langle\nu_{x,t},|\cdot|^2\rangle\,\mathrm{d}x\,\mathrm{d}t<\infty. \tag{11.16}$$

Applying this to our vanishing viscosity sequence $\{u_n\}$ with associated viscosities $\nu(n)\searrow 0$, we see that the uniform L^2 bound on the u_n implies in particular that $\{u_n\}$ itself is equiintegrable, so that $f(z)=z$ is an admissible test function in the Fundamental Theorem. Therefore, we

---

[4] See the next section for a definition.

can easily pass to the limit in the linear terms: indeed, for any $\varphi \in C^1(\mathbb{T}^d \times [0, T]; \mathbb{R}^d)$,

$$\int_0^T \int_{\mathbb{T}^d} u_n(x, t) \cdot \partial_t \varphi(x, t) \, \mathrm{d}x \, \mathrm{d}t \to \int_0^T \int_{\mathbb{T}^d} \langle \nu_{x,t}, \mathrm{id} \rangle \cdot \partial_t \varphi(x, t) \, \mathrm{d}x \, \mathrm{d}t,$$

$$\nu(n) \int_0^T \int_{\mathbb{T}^d} u_n(x, t) \cdot \Delta \varphi(x, t) \, \mathrm{d}x \, \mathrm{d}t \to 0,$$

and, for any $\psi \in C^1(\mathbb{T}^d \times [0, T])$,

$$\int_0^T \int_{\mathbb{T}^d} u_n(x, t) \cdot \nabla \psi(x, t) \, \mathrm{d}x \, \mathrm{d}t \to \int_0^T \int_{\mathbb{T}^d} \langle \nu_{x,t}, \mathrm{id} \rangle \cdot \nabla \psi(x, t) \, \mathrm{d}x \, \mathrm{d}t.$$

The nonlinear term, however, may still lack compactness due to possible concentrations: for the sequence $\{u_n \otimes u_n\}$ we merely have an $L^\infty((0, T); L^1(\mathbb{T}^d))$ bound, which does not guarantee equiintegrability. Hence the difference

$$u_n \otimes u_n - \langle \nu, \mathrm{id} \otimes \mathrm{id} \rangle$$

may not converge to zero. It will, however, converge weakly-$*$ to a bounded measure m on $\mathbb{T}^d \times [0, T]$ by virtue of the $L^\infty((0, T); L^1(\mathbb{T}^d))$ bound and (11.16). By standard measure theory arguments, it is also not difficult to show that the $L^\infty((0, T); L^1(\mathbb{T}^d))$ bound implies the validity of the disintegration

$$m(\mathrm{d}x \, \mathrm{d}t) = m_t(\mathrm{d}x) \otimes \mathrm{d}t \qquad (11.17)$$

for some family $\{m_t\}_{t \in (0,T)}$ of uniformly bounded measures on $\mathbb{R}^d$. We thus obtain for any divergence-free $\varphi \in C^1(\mathbb{T}^d \times [0, T]; \mathbb{R}^d)$ and almost every $\tau \in (0, T)$ the equation

$$\int_0^\tau \int_{\mathbb{T}^d} \partial_t \varphi(x, t) \cdot \langle \nu_{x,t}, \mathrm{id} \rangle + \nabla \varphi(x, t) : (\langle \nu_{x,t}, \mathrm{id} \otimes \mathrm{id} \rangle + m_t) \, \mathrm{d}x \, \mathrm{d}t$$

$$= \int_{\mathbb{T}^d} \langle \nu_{x,\tau}, \mathrm{id} \rangle \cdot \varphi(x, \tau) - u^0(x) \cdot \varphi(x, 0) \, \mathrm{d}x,$$

$$(11.18)$$

provided the initial data u^0 was kept fixed along the viscosity sequence. Moreover, for any $\psi \in C^1(\mathbb{T}^d)$ we have the divergence condition

$$\int_{\mathbb{T}^d} \langle \nu_{x,t}, \mathrm{id} \rangle \cdot \nabla \psi(x) \, \mathrm{d}x = 0 \qquad (11.19)$$

for almost every $t \in (0, T)$.

As a final ingredient in our definition of dissipative measure-valued solutions, we need to impose an analogue of the energy admissibility condition. To this end, define for almost every $\tau \in (0, T)$ the **dissipation defect**

$$D(\tau) = \frac{1}{2} \operatorname{trace} m_\tau(\mathbb{T}^d)$$

and the **measure-valued energy** as

$$E(\tau) = \frac{1}{2} \int_{\mathbb{T}^d} \langle \nu_{x,\tau}, |\cdot|^2 \rangle \, dx + D(\tau).$$

First, since m_τ is uniformly bounded (in the sense of measures) in τ, we find $D \in L^\infty(0,T)$. Second, $D \geq 0$, because for any non-negative $\chi \in C([0,T])$ and any $M > 0$,

$$2 \int_0^T \chi(\tau) D(\tau) \, d\tau = \operatorname{trace} \int_0^T \int_{\mathbb{T}^d} \chi(\tau) \, dm(x, \tau)$$

$$= \lim_{n \to \infty} \int_0^T \int_{T^d} \chi(\tau) \operatorname{trace}[u_n \otimes u_n - \langle \nu, \mathrm{id} \otimes \mathrm{id} \rangle] \, dx \, d\tau$$

$$= \lim_{n \to \infty} \int_0^T \int_{T^d} \chi(\tau)(|u_n|^2 - \langle \nu, |\cdot|^2 \rangle) \, dx \, d\tau$$

$$= \lim_{n \to \infty} \left\{ \int_0^T \int_{T^d} \chi(\tau)(|u_n|^2 \wedge M - \langle \nu, |\cdot|^2 \wedge M \rangle) \, dx \, d\tau \right.$$

$$\left. + \int_0^T \int_{T^d} \chi(\tau)((|u_n|^2 - M) \vee 0 - \langle \nu, (|\cdot|^2 - M) \vee 0 \rangle) \, dx \, d\tau \right\}$$

$$\geq - \int_0^T \int_{T^d} \chi(\tau) \langle \nu, (|\cdot|^2 - M) \vee 0 \rangle \, dx \, d\tau.$$

For the last inequality, we used the Fundamental Theorem for the test function $|\cdot|^2 \wedge M \in C_b(\mathbb{R}^d)$. But the remaining term becomes arbitrarily close to zero for sufficiently large M by the Dominated Convergence Theorem (since $(|\cdot|^2 - M) \vee 0 \leq |\cdot|^2$), whence the claim $D \geq 0$ follows.

Third, for our viscosity sequence $\{u_n\}$ we have the energy bound

$$\frac{1}{2} \int_{\mathbb{T}^d} |u_n(x,\tau)|^2 \, dx \leq \frac{1}{2} \int_{\mathbb{T}^d} |u^0|^2 \, dx$$

for almost every $\tau \in (0,T)$, and passing to the measure-valued limit thus yields

$$E(\tau) \leq \frac{1}{2} \int_{\mathbb{T}^d} |u^0|^2 \, dx \tag{11.20}$$

for almost every $\tau \in (0,T)$.

We condense our discussion into the following definition.

Definition 11.4 Let ν be a Young measure, m a matrix-valued measure on $\mathbb{T}^d \times [0,T]$ satisfying (11.17), and $D \in L^\infty(0,T)$ with $D \geq 0$ such that $|m_t|(\mathbb{T}^d) \leq CD(t)$ for some constant C and almost every $t \in [0,T]$.

The triple (ν, m, D) is called a *dissipative measure-valued solution* of the Euler equations with initial datum u^0 if it satisfies (11.18), (11.19), and (11.20).

Note carefully that our definition does not require the measure-valued solution to arise from a viscosity limit. Indeed, if u^0 is kept fixed along the viscosity limit, then there exist dissipative measure-valued solutions (in fact even admissible weak solutions) that cannot be recovered as a vanishing viscosity limit.

From the previous discussion it is clear that there exists a dissipative measure-valued solution for any $u^0 \in L^2(\mathbb{T}^d)$ (just take a viscosity sequence). But we also have weak-strong uniqueness.

Theorem 11.5 *Let (ν, m, D) be a dissipative measure-valued solution and $U \in C^1(\mathbb{T}^d \times [0,T])$ a strong solution of (11.1), both with initial datum u^0. Then $\nu_{x,\tau} = \delta_{U(x,\tau)}$ for almost every $(x,\tau) \in \mathbb{T}^d \times (0,T)$, and $m = 0$, $D = 0$.*

Proof The proof is achieved essentially by rewriting the proof of Theorem 11.1 in measure-valued notation. Specifically, we estimate for almost every $\tau \in (0,T)$ the **relative energy** E_{rel} as

$$
E_{\text{rel}}(\tau) := \frac{1}{2} \int_{\mathbb{T}^d} \langle \nu_{x,\tau}, |\operatorname{id} - U(x,\tau)|^2 \rangle \, dx + D(\tau)
$$

$$
= \frac{1}{2} \int_{\mathbb{T}^3} |U(x,\tau)|^2 \, dx + \frac{1}{2} \int_{\mathbb{T}^d} \langle \nu_{x,\tau}, |\cdot|^2 \rangle \, dx + D(\tau)
$$

$$
- \int_{\mathbb{T}^d} \langle \nu_{x,\tau}, \operatorname{id} \rangle \cdot U(x,\tau) \, dx
$$

$$
\leq \frac{1}{2} \int_{\mathbb{T}^3} |u^0(x)|^2 \, dx + \frac{1}{2} \int_{\mathbb{T}^d} |u^0(x)|^2 \, dx - \int_{\mathbb{T}^d} \langle \nu_{x,\tau}, \operatorname{id} \rangle \cdot U(x,\tau) \, dx
$$

$$
= \int_{\mathbb{T}^3} |u^0(x)|^2 \, dx - \int_{\mathbb{T}^3} u^0(x) \cdot u^0(x) \, dx
$$

$$
- \int_0^\tau \int_{\mathbb{T}^d} [\partial_t U(x,t) \cdot \langle \nu_{x,t}, \operatorname{id} \rangle + \nabla U(x,t) : \langle \nu_{x,t}, \operatorname{id} \otimes \operatorname{id} \rangle] \, dx \, dt
$$

$$
- \int_0^\tau \int_{\mathbb{T}^d} \nabla U(x,t) \, dm(x,t)
$$

$$= -\int_0^\tau \int_{\mathbb{T}^d} [\partial_t U(x,t) \cdot \langle \nu_{x,t}, \mathrm{id} \rangle + \nabla U(x,t) : \langle \nu_{x,t}, \mathrm{id} \otimes \mathrm{id} \rangle] \, \mathrm{d}x \, \mathrm{d}t$$

$$- \int_0^\tau \int_{\mathbb{T}^d} \nabla_{sym} U(x,t) \, \mathrm{d}m(x,t)$$

$$= \int_0^\tau \int_{\mathbb{T}^d} [\mathrm{div}(U \otimes U)(x,t) \cdot \langle \nu_{x,t}, \mathrm{id} \rangle - \nabla U(x,t) \langle \nu_{x,t}, \mathrm{id} \otimes \mathrm{id} \rangle] \, \mathrm{d}x \, \mathrm{d}t$$

$$- \int_0^\tau \int_{\mathbb{T}^d} \nabla_{sym} U(x,t) \, \mathrm{d}m(x,t)$$

$$= \int_0^\tau \int_{\mathbb{T}^d} \langle \nu_{x,t}, (U(x,t) - \mathrm{id}) \cdot \nabla_{\mathrm{sym}} U(x,t) (\mathrm{id} - U(x,t)) \rangle \, \mathrm{d}x \, \mathrm{d}t$$

$$- \int_0^\tau \int_{\mathbb{T}^d} \nabla_{\mathrm{sym}} U(x,t) \, \mathrm{d}m(x,t)$$

$$\leq \int_0^\tau \| \nabla_{\mathrm{sym}} U(t) \|_{L^\infty(\mathbb{T}^d)} \left[\int_{\mathbb{T}^d} \langle \nu_{x,t}, |\mathrm{id} - U(x,t)|^2 \rangle \, \mathrm{d}x + |m_t|(\mathbb{T}^d) \right] \, \mathrm{d}t$$

$$\leq C \int_0^\tau \| \nabla_{\mathrm{sym}} U(t) \|_{L^\infty(\mathbb{T}^d)} E_{\mathrm{rel}}(t) \, \mathrm{d}t,$$

and from Grönwall's inequality it follows that $E_{\mathrm{rel}}(\tau) = 0$ almost everywhere. But since both terms in the relative energy are non-negative, this implies indeed $\nu_{x,\tau} = \delta_{U(x,\tau)}$ and $D(\tau) = 0$ almost everywhere, and $m = 0$ follows from $|m_\tau|(\mathbb{T}^d) \leq CD(\tau)$ for almost every τ. $\square$

11.4 Dealing with Viscosity

11.4.1 Incompressible Navier–Stokes Equations

For the incompressible Navier–Stokes equations

$$\partial_t u + \mathrm{div}(u \otimes u) + \nabla p = \nu \Delta u,$$
$$\mathrm{div}\, u = 0, \tag{11.21}$$

the improved regularity thanks to the diffusion term enables us to show a much better weak-strong uniqueness result than the one for the Euler equations (Theorem 11.1). "Better" means that the "strong" solution is allowed to be much less regular that Lipschitz. In our discussion of the incompressible Navier–Stokes equations, we always assume $d = 2$ or $d = 3$, as Sobolev embeddings come into play which are no longer valid in higher dimensions.

For the Navier–Stokes equations, we can afford to work in the more realistic setting of bounded domains. Indeed, the no-slip boundary conditions

$$u = 0 \quad \text{on } \partial\Omega$$

that are most commonly imposed on the equations do not conflict with weak-strong uniqueness. The situation for the Euler equations is vastly different, see Section 11.5 below. In what follows, let $\Omega \subset \mathbb{R}^d$ be a bounded Lipschitz domain.

Before we state the result, let us discuss some preliminaries.

Definition 11.6 (Leray, 1934) A vector field

$$u \in L^\infty((0,T); L^2(\Omega)) \cap L^2((0,T); H_0^1(\Omega))$$

is called a *Leray solution* of (11.21) with initial datum u^0 if

$$\int_0^\tau \int_\Omega \partial_t\varphi(x,t) \cdot u(x,t) - \varphi(x,t) \cdot (u(x,t) \cdot \nabla u(x,t))$$

$$- \nu\nabla\varphi(x,t) : \nabla u(x,t)\,\mathrm{d}x\,\mathrm{d}t$$

$$= \int_\Omega u(x,\tau) \cdot \varphi(x,\tau) - u^0(x) \cdot \varphi(x,0)\,\mathrm{d}x$$

for almost every $\tau \in (0,T)$ and for every $\varphi \in C_c^1(\Omega \times [0,T]; \mathbb{R}^d)$ with $\operatorname{div}\varphi = 0$,

$$\int_\Omega u(x,\tau) \cdot \nabla\psi(x)dx = 0$$

for almost every $\tau \in (0,T)$ and every $\psi \in C^1(\Omega)$, and

$$\frac{1}{2}\int_\Omega |u(x,\tau)|^2\,\mathrm{d}x + \nu\int_0^\tau\int_\Omega |\nabla u(x,t)|^2\,\mathrm{d}x\,\mathrm{d}t \le \frac{1}{2}\int_\Omega |u^0(x)|^2\,\mathrm{d}x$$

for almost every $\tau \in (0,T)$.

We will use the following inequality, which can easily be verified via Hölder and Sobolev inequalities: let r, s be such that

$$\frac{d}{s} + \frac{2}{r} = 1, \quad d < s < \infty,$$

and let $u, U \in L^\infty((0,T); L^2(\Omega)) \cap L^2((0,T); H_0^1(\Omega))$ be divergence free. Assume in addition $U \in L^r((0,T); L^s(\Omega))$. Then, for every $\tau \in (0,T)$,

$$\left| \int_0^\tau \int_\Omega \nabla(u - U) : ((u - U) \otimes U)\,\mathrm{d}x\,\mathrm{d}t \right|$$

$$\le C \left(\int_0^\tau \int_\Omega |\nabla(u - U)|^2\,\mathrm{d}x\,\mathrm{d}t \right)^{1-1/r} \left(\int_0^\tau \|U\|_{L_x^s}^r \int_\Omega |u - U|^2\,\mathrm{d}x\,\mathrm{d}t \right)^{1/r}.$$

The classical weak-strong uniqueness theorem of Prodi (1959) and Serrin (1963) reads as follows.

Theorem 11.7 *Let $d \in \{2,3\}$, r,s as above and u, U be two Leray solutions with the same initial datum u^0. If $U \in L^r((0,T); L^s(\mathbb{T}^d))$ then $u = U$ for almost every $(x,\tau) \in \mathbb{T}^d \times (0,T)$.*

Proof For this sketch proof take u and v smooth; the general case is handled by approximation Galdi, 2000, Chapter 4). For a.e. $\tau \in (0,T)$

$$\frac{1}{2} \int_\Omega |u - U|(\tau)^2 \, \mathrm{d}x + \nu \int_0^\tau \int_\Omega |\nabla(u - U)|^2 \, \mathrm{d}x \, \mathrm{d}t$$

$$= \frac{1}{2} \int_\Omega |u(\tau)|^2 \, \mathrm{d}x + \frac{1}{2} \int_\Omega |U(\tau)|^2 \, \mathrm{d}x + \nu \int_0^\tau \int_\Omega |\nabla u|^2 \, \mathrm{d}x \, \mathrm{d}t$$

$$+ \nu \int_0^\tau \int_\Omega |\nabla U|^2 \, \mathrm{d}x \, \mathrm{d}t - \int_\Omega u \cdot U(\tau) \, \mathrm{d}x - 2\nu \int_0^\tau \int_\Omega \nabla u : \nabla U \, \mathrm{d}x \, \mathrm{d}t$$

$$\leq \int_\Omega |u^0|^2 \, \mathrm{d}x - \int_\Omega u \cdot U(\tau) \, \mathrm{d}x - 2\nu \int_0^\tau \int_\Omega \nabla u : \nabla U \, \mathrm{d}x \, \mathrm{d}t$$

$$= \int_0^\tau \int_\Omega -\partial_t U \cdot u + U \cdot (u \cdot \nabla)u + \nu \nabla U : \nabla u \, \mathrm{d}x \, \mathrm{d}t$$

$$- 2\nu \int_0^\tau \int_\Omega \nabla u : \nabla U \, \mathrm{d}x \, \mathrm{d}t$$

$$= \int_0^\tau \int_\Omega -\partial_t U \cdot u + U \cdot (u \cdot \nabla)u - \nu \nabla U : \nabla u \, \mathrm{d}x \, \mathrm{d}t$$

$$= \int_0^\tau \int_\Omega U \cdot \partial_t u + U \cdot (u \cdot \nabla)u - \nu \nabla U : \nabla u \, \mathrm{d}x \, \mathrm{d}t + \int_\Omega |u^0|^2 \, \mathrm{d}x$$

$$- \int_\Omega u \cdot U(\tau) \, \mathrm{d}x$$

$$= \int_0^\tau \int_\Omega -\nabla u : (U \otimes U) + U \cdot (u \cdot \nabla)u \, \mathrm{d}x \, \mathrm{d}t$$

$$= \int_0^\tau \int_\Omega \nabla u : ((u - U) \otimes U) \, \mathrm{d}x \, \mathrm{d}t$$

$$= \int_0^\tau \int_\Omega \nabla(u - U) : ((u - U) \otimes U) \, \mathrm{d}x \, \mathrm{d}t.$$

On the previous page, this is bounded by

$$C \left(\int_0^\tau \int_\Omega |\nabla(u - U)|^2 \, \mathrm{d}x \, \mathrm{d}t \right)^{1-1/r} \left(\int_0^\tau \|U\|_{L_x^s}^r \int_\Omega |u - U|^2 \, \mathrm{d}x \, \mathrm{d}t \right)^{1/r}$$

$$\leq \nu \int_0^\tau \int_\Omega |\nabla(u - U)|^2 \, \mathrm{d}x \, \mathrm{d}t + C(\nu) \int_0^\tau \|U\|_{L_x^s}^r \int_\Omega |u - U|^2 \, \mathrm{d}x \, \mathrm{d}t,$$

where in the last step we used Young's inequality

$$ab \leq \frac{\delta a^p}{p} + \frac{b^q}{\delta^{q/p} q}$$

for any $\delta > 0$, $1/p + 1/q = 1$, and $a, b \geq 0$.

Since the terms involving the factor ν vanish on both sides, we can use Grönwall's inequality to conclude. □

Remark The statement is also true in the endpoint cases $s \in \{d, \infty\}$, with slightly modified proofs. See again Galdi (2000), Chapter 4, for details.

It is worthwhile noting the fundamental differences between this proof and the one for the incompressible Euler equations: for the Navier–Stokes equations, we use both u and U as the test field in the definition of weak solution of the respective other function (cf. lines 3 and 7 in the above estimate); and, most importantly, the compensation of the (insufficient) exponent $1/r$ by the viscosity term in the last step relies crucially on $\nu > 0$ (indeed, the constant $C(\nu)$ blows up as $\nu \searrow 0$).

11.4.2 Compressible Navier–Stokes Equations

In the case of the incompressible Navier–Stokes equations, the Laplacian helped us achieve a much better weak-strong uniqueness result than for the inviscid (Euler) system. The situation is much different for the **isentropic compressible Navier–Stokes equations**, at least as far as we currently know. Here, the viscosity term presents an additional difficulty that needs to be overcome to show weak-strong uniqueness at the same regularity as for the compressible Euler system. It is an interesting open question whether the result can be improved by a more clever treatment of the viscous term.

The equations read

$$\partial_t(\rho u) + \operatorname{div}(\rho u \otimes u) + \nabla \rho^\gamma = \operatorname{div} \mathbb{S}(\nabla u),$$
$$\partial_t \rho + \operatorname{div}(\rho u) = 0, \tag{11.22}$$

where

$$\mathbb{S}(\nabla u) := \mu \left(\nabla u + \nabla^t u - \frac{2}{3}(\operatorname{div} u)I \right) + \eta(\operatorname{div} u)I$$

is the **Newtonian viscous stress** for some given parameters $\mu > 0$ and $\eta \geq 0$.

Similarly as before, a pair (ρ, u) is a **weak solution** of the isentropic

Navier–Stokes system with initial datum (ρ^0, u^0) if $\rho \in L^\gamma(\mathbb{T}^d \times (0,T))$, $\rho|u|^2 \in L^1(\mathbb{T}^d \times (0,T))$, $u \in L^2(0,T; H_0^1(\mathbb{T}^d))$, and

$$\int_0^\tau \int_\Omega \partial_t \psi \rho + \nabla \psi \cdot \rho u \, dx \, dt + \int_\Omega \psi(x,0)\rho^0 - \psi(x,\tau)\rho(x,\tau) \, dx = 0,$$

$$\int_0^\tau \int_\Omega \partial_t \varphi \cdot \rho u + \nabla \varphi : (\rho u \otimes u) + \operatorname{div} \varphi \rho^\gamma - \mathbb{S}(\nabla u) : \nabla \varphi \, dx \, dt$$

$$+ \int_\Omega \varphi(x,0) \cdot \rho^0 u^0 - \varphi(x,\tau) \cdot \rho u(x,\tau) \, dx = 0$$

$$\text{(11.23)}$$

for almost every $\tau \in (0,T)$, every $\psi \in C^1(\Omega \times [0,T])$, and every $\varphi \in C_c^1(\Omega \times [0,T]; \mathbb{R}^d)$. Again, we can impose a weak energy inequality: a solution is **globally admissible** if for almost every $\tau \in (0,T)$

$$E(\tau) + \int_0^\tau \int_\Omega \mathbb{S}(\nabla u) : \nabla u \, dx \, dt \leq E^0, \qquad \text{(11.24)}$$

where the energy E is defined as in the inviscid situation, i.e.

$$E(\tau) = \frac{1}{2} \int_\Omega \rho(x,\tau)|u(x,\tau)|^2 + \frac{1}{\gamma-1}\rho(x,\tau)^\gamma \, dx,$$

and

$$E^0 = \frac{1}{2} \int_\Omega \rho^0(x)|u^0(x)|^2 + \frac{1}{\gamma-1}\rho^0(x)^\gamma \, dx.$$

Theorem 11.8 *Let (ρ, u) be a globally admissible weak solution of (11.22) and (R, U) a strong solution such that*

$$R \in C^1(\Omega \times [0,T]), \quad R > 0, \quad U \in C^1([0,T]; C^2(\Omega)), \quad U \cdot n = 0 \text{ on } \partial\Omega.$$

If both solutions share the same initial data, then they coincide for almost every $(x,t) \in \Omega \times (0,T)$.

Proof Large parts of the proof are very similar to the case of the isentropic Euler equations, see Section 11.2. We give the full argument nevertheless for the reader's convenience.

Again we define for a.e. $\tau \in [0,T]$ the relative energy between (R,U) and (ρ, u) as

$$E_{\mathrm{rel}}(\tau) = \int_\Omega \frac{1}{2}\rho|u - U|^2 + \frac{1}{\gamma-1}\rho^\gamma - \frac{\gamma}{\gamma-1}R^{\gamma-1}\rho + R^\gamma \, dx.$$

Note once more that, since $\gamma > 1$, it holds that $E_{\mathrm{rel}}(\tau) = 0$ for a.e. τ implies Theorem 11.8.

Setting $\varphi = U$ in the momentum equation gives

$$\int_\Omega \rho u \cdot U(\tau) \, \mathrm{d}x = \int_\Omega \rho^0 |u^0|^2 \, \mathrm{d}x$$

$$+ \int_0^\tau \int_\Omega \rho u \cdot \partial_t U + (\rho u \otimes u) : \nabla U \, \mathrm{d}x \, \mathrm{d}t$$

$$+ \int_0^\tau \int_\Omega \rho^\gamma \operatorname{div} U \, \mathrm{d}x \, \mathrm{d}t - \int_0^\tau \int_\Omega \mathbb{S}(\nabla u) : \nabla U \, \mathrm{d}x \, \mathrm{d}t.$$

$$(11.25)$$

Then, setting $\psi = \frac{1}{2}|U|^2$ and $\psi = \gamma R^{\gamma-1}$ in the first equation of (11.3) yields

$$\frac{1}{2} \int_\Omega |U(\tau)|^2 \rho(\tau, x) \, \mathrm{d}x$$

$$= \int_0^\tau \int_\Omega U \cdot \partial_t U \rho + \nabla U U \cdot \rho u \, \mathrm{d}x \, \mathrm{d}t + \int_\Omega \frac{1}{2} |u^0|^2 \rho^0 \, \mathrm{d}x$$

$$(11.26)$$

and

$$\int_\Omega \gamma R^{\gamma-1}(\tau) \rho(\tau) \, \mathrm{d}x$$

$$= \int_0^\tau \int_\Omega \gamma(\gamma-1) R^{\gamma-2} \partial_t R \rho + \gamma(\gamma-1) R^{\gamma-2} \nabla R \cdot \rho u \, \mathrm{d}x \, \mathrm{d}t$$

$$+ \int_\Omega \gamma(\rho^0)^\gamma \, \mathrm{d}x, \qquad (11.27)$$

respectively.

We write the relative energy as

$$E_{\mathrm{rel}}(\tau) = \int_\Omega \frac{1}{2} \rho |u|^2 + \frac{1}{\gamma-1} \rho^\gamma \, \mathrm{d}x + \int_\Omega R^\gamma \, \mathrm{d}x + \frac{1}{2} \int_\Omega |U|^2 \rho \, \mathrm{d}x$$

$$- \int_\Omega U \cdot \rho u \, \mathrm{d}x - \int_\Omega \frac{\gamma}{\gamma-1} R^{\gamma-1} \rho \, \mathrm{d}x$$

$$= E(\tau) + \int_\Omega R^\gamma \, \mathrm{d}x + \frac{1}{2} \int_\Omega |U|^2 \rho \, \mathrm{d}x - \int_\Omega U \cdot \rho u \, \mathrm{d}x$$

$$- \int_\Omega \frac{\gamma}{\gamma-1} R^{\gamma-1} \rho \, \mathrm{d}x.$$

Next, using the balances (11.25), (11.26), (11.27) for the last three integrals, we have

$$
\begin{aligned}
E_{\mathrm{rel}}(\tau) = {}& E(\tau) + \int_\Omega R^\gamma \, \mathrm{d}x \\
&+ \int_0^\tau \int_\Omega U \cdot \partial_t U \rho + \nabla U U \cdot \rho u \, \mathrm{d}x \, \mathrm{d}t + \int_\Omega \frac{1}{2}|u^0|^2 \rho^0 \, \mathrm{d}x \\
&- \int_\Omega \rho^0 |u^0|^2 \, \mathrm{d}x - \int_0^\tau \int_\Omega \rho u \cdot \partial_t U + (\rho u \otimes u) : \nabla U \, \mathrm{d}x \, \mathrm{d}t \\
&- \int_0^\tau \int_\Omega \rho^\gamma \operatorname{div} U \, \mathrm{d}x \, \mathrm{d}t + \int_0^\tau \int_\Omega \mathbb{S}(\nabla u) : \nabla U \, \mathrm{d}x \, \mathrm{d}t \\
&- \int_0^\tau \int_\Omega (\gamma R^{\gamma-2} \partial_t R \rho + \gamma R^{\gamma-2} \nabla R \cdot \rho u) \, \mathrm{d}x \, \mathrm{d}t \\
&- \int_\Omega \frac{\gamma}{\gamma-1} (\rho^0)^\gamma \, \mathrm{d}x,
\end{aligned}
$$

and using (11.24) we have, for a.e. τ,

$$
\begin{aligned}
E_{\mathrm{rel}}(\tau) + & \int_0^\tau \int_\Omega \mathbb{S}(\nabla u) : (\nabla u - \nabla U) \, \mathrm{d}x \, \mathrm{d}t \\
\leq {}& -\int_\Omega (\rho^0)^\gamma \, \mathrm{d}x + \int_0^\tau \int_\Omega R^\gamma \, \mathrm{d}x \\
&+ \int_0^\tau \int_\Omega U \cdot \partial_t U \rho + \nabla U U \cdot \rho u \, \mathrm{d}x \, \mathrm{d}t \\
&- \int_0^\tau \int_\Omega \rho u \cdot \partial_t U + (\rho u \otimes u) : \nabla U \, \mathrm{d}x \, \mathrm{d}t \\
&- \int_0^\tau \int_\Omega \rho^\gamma \operatorname{div} U \, \mathrm{d}x \, \mathrm{d}t \\
&- \int_0^\tau \int_\Omega (\gamma R^{\gamma-2} \partial_t R \rho + \gamma R^{\gamma-2} \nabla R \cdot \rho u) \, \mathrm{d}x \, \mathrm{d}t.
\end{aligned}
\tag{11.28}
$$

Next, we collect some terms:

$$
\begin{aligned}
& \int_\Omega R^\gamma \, \mathrm{d}x - \int_\Omega (\rho^0)^\gamma \, \mathrm{d}x - \int_0^\tau \int_\Omega \gamma R^{\gamma-2} \partial_t R \rho \, \mathrm{d}x \, \mathrm{d}t \\
&= \int_0^\tau \int_\Omega \frac{\mathrm{d}}{\mathrm{d}t} R^\gamma - \gamma R^{\gamma-2} \partial_t R \rho \, \mathrm{d}x \, \mathrm{d}t \\
&= \int_0^\tau \int_\Omega \gamma R^{\gamma-1} \partial_t R - \gamma R^{\gamma-2} \partial_t R \rho \, \mathrm{d}x \, \mathrm{d}t \\
&= \int_0^\tau \int_\Omega \gamma R^{\gamma-2} \partial_t R (R - \rho) \, \mathrm{d}x \, \mathrm{d}t
\end{aligned}
\tag{11.29}
$$

and

$$\int_0^\tau \int_\Omega U \cdot \partial_t U \rho + \nabla U U \cdot \rho u - \rho u \cdot \partial_t U - (\rho u \otimes u) : \nabla U \, dx \, dt$$

$$= \int_0^\tau \int_\Omega \partial_t U \cdot \rho (U - u) + \nabla U : (\rho u \otimes (U - u)) \, dx \, dt. \tag{11.30}$$

Plugging equalities (11.29) and (11.30) into (11.28), we obtain

$$E_{\mathrm{rel}}(\tau) + \int_0^\tau \int_\Omega \mathbb{S}(\nabla u) : (\nabla u - \nabla U) \, dx \, dt$$

$$\leq \int_0^\tau \int_\Omega \gamma R^{\gamma-2} \partial_t R(R - \rho) \, dx \, dt$$

$$+ \int_0^\tau \int_\Omega \partial_t U \cdot \rho(U - u) + \nabla U : (\rho u \otimes (U - u)) \, dx \, dt \tag{11.31}$$

$$- \int_0^\tau \int_\Omega \rho^\gamma \operatorname{div} U \, dx \, dt - \int_0^\tau \int_\Omega \gamma R^{\gamma-2} \nabla R \cdot \rho u \, dx \, dt.$$

For the last two integrals, we have

$$- \int_0^\tau \int_\Omega \rho^\gamma \operatorname{div} U \, dx \, dt - \int_0^\tau \int_\Omega \gamma R^{\gamma-2} \nabla R \cdot \rho u \, dx \, dt$$

$$= \int_0^\tau \int_\Omega -\rho^\gamma \operatorname{div} U + \gamma R^{\gamma-2} \nabla R \cdot (RU - \rho u)$$

$$- \gamma R^{\gamma-2} \nabla R \cdot RU \, dx \, dt$$

$$= \int_0^\tau \int_\Omega (R^\gamma - \rho^\gamma) \operatorname{div} U + \gamma R^{\gamma-2} \nabla R \cdot (RU - \rho u) \, dx \, dt.$$

Inserting this back into (11.31) and observing that, owing to the mass equation for (R, U),

$$\gamma R^{\gamma-2} \partial_t R(R - \rho) + \gamma R^{\gamma-2} \operatorname{div} U R(R - \rho) + \gamma R^{\gamma-2} \nabla R \cdot RU$$

$$= \gamma R^{\gamma-2} U \cdot \nabla R \rho,$$

we get

$$E_{\mathrm{rel}}(\tau) + \int_0^\tau \int_\Omega \mathbb{S}(\nabla u) : (\nabla u - \nabla U) \, dx \, dt$$

$$\leq \int_0^\tau \int_\Omega \gamma R^{\gamma-2} \cdot \nabla R \rho(U - u) \, dx \, dt$$

$$+ \int_0^\tau \int_\Omega \partial_t U \cdot \rho(U - u) + \nabla U : (\rho u \otimes (U - u)) \, dx \, dt$$

$$- \int_0^\tau \int_\Omega \gamma R^{\gamma-1} \operatorname{div} U(R - \rho) \, dx \, dt + \int_0^\tau \int_\Omega (R^\gamma - \rho^\gamma) \operatorname{div} U. \tag{11.32}$$

The integrand in the third line can be rewritten pointwise as

$$\partial_t U \cdot \rho(U - u) + \nabla U : (\rho u \otimes (U - u))$$
$$= \partial_t U \cdot \rho(U - u) + \nabla U : (\rho U \otimes (U - u)) + \nabla U : (\rho(u - U) \otimes (U - u)),$$
(11.33)

and the integral of the last term and the one from the last line in (11.32) can be estimated by

$$C\|U\|_{C^1} \int_0^\tau E_{\mathrm{rel}}(t)\,\mathrm{d}t. \tag{11.34}$$

For the remaining terms in (11.33) we obtain by the momentum equation for (R, U)

$$\partial_t U \cdot \rho(U - u) + \nabla U : U \otimes \rho(U - u)$$
$$= \frac{1}{R}(\partial_t(RU) + \mathrm{div}(RU \otimes U)) \cdot \rho(U - u) \tag{11.35}$$
$$= -\gamma R^{\gamma-2}\nabla R \cdot \rho(U - u) + \frac{1}{R}\,\mathrm{div}\,\mathbb{S}(\nabla U) \cdot \rho(U - u).$$

Combining (11.32), (11.34), and (11.35), we obtain

$$E_{\mathrm{rel}}(\tau) + \int_0^\tau \int_\Omega \mathbb{S}(\nabla u) : (\nabla u - \nabla U)\,\mathrm{d}x\,\mathrm{d}t$$
$$\leq C\|U\|_{C^1} \int_0^\tau E_{\mathrm{rel}}(t)\,\mathrm{d}t + \int_0^\tau \int_\Omega \frac{1}{R}\,\mathrm{div}\,\mathbb{S}(\nabla U) \cdot \rho(U - u)\,\mathrm{d}x\,\mathrm{d}t$$

and thus

$$E_{\mathrm{rel}}(\tau) + \int_0^\tau \int_\Omega (\mathbb{S}(\nabla u) - \mathbb{S}(\nabla U)) : (\nabla u - \nabla U)\,\mathrm{d}x\,\mathrm{d}t$$
$$\leq C\|U\|_{C^1} \int_0^\tau E_{\mathrm{rel}}(t)\,\mathrm{d}t$$
$$+ \int_0^\tau \int_\Omega \frac{1}{R}\,\mathrm{div}\,\mathbb{S}(\nabla U) \cdot \rho(U - u) - \mathbb{S}(\nabla U) : (\nabla u - \nabla U)\,\mathrm{d}x\,\mathrm{d}t.$$

The desired result will follow from Grönwall's inequality if we can absorb the remaining viscosity term

$$\int_0^\tau \int_\Omega \frac{1}{R}\,\mathrm{div}\,\mathbb{S}(\nabla U) \cdot \rho(U - u) - \mathbb{S}(\nabla U) : (\nabla u - \nabla U)\,\mathrm{d}x\,\mathrm{d}t$$

into the relative energy. First observe that, by the divergence theorem,

$$\int_0^\tau \int_\Omega \frac{1}{R}\,\mathrm{div}\,\mathbb{S}(\nabla U) \cdot \rho(U - u) - \mathbb{S}(\nabla U) : (\nabla u - \nabla U)\,\mathrm{d}x\,\mathrm{d}t$$
$$= \int_0^\tau \int_\Omega \frac{1}{R}\,\mathrm{div}\,\mathbb{S}(\nabla U) \cdot (\rho - R)(U - u)\,\mathrm{d}x\,\mathrm{d}t.$$

By our assumptions on U and R, the factor $R^{-1} \operatorname{div} \mathbb{S}(\nabla U)$ is bounded in L^∞, so it remains to estimate $\int \int |\rho - R||U - u| \, dx \, dt$.

To this end, we split the domain of integration into the three parts

$$\Omega_< \cup \Omega_\sim \cup \Omega_> := \{\rho \le \underline{R}/2\} \cup \{\underline{R}/2 < \rho < 2\overline{R}\} \cup \{\rho \ge 2\overline{R}\},$$

where $\underline{R} := \inf_{\bar\Omega \times [0,T]} R(x,t) > 0$ and $\overline{R} := \sup_{\bar\Omega \times [0,T]} R(x,t) < \infty$.

Since the map $s \mapsto s^\gamma$ is smooth and strictly convex in $(0, \infty)$ and the range of R is compact, there exists a constant (depending only on R, Ω, d, and γ) such that on $\Omega_< \cup \Omega_\sim$,

$$(\rho - R)^2 \le C \left(\frac{1}{\gamma - 1} \rho^\gamma - \frac{\gamma}{\gamma - 1} R^{\gamma - 1} \rho + R^\gamma \right).$$

On the other hand, on $\Omega_>$ the term ρ^γ dominates, and therefore there exists another constant (also denoted C) such that on $\Omega_>$,

$$\rho \le C \left(\frac{1}{\gamma - 1} \rho^\gamma - \frac{\gamma}{\gamma - 1} R^{\gamma - 1} \rho + R^\gamma \right).$$

Armed with these estimates, we find

$$\int \int_{\Omega_<} |\rho - R||U - u| \, dx \, dt$$

$$\le C(\delta) \int_0^\tau \int_\Omega (\rho - R)^2 \, dx \, dt + \delta \int_0^\tau \int_\Omega |U - u|^2 \, dx \, dt$$

$$\le C(\delta) \int_0^\tau E_{\text{rel}}(t) \, dt + \int_0^\tau \int_\Omega (\mathbb{S}(\nabla u) - \mathbb{S}(\nabla U)) : (\nabla u - \nabla U) \, dx \, dt,$$

where we chose a sufficiently small number $\delta > 0$ and invoked the Poincaré–Korn inequality

$$\int_\Omega |w|^2 \le C \int_\Omega \mathbb{S}(\nabla w) : \nabla w \, dx$$

for all $w \in H_0^1(\Omega)$.

Further, we have

$$\int \int_{\Omega_\sim} |\rho - R||U - u| \, dx \, dt$$

$$\le \frac{1}{2} \int \int_{\Omega_\sim} \frac{(\rho - R)^2}{\sqrt{\rho}} \, dx \, dt + \frac{1}{2} \int \int_{\Omega_\sim} \frac{1}{\sqrt{\rho}} \rho |U - u|^2 \, dx \, dt$$

$$\le C \int_0^\tau E_{\text{rel}}(t) \, dt,$$

since ρ is bounded away from zero on $\Omega_\sim$. $\qquad\square$

Finally, we estimate (keeping in mind $\rho > R$ on $\Omega_>$)

$$\int\int_{\Omega_>} |\rho - R||U - u| \, \mathrm{d}x \, \mathrm{d}t$$

$$\leq \int\int_{\Omega_>} \rho|U - u| \, \mathrm{d}x \, \mathrm{d}t$$

$$\leq \frac{1}{2}\int\int_{\Omega_>} \rho \, \mathrm{d}x \, \mathrm{d}t + \frac{1}{2}\int\int_{\Omega_>} \rho|U - u|^2 \, \mathrm{d}x \, \mathrm{d}t$$

$$\leq C\int_0^\tau E_{\mathrm{rel}}(t) \, \mathrm{d}t.$$

Putting everything together, we obtain $E_{\mathrm{rel}}(\tau) \leq C \int_0^\tau E_{\mathrm{rel}}(t) \, \mathrm{d}t$, and the claim follows.

11.5 Physical Boundaries in the Inviscid Situation

Choosing $\mathbb{T}^d$ as the spatial domain (thus imposing periodic boundary conditions), as we did for the incompressible and compressible Euler equations, often comes in handy, because the torus is compact but has no boundary. All the arguments so far carry over to the whole space $\mathbb{R}^d$ with only a little more effort.

The story becomes very different upon the appearance of physical boundaries: weak-strong uniqueness can fail in this situation due to the possible formation of a boundary layer in inviscid models. Let us discuss again the incompressible Euler equations, for which the usual slip boundary conditions are required:

$$u \cdot n = 0 \quad \text{on } \partial\Omega,$$

where n denotes the outward unit normal to the boundary of Ω. For weak solutions, in general we only have $u \in L^2_{\mathrm{loc}}(\Omega \times (0,T))$, and so it makes no sense to evaluate u on the boundary. Instead one defines the space $H(\Omega)$ of **solenoidal vectorfields**, which is the completion in the norm of $L^2(\Omega)$ of the space

$$\left\{ w \in C_c^\infty(\Omega; \mathbb{R}^d) : \operatorname{div} w = 0 \right\}.$$

A solenoidal vector field thus satisfies the divergence-free and slip boundary conditions in a weak sense. This allows us to define the notion of admissible weak solution on $\Omega \times [0, T]$ as follows.

Definition 11.9 A vector field $u \in L^\infty((0,T); H(\Omega))$ is called an *admissible weak solution* of the incompressible Euler equations with initial data $u^0 \in H(\Omega)$ if

$$\int_0^T \int_\Omega \partial_t \varphi(x,t) \cdot u(x,t) + \nabla\varphi(x,t) : (u \otimes u)(x,t) \, \mathrm{d}x \, \mathrm{d}t$$

$$= \int_\Omega u(x,\tau) \cdot \varphi(x,\tau) - u^0(x) \cdot \varphi(x,0) \, \mathrm{d}x$$

(11.36)

for almost every $\tau \in (0,T)$ and for every $\varphi \in C_c^1(\Omega \times [0,T]; \mathbb{R}^d)$ with $\mathrm{div}\,\varphi = 0$, and if for almost every $\tau \in (0,T)$ we have

$$\frac{1}{2} \int_\Omega |u(x,\tau)|^2 \, \mathrm{d}x \leq \frac{1}{2} \int_\Omega |u^0(x)|^2 \, \mathrm{d}x.$$

In order to show weak-strong uniqueness, we would like to use the relative energy argument as for Theorem 11.1. To do this, we would have to use the strong solution U as a test function in (11.36). However, even a classical smooth solution of the Euler equations is only required to satisfy $v \cdot n = 0$ on the boundary, whereas the tangential component may be nonzero. But the test function in (11.36) has to be zero at the boundary in *every* direction. And indeed it is possible to produce an example of the failure of weak-strong uniqueness that exploits exactly this discrepancy. Before we state it, we remark[5] that a possible remedy for the problem at hand would be to demand that (11.36) hold for all $\varphi \in C^1(\Omega \times [0,T]; \mathbb{R}^d)$ with $\mathrm{div}\,\varphi = 0$ and $\varphi \cdot n = 0$ on $\partial\Omega$ (thus admitting a non-vanishing tangential component for the test function). Then weak-strong uniqueness would follow exactly as in Theorem 11.1. We argue however that this ad hoc extension of the class of test functions would amount to begging the question in favour of weak-strong uniqueness: a boundary-value problem for a system of partial differential equations imposes the equations *in the interior* of the domain, which is guaranteed by testing only against compactly supported test fields, and by requiring the boundary condition, which is encoded in the assumption $u \in H(\Omega)$ for almost every time. The further requirement that (11.36) hold additionally for test fields with non-vanishing tangential component, however, lacks justification from the modelling perspective.

Theorem 11.10 (Bardos, Székelyhidi, Jr., & Wiedemann, 2014) *There exists a smooth bounded domain $\Omega \subset \mathbb{R}^2$, a time $T > 0$, smooth data $u^0 \in C^\infty(\Omega)$ with $\mathrm{div}(u^0) = 0$ and $u^0 \cdot n = 0$ on $\partial\Omega$, a smooth solution*

---

[5] I would like to thank Eduard Feireisl for pointing this out.

$U \in C^\infty(\Omega \times [0, T])$ with data u^0, and infinitely many admissible weak solutions with the same data.

We refer to Bardos et al. (2014) for the proof, which is based on convex integration techniques, in particular on a construction of Székelyhidi (2011).

The good news is that weak-strong uniqueness can be restored by requiring the weak solution not to behave "wildly" near the boundary:

Theorem 11.11 *Suppose that $\Omega \in \mathbb{R}^2$ is a smooth bounded domain. Let $U \in C^1(\bar\Omega \times [0, T])$ be a solution of the Euler equations with $U(\cdot, 0) = u^0$, and $u \in L^\infty((0, T); H(\Omega))$ an admissible weak solution with initial data u^0 such that u is continuous[6] in a neighbourhood of $\partial\Omega \times [0, T]$. Then, $u = U$ for almost every $(x, t) \in \Omega \times (0, T)$.*

Proof We omit some details in the sequel; for the full proof, see Bardos et al. (2014).

For simplicity we work only on the upper half-plane[7]

$$\Omega = \{(x_1, x_2) \in \mathbb{R} : x_2 > 0\}.$$

Let $\varepsilon > 0$ be so small that u is continuous on

$$\Gamma^\varepsilon := \{(x_1, x_2) \in \bar\Omega : x_2 < \varepsilon\}.$$

Moreover let $\chi \in C_c^\infty((0, \infty]; \mathbb{R})$ be a smooth function with $0 \leq \chi \leq 1$ and $\chi(s) = 1$ for $s \geq 1$, and set $\chi^\varepsilon(s) = \chi(s/\varepsilon)$. Since U is divergence free and satisfies the slip boundary condition, there exists (by Poincaré's Lemma) a scalar potential $\psi \in C([0, T]; C^2(\bar\Omega)) \cap C^1(\bar\Omega \times [0, T])$ with $\psi = 0$ on $\partial\Omega$ such that $U = \nabla^\perp \psi$. Inspired by Kato's famous paper (Kato, 1984), we now set

$$U^\varepsilon(x_1, x_2, t) := \nabla^\perp \left[\chi^\varepsilon(x_2)\, \psi(x_1, x_2, t) \right].$$

Then U^ε is C^1, compactly supported and divergence free, and thus qualifies as a test function in (11.36).

---

[6] In the original paper Bardos et al. (2014), Hölder continuity is required. However, the Hölder condition is not necessary in the proof, which works just fine even if u is merely continuous.

[7] This domain is of course not bounded, but we will ignore this issue since our problem is of a local nature.

Hence we can estimate, for almost every $\tau \in (0, T)$,

$$
\frac{1}{2} \int_\Omega |u(\tau) - U^\varepsilon(\tau)|^2 \, \mathrm{d}x
$$

$$
= \frac{1}{2} \int_\Omega |u(\tau)|^2 \, \mathrm{d}x + \frac{1}{2} \int_\Omega |U^\varepsilon(\tau)|^2 \, \mathrm{d}x - \int_\Omega u(\tau) \cdot U^\varepsilon(\tau) \, \mathrm{d}x
$$

$$
\leq \frac{1}{2} \int_\Omega |U^\varepsilon(\tau)|^2 \, \mathrm{d}x - \frac{1}{2} \int_\Omega |u^0|^2 \, \mathrm{d}x
$$

$$
- \int_0^\tau \int_\Omega (\partial_t U^\varepsilon \cdot u + \nabla U^\varepsilon : (u \otimes u)) \, \mathrm{d}x \, \mathrm{d}t.
$$

$$(11.37)$$

We now want to let $\varepsilon \searrow 0$ in order to obtain an estimate for the relative energy between u and U. To this end, observe that

$$
U^\varepsilon(x, \tau) = \nabla^\perp \left[\chi^\varepsilon(x_2) \, \psi(x_1, x_2, \tau) \right]
$$

$$
= \chi^\varepsilon(x_2) \nabla^\perp \psi(x_1, x_2, t) + \frac{1}{\varepsilon} \psi(x_1, x_2, \tau) \chi' \left(\frac{x_2}{\varepsilon} \right) e_1 \quad (11.38)
$$

$$
= \chi^\varepsilon(x_2) U(x, \tau) + \frac{1}{\varepsilon} \psi(x_1, x_2, \tau) \chi' \left(\frac{x_2}{\varepsilon} \right) e_1.
$$

The first term clearly converges strongly in $L^2(\Omega)$ to U, uniformly in time, as $\varepsilon \searrow 0$. For the second term, recall that ψ is twice continuously differentiable in space, uniformly in time, and assumes the value zero at $x_2 = 0$, so that there exists a constant C such that

$$
|\psi(x, \tau)| \leq C x_2 \leq C \varepsilon
$$

in the boundary region Γ^ε. Together with the observation that $\chi'(\cdot/\varepsilon)$ vanishes outside Γ^ε, we obtain convergence in $L^2(\Omega)$ of the second term of (11.38) to zero, uniformly in time. Hence we get $U^\varepsilon \to U$ in $L^2(\Omega)$ as $\varepsilon \searrow 0$, uniformly in time. In a similar way one can see $\partial_t U^\varepsilon \to \partial_t U$ in $L^2(\Omega)$ as $\varepsilon \searrow 0$, uniformly in time.

The more problematic term is $\int_0^\tau \int_\Omega \nabla U^\varepsilon : (u \otimes u) \, \mathrm{d}x \, \mathrm{d}t$. We split it as

$$
\int_0^\tau \int_\Omega \partial_1 U_1^\varepsilon u_1^2 \, \mathrm{d}x \, \mathrm{d}t + \int_0^\tau \int_\Omega \partial_1 U_2^\varepsilon u_1 u_2 \, \mathrm{d}x \, \mathrm{d}t
$$

$$
+ \int_0^\tau \int_\Omega \partial_2 U_1^\varepsilon u_1 u_2 \, \mathrm{d}x \, \mathrm{d}t + \int_0^\tau \int_\Omega \partial_2 U_2^\varepsilon u_2^2 \, \mathrm{d}x \, \mathrm{d}t.
$$

In view of (11.38), we expect the third integral to behave worst, so we skip the other integrals. We estimate it in the following way, again using

(11.38):

$$\left| \int_0^\tau \int_\Omega \partial_2 (U_1^\varepsilon - U_1) u_1 u_2 \, dx \, dt \right|$$

$$\leq \left| \int_0^\tau \int_\Omega (\chi^\varepsilon(x_2) - 1) \partial_2 U_1 u_1 u_2 \, dx \, dt \right|$$

$$+ \frac{1}{\varepsilon} \left| \int_0^\tau \int_\Omega \partial_2 \psi(x,\tau) \chi' \left(\frac{x_2}{\varepsilon} \right) u_1 u_2 \, dx \, dt \right|$$

$$+ \frac{1}{\varepsilon^2} \left| \int_0^\tau \int_\Omega \psi(x,\tau) \chi'' \left(\frac{x_2}{\varepsilon} \right) u_1 u_2 \, dx \, dt \right|.$$

The first term clearly converges to zero. For the second term, observe that χ' is supported on Γ^ε, where by assumption u is continuous; therefore, u_1, $\nabla \psi$, and $\chi'(\cdot/\varepsilon)$ are bounded on the domain of integration, and by the slip boundary condition $u(x_1, x_2, t) \to 0$ as $x_2 \to 0$, uniformly in x_1, t. It follows that the second integral is bounded by

$$C \frac{1}{\varepsilon} \int_{\Gamma^\varepsilon} u_2(x_1, x_2, t) \, dx \, dt \to 0.$$

Finally, the third integral can be estimated similarly, because $\chi''(\cdot/\varepsilon)$ is again supported on Γ^ε, and there we have as above $|\psi(x_1, x_2, t)| \leq C\varepsilon$.

We have thus shown that taking the limit $\varepsilon \searrow 0$ in (11.37) yields

$$\frac{1}{2} \int_\Omega |u(\tau) - U(\tau)|^2 \, dx \leq - \int_0^\tau \int_\Omega (\partial_t U \cdot u + \nabla U : (u \otimes u)) \, dx \, dt,$$

which allows us to continue exactly as in the proof of Theorem 11.1. $\quad\square$

11.6 An Alternative Approach

We present here an alternative proof of weak-strong-uniqueness for the incompressible Euler equations, inspired by arguments used in the context of so-called statistical solutions (Fjordholm et al., 2017)[8]. This approach may be of some interest, as it demonstrates the usefulness of doubling of variables techniques for the Euler equations, and avoids the assumption of time differentiability of the strong solution. We work again on the torus $\mathbb{T}^d$.

---

[8] I would like to thank U.S. Fjordholm and S. Mishra for valuable discussions on this subject.

Write the Euler equations in components,

$$\partial_t u^i(x) + \sum_{j=1}^{d} \partial_j u^i(x) u^j(x) + \partial_i p(x) = 0.$$

If U is the strong solution, we have similarly (but evaluated at another point $y \in \mathbb{T}^d$)

$$\partial_t U^i(y) + \sum_{j=1}^{d} \partial_j U^i(y) U^j(y) + \partial_i P(y) = 0.$$

Multiplying the first equation by $U^i(y)$, the second by $u^i(x)$, and adding everything together, we arrive at

$$\partial_t(u(x) \cdot U(y)) + \mathrm{div}_x((u(x) \cdot U(y))u(x)) + \mathrm{div}_y((u(x) \cdot U(y))U(y))$$
$$+ \mathrm{div}_x(p(x)U(y)) + \mathrm{div}_y(P(y)u(x)) = 0.$$
$$(11.39)$$

Thanks to its divergence structure, this last equation admits a formulation in the sense of distributions, and it can be shown rigorously (as in Lemma 3.1 in Fjordholm et al., 2017) that (11.39) holds in such a weak sense if u, U are weak solutions of the Euler equations.

Therefore (omitting $dx\,dy\,dt$ from every integral expression) we have for any $\varphi \in C^1(\mathbb{T}^d \times \mathbb{T}^d \times [0, T])$

$$\frac{1}{2} \int_0^T \int_{\mathbb{T}^d} \int_{\mathbb{T}^d} \partial_t \varphi(x, y, t) |u(x) - U(y)|^2$$

$$= \frac{1}{2} \int_0^T \int_{\mathbb{T}^d} \int_{\mathbb{T}^d} \partial_t \varphi(x, y, t) |u(x)|^2 + \frac{1}{2} \int_0^T \int_{\mathbb{T}^d} \int_{\mathbb{T}^d} \partial_t \varphi(x, y, t) |U(y)|^2$$

$$- \int_0^T \int_{\mathbb{T}^d} \int_{\mathbb{T}^d} \partial_t \varphi(x, y, t) u(x) \cdot U(y)$$

$$= \frac{1}{2} \int_0^T \int_{\mathbb{T}^d} \int_{\mathbb{T}^d} \partial_t \varphi(x, y, t) |u(x)|^2 + \frac{1}{2} \int_0^T \int_{\mathbb{T}^d} \int_{\mathbb{T}^d} \partial_t \varphi(x, y, t) |U(y)|^2$$

$$+ \int_0^T \int_{\mathbb{T}^d} \int_{\mathbb{T}^d} (\nabla_x \varphi(x, y, t) \cdot u(x))(u(x) \cdot U(y))$$

$$+ \int_0^T \int_{\mathbb{T}^d} \int_{\mathbb{T}^d} (\nabla_y \varphi(x, y, t) \cdot U(y))(U(y) \cdot u(x))$$

$$- \int_0^T \int_{\mathbb{T}^d} \int_{\mathbb{T}^d} \nabla_x \varphi(x, y, t) \cdot p(x) U(y)$$

$$- \int_0^T \int_{\mathbb{T}^d} \int_{\mathbb{T}^d} \nabla_y \varphi(x, y, t) \cdot P(y) u(x).$$
$$(11.40)$$

For the rest of the argument, assume in addition that U is C^1 in space and that u is globally admissible. Choose

$$\varphi(x, y, t) = \eta^\varepsilon \left(\frac{x - y}{2} \right) \chi(t) \qquad (11.41)$$

for a standard mollifier η, $\eta^\varepsilon(x) = \varepsilon^{-d}\eta(x/\varepsilon)$, and some $\chi \in C^1([0, T])$. It is easy to see (by the Lebesgue Differentiation Theorem) that the left hand side of (11.40) converges to

$$\frac{1}{2} \int_0^T \int_{\mathbb{T}^d} \chi'(t) |u(x) - U(x)|^2 \, dx \, dt$$

as $\varepsilon \searrow 0$.

For the second-to-last line of (11.40), observe that

$$\nabla_x \eta^\varepsilon \left(\frac{x - y}{2} \right) = -\nabla_y \eta^\varepsilon \left(\frac{x - y}{2} \right)$$

and therefore, by the divergence-free condition on U, for the particular choice (11.41),

$$\int_0^T \int_{\mathbb{T}^d} \int_{\mathbb{T}^d} \nabla_x \varphi(x, y, t) \cdot p(x) U(y) \, dx \, dy \, dt = 0,$$

and similarly

$$\int_0^T \int_{\mathbb{T}^d} \int_{\mathbb{T}^d} \nabla_y \varphi(x, y, t) \cdot P(y) u(x) \, dx \, dy \, dt = 0.$$

For the second and third integrals of (11.40), we compute

$$\int_0^T \int_{\mathbb{T}^d} \int_{\mathbb{T}^d} \nabla_x \varphi(x, y, t) \cdot u(x))(u(x) \cdot U(y)$$

$$+ \nabla_y \varphi(x, y, t) \cdot U(y))(U(y) \cdot u(x) \, dx \, dy \, dt$$

$$= \int_0^T \chi(t) \int_{\mathbb{T}^d} \int_{\mathbb{T}^d} (u(x) \cdot U(y)) \nabla_y \eta^\varepsilon \left(\frac{x - y}{2} \right) \cdot (U(y) - u(x)) \, dx \, dy \, dt.$$

$$(11.42)$$

Now we obtain on the one hand (owing to the divergence condition)

$$\int_0^T \chi(t) \int_{\mathbb{T}^d} \int_{\mathbb{T}^d} (u(x) \cdot U(y)) \nabla_y \eta^\varepsilon \left(\frac{x - y}{2} \right) \cdot U(y) \, dx \, dy \, dt$$

$$= -\int_0^T \chi(t) \int_{\mathbb{T}^d} \int_{\mathbb{T}^d} \eta^\varepsilon \left(\frac{x - y}{2} \right) u(x) \cdot \nabla U(y) U(y) \, dx \, dy \, dt$$

and on the other hand

$$- \int_0^T \chi(t) \int_{\mathbb{T}^d} \int_{\mathbb{T}^d} (u(x) \cdot U(y)) \nabla_y \eta^\varepsilon \left(\frac{x-y}{2} \right) \cdot u(x) \, dx \, dy \, dt$$

$$= \int_0^T \chi(t) \int_{\mathbb{T}^d} \int_{\mathbb{T}^d} \eta^\varepsilon \left(\frac{x-y}{2} \right) u(x) \cdot \nabla U(y) u(x) \, dx \, dy \, dt,$$

so that (11.42) turns into

$$\int_0^T \chi(t) \int_{\mathbb{T}^d} \int_{\mathbb{T}^d} \eta^\varepsilon \left(\frac{x-y}{2} \right) u(x) \cdot \nabla U(y) (u(x) - U(y)) \, dx \, dy \, dt$$

and, after another invocation of the identity $\int U \cdot \nabla U w \, dy = 0$ for divergence-free w, into

$$\int_0^T \chi(t) \int_{\mathbb{T}^d} \int_{\mathbb{T}^d} \eta^\varepsilon \left(\frac{x-y}{2} \right) (u(x) - U(y)) \cdot \nabla_{\mathrm{sym}} U(y) (u(x) - U(y)) \, dx \, dy \, dt.$$

Now again we can apply Lebesgue's Differentiation Theorem to obtain in the limit $\varepsilon \searrow 0$

$$\int_0^T \chi(t) \int_{\mathbb{T}^d} \int_{\mathbb{T}^d} (u(x) - U(x)) \cdot \nabla U(x) (u(x) - U(x)) \, dx \, dt.$$

To conclude the analysis of equality (11.40), we observe that the energy admissibility of u and U implies that the first line on the right hand side of (11.40) is non-negative for any choice of $\chi \geq 0$.

Putting everything together, we get from (11.40)

$$\frac{\mathrm{d}}{\mathrm{d}t} E_{\mathrm{rel}} \leq \left| \int_{\mathbb{T}^d} \int_{\mathbb{T}^d} (u(x) - U(x)) \cdot \nabla U(x) (u(x) - U(x)) \, dx \right|$$

$$\leq \|\nabla_{\mathrm{sym}} U\|_\infty E_{\mathrm{rel}},$$

where the relative energy E_{rel} is defined as in the proof of Theorem 11.1. Hence, Grönwall's inequality yields once more weak-strong uniqueness as long as U is C^1 in space for almost every time, and $\nabla_{\mathrm{sym}} U \in L^1((0,T); L^\infty(\mathbb{T}^d))$. Note that no further time regularity is required.

Remark In both proofs of weak-strong uniqueness for the Euler equations, the assumption on $\nabla_{\mathrm{sym}} U$ is crucial. It is a major open problem whether uniqueness holds for the Euler equations when the solutions have less than one derivative. In the light of recent progress on Onsager's Conjecture, one conjecture is that weak solutions are unique if they belong to C^α with $\alpha > 1/3$. However the regime $1/3 < \alpha < 1$ is currently completely open in terms of uniqueness.

Acknowledgements. The author would like to thank Eduard Feireisl, Ulrik Skre Fjordholm, and Siddharta Mishra for very helpful discussions on the topic.

References

Alibert, J.-J. & Bouchitté, G. (1997) Non-uniform integrability and generalized Young measure. *J. Convex Anal.* **4**, 129–148.

Bardos, C. & Titi, E.S. (2010) Loss of smoothness and energy conserving rough weak solutions for the 3*d* Euler equations. *Discrete Contin. Dyn. Syst. Ser. S* **3**(2), 185–197.

Bardos, L., Székelyhidi, Jr., L., & Wiedemann, E. (2014) On the absence of uniqueness for the Euler equations: the effect of the boundary. *Uspekhi Mat. Nauk* **69**(2(416)), 3–22.

Bourgain, J. & Li, D. (2015) Strong ill-posedness of the incompressible Euler equation in borderline Sobolev spaces. *Invent. Math.* **201**(1), 97–157.

Brenier, Y., De Lellis, C., & Székelyhidi, Jr., L. (2011) Weak-strong uniqueness for measure-valued solutions. *Comm. Math. Phys.* **305**(2), 351–361.

Březina, J. & Mácha, V. (2017) Inviscid limit for the compressible Euler system with non-local interactions. *Preprint.*

Březina, J. & Feireisl, E. (2017) Measure-valued solutions to the complete Euler system. *J. Math. Soc. Japan*, to appear.

Buckmaster, T., & Vicol, V. (2017) Nonuniqueness of weak solutions to the Navier–Stokes equation. arXiv: 1709.10033.

Buckmaster, T., De Lellis, C., Székelyhidi, Jr., L., & Vicol, V. (2017) Onsager's Conjecture for admissible weak solutions. *Preprint.*

Chiodaroli, E. (2014) A counterexample to well-posedness of entropy solutions to the compressible isentropic Euler system. *J. Hyperbolic Differ. Equ.* **11**(3), 493–519.

Chiodaroli, E., De Lellis, C., & Kreml, O. (2015) Global ill-posedness of the isentropic system of gas dynamics. *Comm. Pure Appl. Math.* **68**(7), 1157–1190.

Constantin, P. & Foias, C. (1988) *Navier–Stokes equations.* Chicago Lectures in Mathematics. University of Chicago Press, Chicago, IL.

Dafermos, C. (2010) *Hyperbolic conservation laws in continuum physics.* Third edition. Grundlehren der Mathematischen Wissenschaften [Fundamental Principles of Mathematical Sciences], 325. Springer-Verlag, Berlin.

Daneri, S. & Székelyhidi, Jr., L. (2017) Non-uniqueness and h-principle for Hölder-continuous weak solutions of the Euler equations. *Arch. Ration. Mech. Anal.* **224**(2), 471–514.

Demoulini, S., Stuart, D., & Tzavaras, A. (2012) Weak-strong uniqueness of dissipative measure-valued solutions for polyconvex elastodynamics. *Arch. Ration. Mech. Anal.* **205**(3), 927–961.

DiPerna, R.J. & Majda, A.J. (1987). Oscillations and concentrations in weak solutions of the incompressible fluid equations. *Comm. Math. Phys.* **108**(4), 667–689.

De Lellis, C. & Székelyhidi, Jr., L. (2009) The Euler equations as a differential inclusion. *Ann. of Math. (2)* **170**(3), 1417–1436.

De Lellis, C. & Székelyhidi, Jr., L. (2010) On admissibility criteria for weak solutions of the Euler equations. *Arch. Ration. Mech. Anal.* **195**(1), 225–260.

De Lellis, C. & Székelyhidi, Jr., L. (2013) Dissipative continuous Euler flows. *Invent. Math.* **193**(2), 377–407.

Feireisl, E., Novotný, A., & Petzeltová, H. (2001) On the existence of globally defined weak solutions to the Navier–Stokes equations. *J. Math. Fluid Mech.* **3**(4), 358–392.

Feireisl, E. (2004) *Dynamics of viscous compressible fluids.* Oxford Lecture Series in Mathematics and its Applications, 26. Oxford University Press, Oxford.

Feireisl, E., Jin, B.J., & Novotný, A. (2012) Relative entropies, suitable weak solutions, and weak-strong uniqueness for the compressible Navier-Stokes system. *J. Math. Fluid Mech.* **14**, 712–730.

Feireisl, E., Gwiazda, P., Świerczewska-Gwiazda, A., & Wiedemann, E. (2016) Dissipative measure-valued solutions to the compressible Navier–Stokes system. *Calc. Var. Partial Differential Equations* **55**(6):Art. 141.

Feireisl, E. & Lukáčová-Medvid'ová, M. (2017) Convergence of a mixed finite element finite volume scheme for the isentropic Navier–Stokes system via dissipative measure-valued solutions. *Found. Comput. Math.*, to appear. DOI:10.1007/s10208-017-9351-2.

Fjordholm, U.S., Lanthaler, S., & Mishra, S. (2017) Statistical solutions of hyperbolic conservation laws I: Foundations. *Arch. Rational Mech. Anal.* **226**, 809–849.

Galdi, G. (2000) An introduction to the Navier–Stokes initial-boundary value problem. *Fundamental directions in mathematical fluid mechanics*, 1–70, Adv. Math. Fluid Mech. Birkhäuser, Basel.

Guillod, J. & Šverák, V. (2017) Numerical investigations of non-uniqueness for the Navier–Stokes initial value problem in borderline spaces. *Preprint.*

Gwiazda, P., Świerczewska-Gwiazda, A., & Wiedemann, E. (2015) Weak-strong uniqueness for measure-valued solutions of some compressible fluid models. *Nonlinearity* **28**(11), 3873–3890.

Isett, P. (2017) A proof of Onsager's Conjecture. *Preprint.*

Kato, T. (1984) Remarks on zero viscosity limit for nonstationary Navier–Stokes flows with boundary. *Seminar on nonlinear partial differential equations (Berkeley, Calif., 1983)*, 85–98, Math. Sci. Res. Inst. Publ., 2, Springer, New York.

Leray, J. (1934) Sur le mouvement d'un liquide visqueux emplissant l'espace. *Acta Math.* **63**(1), 193–248.

Lions, P.-L. (1996) *Mathematical topics in fluid dynamics. Vol. 1. Incompressible models.* Oxford Lecture Series in Mathematics and its Applications,

3. Oxford Science Publications. The Clarendon Press, Oxford University Press, New York.

Lions, P.-L. (1998) *Mathematical topics in fluid dynamics. Vol. 2. Compressible models.* Oxford Lecture Series in Mathematics and its Applications, 10. Oxford Science Publications. The Clarendon Press, Oxford University Press, New York.

Marchioro, C. & Pulvirenti, M. (1994) *Mathematical theory of incompressible nonviscous fluids.* Applied Mathematical Sciences, 96. Springer-Verlag, New York.

Majda, A. & Bertozzi, A. (2002) *Vorticity and incompressible flow.* Cambridge Texts in Applied Mathematics, 27. Cambridge University Press, Cambridge.

Onsager, L. (1949) Statistical hydrodynamics. *Nuovo Cimento (9)*, 6 (Supplemento, 2 (Convegno Internazionale di Meccanica Statistica)), 279–287.

Prodi, G. (1959) Un teorema di unicità per le equazioni di Navier-Stokes. *Ann. Mat. Pura Appl.* **48**, 173–182.

Robinson, J.C., Rodrigo, J.L., & Sadowski, W. (2016) *The three-dimensional Navier–Stokes equations. Classical theory.* Cambridge Studies in Advanced Mathematics, 157. Cambridge University Press, Cambridge.

Saint-Raymond, L. (2009) *Hydrodynamic limits of the Boltzmann equation.* Lecture Notes in Mathematics, 1971. Springer-Verlag, Berlin.

Scheffer, V. (1993) An inviscid flow with compact support in space-time. *J. Geom. Anal.* **3**(4), 343–401.

Serrin, J. (1963) The initial value problem for the Navier-Stokes equations. In *Nonlinear Problems (Proc. Sympos., Madison, Wis., 1962)*, 69–98. Univ. of Wisconsin Press, Madison, Wisconsin.

Shnirelman, A. (1997) On the nonuniqueness of weak solution of the Euler equation. *Comm. Pure. Appl. Math.* **50**(12), 1261–1286.

Székelyhidi, Jr., L. (2011) Weak solutions to the incompressible Euler equations with vortex sheet initial data. *C. R. Math. Acad. Sci. Paris* **349**(19-20), 1063–1066.

Székelyhidi, Jr., L. & Wiedemann, E. (2012) Young measures generated by ideal incompressible fluid flows. *Arch. Ration. Mech. Anal.* **206**(1), 333–366.

Temam, R. (2001) *Navier–Stokes equations. Theory and numerical analysis.* Reprint of the 1984 edition. AMS Chelsea Publishing, Providence, RI.

Wiedemann, E. (2011) Existence of weak solutions for the incompressible Euler equations. *Ann. Inst. H. Poincaré Anal. Non Linéaire* **28**(5), 727–730.

Printed in the United States
By Bookmasters